Studies in Computational Intelligence

Volume 800

Series editor

Janusz Kacprzyk, Polish Academy of Sciences, Warsaw, Poland
e-mail: kacprzyk@ibspan.waw.pl

The series "Studies in Computational Intelligence" (SCI) publishes new developments and advances in the various areas of computational intelligence—quickly and with a high quality. The intent is to cover the theory, applications, and design methods of computational intelligence, as embedded in the fields of engineering, computer science, physics and life sciences, as well as the methodologies behind them. The series contains monographs, lecture notes and edited volumes in computational intelligence spanning the areas of neural networks, connectionist systems, genetic algorithms, evolutionary computation, artificial intelligence, cellular automata, self-organizing systems, soft computing, fuzzy systems, and hybrid intelligent systems. Of particular value to both the contributors and the readership are the short publication timeframe and the world-wide distribution, which enable both wide and rapid dissemination of research output.

More information about this series at http://www.springer.com/series/7092

Plamen P. Angelov · Xiaowei Gu

Empirical Approach to Machine Learning

Plamen P. Angelov
School of Computing
and Communications
Lancaster University
Lancaster, UK

Xiaowei Gu
School of Computing
and Communications
Lancaster University
Lancaster, UK

Additional material to this book can be downloaded from http://extras.springer.com.

ISSN 1860-949X ISSN 1860-9503 (electronic)
Studies in Computational Intelligence
ISBN 978-3-030-13209-5 ISBN 978-3-030-02384-3 (eBook)
https://doi.org/10.1007/978-3-030-02384-3

This Springer imprint is published by the registered company Springer Nature Switzerland AG
The registered company address is: Gewerbestrasse 11, 6330 Cham, Switzerland

To Rositsa, Mariela and Lachezar

Plamen P. Angelov

To my family and friends

Xiaowei Gu

Forewords

Dimitar Filev

Empirical Approach to Machine Learning by Plamen Angelov and Xiaowei Gu is an original and comprehensive contribution to the modern data analytic and machine learning methods and techniques. This book enthusiastically introduces a data-driven nonparametric "empirical" approach to machine learning that is focused on the data and that minimizes the need for "handcrafting"—making subjective assumptions about the data properties, consideration of models with predefined structure, user-defined parameters, and thresholds.

The idea of empirical learning is a step further towards the concept of evolving systems—a spinoff of the fields of statistical learning, probability theory, fuzzy theory, and artificial neural networks that is receiving growing attention in the recent years. The lead author, Prof. Angelov, with whose work I have been acquainted since his graduate study years, is one of the pioneers of the evolving system area. His long term contributions are extended and further advanced in the text by introducing the notions of cumulative proximity, typicality, and eccentricity as non-parametric characteristics of the data that form the foundation of the empirical learning approach. Some additional original concepts include effective methodology and algorithms for recursive density estimation, generalization of the multidimensional fuzzy sets to fuzzy clouds of general shape, critical analysis of some of the drawbacks of the deep learning methods, introduction of deep rule-based classifier with multilayer structure, enhancements of the unsupervised and semi-supervised learning methods, etc.

The outcome is an efficient data analytic methodology with high level of autonomy and potential for increasing the productivity and automation of wide range of information processing tasks. One additional distinctive characteristic of the proposed empirical approach to machine learning is its ability to operate online by applying recursive learning process without the need for an explicit moving window buffer. This one-pass information processing provides abundant opportunities for fast and highly efficient model learning with minimal computational requirements and applications to signal processing, machine vision, anomaly detection, etc.

In addition, the learned models have hierarchical logical rule-like structure that enables a straightforward interpretability of the information summarized during the learning process.

The emphasis of this work is on a novel empirical approach to machine learning. It also provides a comprehensive set of systematic references and can be a valuable reading for those interested in the fundamentals and advances of machine learning and its applications to pattern recognition, system modeling, anomaly detection, diagnostics, and prognostics. It is further enriched by the numerous applications, examples, and downloadable scripts supporting and illustrating the methods and algorithms presented in this book.

The book *Empirical Approach to Machine Learning* opens new horizons to automated and efficient data processing. In the current reality of exponentially growing interest in AI technology, this book is an indispensable reading for professionals in the field of data science and machine learning, including graduate students, researchers, and practitioners.

Dearborn, MI, USA

Dimitar Filev
Henry Ford Technical Fellow
Research and Innovations Center
Ford Motor Company

and

Washington, DC, USA

Member of the National Academy of Engineering

Paul J. Werbos

I owe great thanks to Professor Plamen Angelov for making this important material available to the community just as I see great practical needs for it, in the new area of making real sense of high-speed data from the brain.[1] There are huge areas of industry and of data analysis, which have simply not been touched yet by the latest reincarnation of neural networks in computer science, which even now does not yet appreciate the great capabilities, which have been developed over many decades in the field of computational intelligence aimed at applications in engineering.[2]

One of these capabilities is clustering analysis, one of the main themes developed in this book, with incredible clarity and coherence, focused strongly on the needs of application, which Professor Angelov has worked to improve and strengthen for many decades, informed by all the best of what others have done as

[1] P. J. Werbos and J. J. J. Davis, "Regular cycles of forward and backward signal propagation in prefrontal cortex and in consciousness," *Front. Syst. Neurosci.*, vol. 10, p. 97, 2016.
[2] https://cis.ieee.org/.

well. It is common in many areas of application for users to grab onto the first old method they find on the web or in their undergraduate textbooks, and not put much energy into trying to do better. However, since the quality of clustering is crucial to the quality of results in many areas, it is critically important to be very careful how to proceed. This book is certainly on the short list of two things I would want to read, before doing anything practical with clustering—as I am very much thinking of doing soon. This is in part because it is so clear and so logical, and because the tools are ready to use.

At the end of the day, clustering is one of the crucial capabilities, which allow the brain itself to work. Deep learning is an important part of it, but the combination of highly evolved clustering and memory based on clustering together with deep learning still gives the brain capabilities far beyond any of the simpler forms of deep learning available today.

This book also addresses many of the challenges in nonparametric statistics, which pioneers like Mosteller and Tukey explained to the world in the 1970s, which go far beyond the simplified versions of it which were popularized by Vapnik and his followers in recent years. These, too, have wide-ranging applications. There are important things to build on starting from this book.

Arlington, VA, USA

Paul J. Werbos
Inventor of the back-propagation method
Pioneer of neural networks research

Jose Principe

Our colleagues working in statistical data analysis see the world through the lens of a model. Statistical modeling has many advantages mostly when data sets are small, but in the current age of big data, data-driven alternatives have a role to play, i.e., data should speak by itself. One knows from information theory that the only way to improve information is to perform an experiment to collect new data, so one can attempt to compensate modeling by large data sets, and in the process simplify the algorithms and the mathematics required for data analytics.

This book addresses data analysis from an empirical perspective, i.e., it seeks to quantify data structure by exploring local to global properties of the sample set. The only mathematical requirement is the existence of a metric space where the data exists, i.e., it can be applied to both deterministic and stochastic data sources. Because of this viewpoint, only nonparametric tests of performance should be applied and optimality is hard to quantify. However, as an advantage, there is no need for a priori statistical assumptions of data distributions nor to quantify the properties of estimators that is a difficult and often forgotten step in statistical modeling.

The book is divided into three parts. Part I have three chapters and explain the basic approach and surveys at an introductory level the conventional pattern recognition and fuzzy analytics' tools that will be used as a comparison for the empirical approach. Part II explains in detail the empirical approach and the algorithms to quantify local structure from data. Chapter 4 starts to review the concept of distance and different metrics, and then presents descriptors for data concentration (or density), eccentricity, typicality, which lead to the concept of a data cloud, and their multi-modal extensions. In order to address inference for new data sets, these descriptors are extended to their continuous versions. Chapter 5 addresses the application of these empirical descriptors to substitute the mathematically based membership function in fuzzy sets. Chapters 6–9 propose architectures to solve practical problems using the descriptors outlined in Chaps. 4 and 5, exploiting the advantages of the empirical approach, which requires fewer a priori parameters that have to be supplied by the designers' experience. They have been named autonomous because of this feature. Chapter 6 addresses the important practical problem of autonomous anomaly detection. It starts with a brief description of the different types of anomalies and the different classes of methods (unsupervised or supervised techniques), and then shows how the empirical descriptors of centrality and eccentricity can be used to create effective anomaly detectors. Chapter 7 develops an autonomous data partition methodology that does not require the a priori selection of number of clusters, using the empirical estimators of multi-modal data density and eccentricity. Chapter 8 aims at an autonomous learning of multi-models, which the authors call ALMMo, which is an empiric version of the neuro-fuzzy approach. This chapter only covers the type 0 (data partition based) and the type 1 (linear based modeling), and shows how it is possible to learn directly from data multiple models that exploit the local properties of the data structures. Finally, Chapter 9 introduces deep architectures based on the ALMMo framework, but relying on the peaks of the data clouds, which are called prototypes. This sparsification is efficient and also preserves the most information about the local data structures. The authors call this methodology transparent rule based classifiers, which has a role to play in medical diagnostics and data analytics. Part III of the book comprises six chapters and presents comparative tests on autonomous anomaly detection, autonomous data partitioning, autonomous learning multiple models, deep rule based classification, and semi-supervised classification, followed by an epilogue. The importance of Part III is that the reader can see how the methodology is applied in practical problems, and pseudocodes (and Matlab calls) are presented in the appendices to facilitate porting the ideas for user-specific applications. As demonstrated in these chapters, the empirical data analysis developed in the book solves efficiently problems and the solutions are computational lighter than alternate statistical methods, while the performance is on par with them. Therefore, the book is a welcome addition to the literature on data analytics presenting a complementary view for the design of practical data processing algorithms. I also hope that this book will be useful for young researchers to further develop the methodologies. There are many interesting avenues of research that can be pursued from the presented set of core results.

I hope you enjoy the reading as much as I have.

Gainesville, FL, USA

Jose Principe
Distinguished Professor
Eckis Professor
Director of Computational NeuroEngineering Lab
Electrical and Computer Engineering Department
University of Florida

Edward Tunstel

The lead author and I have been active for years in the intelligent systems related to technical activities of the IEEE. Over the years, I have become increasingly aware of and interested in his work on evolving intelligent systems and the realization of systems that can learn autonomously from streaming data. His work has a persistent presence at the cutting edge of these technologies, and this book is the latest in a productive and steady stream of discovery from the research he and colleagues have done. While his work has relevance to my work in intelligent robotic systems, what I value most is his typical broad treatment expressing how the technology can be applied to many different problems. This book continues that trend.

As my colleagues and I have sought to develop and apply computational techniques that enable intelligent systems solutions to real-world problems, a common occurrence has been encountered with practical limitations of the techniques. Where limitations may have been of low consequence, we have often had to reluctantly adopt inaccurate or unrealistic assumptions about the problem in order to get the technology to work as algorithmically designed. Working as designed is not always equivalent to working as needed for a given problem. So, it is refreshing to learn of techniques that do away with or substantially relax constraints on problem formulation and accommodate the flexibility warranted by real-world problems. This book does just that while offering a fresh viewpoint on the latest machine learning methods.

From a perspective of keen insight and understanding of what limits current state of the art, the book delivers a new empirical approach to realizing intelligent machines and systems with greater capacity for learning and with greater potential for more effective operation in the real-world. With an elevated approach to realizing smarter devices, machines, and more autonomous systems, empirical machine learning forges a path beyond limitations that are stalling prior approaches. At a time when advances seem to be occurring at an increasingly rapid pace, new manuscripts that build upon cognizance of what limits the latest technology are most valuable. This work is timely in its delivery of new ideas needed to advance machine learning to the next level, particularly in its relationship to data.

The authors present the material in a systematic manner in which researchers who are new to the area and those who are current practitioners can both appreciate. Setting the stage for the path that the new empirical approach trail blazes, the book opens with discussion of its visionary transformation in thinking relative to recent prior approaches. Foundational material is reviewed in early chapters, instructive to new researchers and a familiar refresher to more seasoned researchers. Subsequent chapters reveal the manner in which empirical machine learning extends flexibility with the more powerful capacity of continuously learning. The latter chapters reveal the breadth of applicability across a range of problems, bringing increased autonomy to machine learning and deep classification solutions. Those application chapters serve to expose only the tip of the iceberg of the true potential of empirical machine learning. This leaves plenty of room for readers to appreciate that potential and reap the benefits through application to their real-world problems.

Empirical Approach to Machine Learning provides an insightful and visionary boost of progress in the evolution of computational learning capabilities yielding interpretable and transparent implementations. I am anxious to consider how I can apply the new approach to develop capabilities such as autonomous, anytime, and lifelong learning for intelligent robots operating in unstructured field and service environments on Earth as well as unknown environments on other planets. Truly exciting!

Canton, CT, USA

Edward Tunstel
President of IEEE Systems, Man, Cybernetics Society

Vladik Kreinovich

We are all fascinated by the recent successes of deep learning. Deep learning has performed wonders in image processing, in natural language processing, in bioinformatics, and in many other application areas. Deep learning has fulfilled a long-time dream of Artificial Intelligence community: to design a computer program that excels in Go. These successes are well known, but it is also well known that sometimes deep learning makes mistakes. For example, small changes in a picture—changes invisible to a human eye—can lead to a gross misclassification, like interpreting a coffee machine for a cobra. Because of these mistakes, we cannot always trust the results of these exciting machine learning algorithms.

The problem is not so much the complexity of these algorithms: we often trust the results of very complex algorithms. For example, we trust the planes when they are controlled by very complex algorithm implementing the autopilot, and we trust the X-ray and MRI machines where algorithms are also very complex. We trust them because in all these situations, while the algorithms are complex, we—or at least experts whom we trust—have an intuitive understanding of these algorithms.

Not only we have the results of the complex algorithms, we also have a clear qualitative explanation of these results.

In contrast, many modern machine learning algorithms are, to a large extent, black boxes—they provide predictions and recommendations, but they do not explain in clear terms why. This is a known serious problem of machine learning. Many researchers are working on it, and some progress has been achieved—but the challenge largely remains. This book describes a new innovative approach to making machine learning intuitively interpretable—an approach whose success has been proved by many successful applications. To fully understand this approach, one needs to read this very interesting book.

What do we mean by intuitive explanation? In a nutshell, it means that, a simple IF…THEN set of rules using prototypes (actual data, which may be images or else) is automatically derived. Thus, we have simple commonsense explanations described by using words from natural language such as IF… is like…OR… is like…THEN.

To solve this new task, Dr. Angelov comes up with a systematic way to generate powerful models (classifiers, predictors, anomaly detectors, even controllers) which can be expressed through fuzzy sets based on data only. Expert knowledge can optionally be used (if available and convenient) and, equally, experts can interrogate and analyze the automatically extracted from data models.

Summarizing: this book is an important step in Computational Intelligence. It provides a new way to make machine learning results interpretable, and it also provides a new innovative application of fuzzy techniques. This book will definitely inspire new ideas, new techniques, and new applications.

El Paso, TX, USA

Vladik Kreinovich
Vice President of International Fuzzy Systems Association
Professor of Computer Science
University of Texas at El Paso

Preface

This book is a product of focused research over the last several years and brings in one place the fundamentals of a new methodological approach to machine learning that is centered entirely at the actual data. We call this approach "*empirical*" to distinguish it from the traditional approach that is heavily restricted and driven by the *prior* assumptions about data distribution and data generation model. This new approach is not only liberating (no need to make *prior* assumptions about the type of data distribution, amount of the data and even their nature—random or deterministic), but it is also fundamentally different—it places the mutual position of each data sample in the data space at the center of the analysis. It is also closely related to the concepts of *data density* and has close resemblance to *centrality* (known from the network theory) and *inverse square distance* rule/law (known form physics/astronomy).

Furthermore, this approach has anthropomorphic characteristics. For example, unlike the vast majority of the existing machine learning methods which require a large amount of training data the proposed approach allows to learn from a handful or even a single example, that is to start "from scratch" and also to learn continuously even after training/deployment. That is, the machine can learn lifelong or continuously without or with very little human intervention or supervision. Critically, the proposed approach is not "*black box*" unlike many of the existing competitors, e.g., most of the neural networks (NN), the celebrated *deep learning*, etc. On the contrary, it is fully interpretable, transparent, has a clear and logical internal model structure, can carry semantic meaning and is, thus, much more human-like.

Traditional machine learning is statistical and is based on the classical probability theory which guarantees, due to its solid mathematical foundation, the properties of these learning algorithms when the amount of the data tends to infinity and all the data come from the same distribution. Nonetheless, the presumed random nature and same distribution imposed on the data generation model are too strong and impractical to be held true in real situations. In addition, the predefined parameters of machine learning algorithms usually require a certain amount of *prior* knowledge of the problem, which, in reality, is often unavailable. Thus, these parameters are impossible to be correctly defined in real applications, and the performance of the algorithms can be largely influenced by the improper choice. Importantly, even though the newly proposed concept is centered at experimental data, it leads to a theoretically sound closed-form model of the data distribution and has theoretically proven convergence (in the mean), stability, and local optimality.

Despite the seeming similarity of the end result to the traditional approach, this data distribution is extracted from the data and not assumed *a priori*. This new quantity that represents the likelihood also integrates to *1* and is represented as a continuous function; however, unlike the traditional pdf (probability density function), it does not suffer from obvious paradoxes. We call this new quantity "*typicality*". We also introduce another new quantity, called "*eccentricity*" which is inverse of the *data density* and is very convenient for the analysis of the anomalies/outliers/faults simplifying the Chebyshev inequality expression and analysis. *Eccentricity* is a new measure of the tail of the distribution, which is introduced for the first time by the lead author in his recent previous works.

Based on the *typicality* (used instead of the probability density function) and *eccentricity* as new measures derived directly and entirely from the experimental data, we develop a new methodological basis for data analysis called *empirical*. We also redefine and simplify the fuzzy sets and systems definition. Traditionally, fuzzy sets are defined through their membership functions. This is often a problem because to define a suitable membership function may not be easy or convenient and is based on *prior* assumptions and approximations. Instead, we propose to only select *prototypes*. These *prototypes* can be actual data samples selected autonomously due to their high descriptive/representative power (having high local *typicality* and density) or pointed by an expert (for the fuzzy sets the role of the human expert and user is desirable and natural). Even when these prototypes are identified by the human expert and not autonomously from the experimental data, the benefit is significant because the cumbersome, possibly prohibitive and potentially controversial problem of defining potentially a huge number of membership functions, can be circumvented and tremendously simplified.

Based on the new, *empirical* (based on the actual/real data and not on the assumptions made about the data generation model) methodology, we further analyze and redefine the main elements of the machine learning/pattern recognition/data mining, deep learning as well as anomaly detection, fault detection and identification. We start with the data pre-processing and anomaly detection. This problem is the basis of multiple and various applications to fault detection in engineering systems, intruder and insider detection in cybersecurity systems, outlier detection in data mining, etc. *Eccentricity* is a new, more convenient form for analysis of such properties of the data. *Data density* and, especially, its recursive form of update, which we call RDE (recursive density estimation), makes the analysis of anomalies very convenient, as it will be illustrated in the book.

We further introduce a new method for fully autonomous (and, thus, not based on handcrafting, selecting thresholds, parameters, and coefficients by the user or "tailoring" these to the problem) data partitioning. In essence, this is a new method for clustering, which is, however, fully data-driven. It combines rank-ordering (in terms of *data density*) with the distance between any point and the point with the maximum *typicality*. We also introduce a number of autonomous clustering methods (online, evolving, taking into account local anomalies, etc.) and compare these with the currently existing alternatives. In this sense, this book builds upon the previous research monograph by the lead author entitled *Autonomous Learning Systems: From Data Streams to Knowledge in Real time, Willey, 2012, ISBN 978-1-119-95152-0.*

We then move to supervised learning starting with the classifiers. We focus on fuzzy rule-based (FRB) systems as classifiers, but it is important to stress that since FRBs and artificial neural networks (ANN) were demonstrated to be dual (the term neuro-fuzzy is widely used to indicate their close relation), everything presented in this book can also be interpreted as NNs. Using FRB and, respectively, ANNs as classifiers is not a new concept. In this book, we introduce the interpretable deep rule-based (DRB) classifiers as a new powerful form of machine learning, specifically effective for image classification, which has anthropomorphic characteristics as described earlier. The importance of DRB is multifaceted. It concerns not only the efficiency (very low training time, low computing resources required—no graphic processing units (GPU), for example), high precision (classification rate) competing, and surpassing the best published results and the human abilities, but also high interpretability/transparency, repeatability, proven convergence, optimality, non-parametric, non-iterative nature, and self-evolving capability. This new method is compared thoroughly with the best existing alternatives. It can start learning from the very first image presented (very much like humans can). The DRB method can be considered as neuro-fuzzy. We pioneer the deep FRB as highly parallel multi-layer classifiers which offer the high interpretability/transparency typical for the FRB. Indeed, up until now the so-called *deep learning* method proved its efficiency and high potential as a type of artificial/computational NN, but it was not combined with fuzzy rules to benefit from their semantic clarity.

Another important supervised learning constructs are the predictive models, which can be of regression or time series type. These are traditionally being approached in the same way—starting with the *prior* assumptions about inputs/features, cause-effect relations, data generation model, and density distributions and the actual experimental data are only used to confirm or correct these assumptions made *a priori*. The proposed *empirical* approach, on the contrary, starts with the data and their mutual position in the data space and extracts all internal dependencies in a convenient form from these. It self-evolves from data complex non-linear predictive models. These can be interpreted as IF…THEN FRB of a particular type, called *AnYa* or, equally, as self-evolving computational ANN. In this sense, this book builds upon the first research monograph by the lead author entitled *Evolving Rule-based Models: A Tool for Design of Flexible Adaptive Systems, Springer, 2002, ISBN 978-3-7908-1794-2.*

In this book, we use the fully autonomous data partitioning (ADP) method introduced in earlier chapters to form the model structure (the premise/IF part). These are the local peaks (modes) of the multi-modal (mountain-like) *typicality* distribution, which is automatically extracted from the actual/observable data. In this book, we offer locally optimal method for ADP (satisfying Karush-Kuhn-Tucker conditions). The consequent/THEN part of the self-evolving FRB based predictive models is linear and fuzzily weighted. In this book, we provide theoretical proof of the convergence of the error (in the mean) using Lyapunov functions. In this way, this presents the first FRB with self-evolving nature with theoretically proven convergence (in the mean) in training (including online, during the use), stability as well as local optimality of the premise (IF) part of the model structure. The properties of local optimality, convergence, and stability are illustrated on a set of benchmark experimental data sets and streams.

Last, but not least, the authors would like to express their gratitude for the close collaboration on some aspects of this new concept with Prof. Jose Principe (University of Florida, USA, in the framework of The Royal society grant "*Novel Machine Learning Methods for Big Data Streams*"), Dr. Dmitry Kangin (former Ph.D. student at Lancaster University with the lead author, currently Postdoc Researcher at Exeter University, UK), Dr. Bruno Sielly Jales Costa (visiting Ph.D. student at Lancaster University with the lead author, now with Ford R&D, Palo Alto, USA), Dr. Dimitar Filev (former PhD advisor of the lead author in early 1990s, now Henry Ford Technical Fellow at Ford R&D, Dearborn, MI, USA), and Prof. Ronald Yager (Iona College, NY, USA), Dr. Hai-Jun Rong (visiting scholar at Lancaster University with the lead author, Associate Professor at Xi'an Jiaotong University, Xi'an, China).

Lancaster, UK
November 2017–August 2018

Plamen P. Angelov
Xiaowei Gu

Contents

Glossary

Abbreviations

AAD	Autonomous anomaly detection algorithm
ADP	Autonomous data partitioning algorithm
AI	Artificial intelligence
ALMMo	Autonomous learning multi-model system
ALMMo-0	Zero-order autonomous learning multi-model system
ALMMo-1	First-order autonomous learning multi-model system
ANFIS	Adaptive-network-based fuzzy inference system
ANN	Artificial neural network
ASR	Adaptive sparse representation algorithm
BOVW	Bag of visual words algorithm
BPTT	Backpropagation through time
CAFFE	Convolutional architecture for fast feature embedding model
CBDN	Convolutional deep belief network
CDF	Cumulative distribution function
CENFS	Correntropy-based evolving fuzzy neural system
CH	Color histogram feature
CLT	Central Limit Theorem
CNN	Convolution neural network
CSAE	Convolutional sparse autoencoders
DBSCAN	Density-based spatial clustering of applications with noise
DCNN	Deep convolutional neural network
DENFIS	Dynamic evolving neural-fuzzy inference system
DL	Deep learning
DLNN	Deep learning neural network
DRB	Deep rule-based system
DRNN	Deep recurrent neural network
DT	Decision tree classifier
EBP	Error back-propagation algorithm

ECM	Evolving clustering method
EFS	Evolving fuzzy system
EFuNN	Evolving fuzzy neural network
ELM	Evolving local means algorithm
EM	Expectation maximization algorithm
eTS	Evolving Takagi-Sugeno type fuzzy rule-based system
FCM	Fuzzy c-means algorithm
FCMM	Fuzzily connected multi-model system
FD	Fault detection
FDI	Fault detection and isolation
FI	Fault isolation
FLEXFIS	Flexible fuzzy inference system
FRB	Fuzzy rule-based system
FWRLS	Fuzzily weighted recursive least squares algorithm
GENFIS	Generic evolving neuro-fuzzy inference system
GGMC	Greedy gradient Max-Cut based semi-supervised classifier
GMM	Gaussian mixture model
GPU	Graphic processing unit
HFT	High frequency trading
HOG	Histogram of oriented gradient feature
HPC	High performance computing
IBeGS	Interval-based evolving granular system
iid	Independent and identically distributed
KDE	Kernel density estimation
KL	Kullback–Leibler divergence
kNN	k-nearest neighbor algorithm
KPCA	Kernel principle component analysis
LapSVM	Laplacian support vector machine
LCFH	Learning convolutional feature hierarchies algorithm
LCLC	Local-constrained linear coding algorithm
LGC	Local and global consistency based semi-supervised classifier
LMS	Least mean squares algorithm
LS	Least square algorithm
LSLR	Least square linear regression algorithm
LSPM	Linear spatial pyramid matching algorithm
LSTM	Long short term memory
LVQ	Learning vector quantization
MLP	Multi-layer perceptron
MNIST	Modified National Institute of Standards and Technology
MS	Mean-shift algorithm
MUFL	Multipath unsupervised feature learning algorithm
NDEI	Non-dimensional error index
NFS	Neuro-fuzzy system
NLP	Natural language processing
NMI	Nonparametric mode identification algorithm

NMM	Nonparametric mixture model algorithm
NN	Neural network
ODRW	Outlier detection using random walks
PANFIS	Parsimonious network based fuzzy inference system
PC	Principal component
PCA	Principle component analysis
PDF	Probability density function
PMF	Probability mass function
RBF	Radial basis function
RBM	Restricted Boltzmann machine
RCN	Random convolutional network
RDE	Recursive density estimation
RLS	Recursive least squares
RNN	Recurrent neural network
$SDAL_{21}M$	Sparse discriminant analysis via joint $L_{2,1}$-norm minimization algorithm
SFC	Sparse fingerprint classification algorithm
SIFT	Scale-invariant feature transform
SIFTSC	Scale-invariant feature transform with sparse coding transform algorithm
SOM	Self-organizing map
SPM	Spatial pyramid matching algorithm
SPMK	Spatial pyramid matching kernel algorithm
SS_DRB	Semi-supervised deep rule-based system
SURF	Speeded up robust feature
SV	Support vector
SVD	Singular-value decomposition
SVM	Support vector machine
TEDA	Typicality and eccentricity-based data analytics
TLFP	Two-level feature representation algorithm
VGG-VD	Visual geometry group-very deep model

Synonyms

- Data Point/Data Sample/Data Vector
- Data Space/Feature Space
- Feature/Attribute
- Distance/Proximity/Dissimilarity
- Activation Level/Degree of Closeness/Degree of Membership/Degree of Confidence/Firing Strength/Score of Confidence

Website

http://angeloventelsensys.wixsite.com/plamenangelov/software-downloads

This book is accompanied by the above website. This website provides the source code (in MATLAB) with extensive comments and example problems, lecture notes that can form a standalone postgraduate course or be used in advanced machine learning undergraduate courses, and other supplementary materials of this book.

Chapter 1
Introduction

1.1 Motivation

Today we live in a data-rich environment. This is dramatically different from the last century when the fundamentals of machine learning, control theory and related subjects were established. Nowadays, vast and exponentially increasing data sets and streams which are often non-linear, non-stationary and increasingly more multi-modal/heterogeneous (combining various physical variables, signals with images/videos as well as text) are being generated, transmitted and recorded as a result of our everyday live. This is drastically different from the reality when the fundamental results of the probability theory, statistics and statistical learning where developed few centuries ago. The famous Moore's law [1] now applies not to the hardware capacity but to the data. The obvious reasons mentioned in the previous sentence are related to the miniaturization of semiconductor chips which cannot go infinite using the same physical principles because now the size of transistors reaches nanoscale comparable to molecules and atoms. So, now we can say that we observe a new law: the law of exponential growth of the amount and complexity of the data (streams). Indeed, each one of us carries literally in our pockets gigabytes of data in the form of flash memory USB sticks, smartphones and other gadgets. We are also often involved (especially, the younger generation) in consuming and/or producing vast amounts of data through video, images, text as well as increasingly also signals through smart devices (such as watches, positioning systems in cars and smartphones, etc.).

This reality calls for new approaches, it brings some specific new challenges, but also offers new opportunities:

- **Increase the autonomy** of the whole process (reduce the human involvement in selecting *prior* model structures, assumptions, selection of problem and user-specific parameters to the bare minimum or remove it all together wherever possible);

P. P. Angelov and X. Gu, *Empirical Approach to Machine Learning*, Studies in Computational Intelligence 800,
https://doi.org/10.1007/978-3-030-02384-3_1

- Take into account the **dynamically evolving nature** of data streams, non-stationarity and non-linearity of most of the real problems;
- Be able to learn from a single data sample/example or from a handful (few) examples—that means, to **learn "from scratch"**;
- To be able to learn continuously; even after training and deployment; that is, **lifelong learning, adapting and improving** to the possibly new situations, environment, data;
- To be transparent, interpretable, understandable by a human (**not a "black box"**); ideally, it will also have anthropomorphic characteristics; that is, it will **learn** more **like a human** and less like a machine;
- To **have internal hierarchical representation**, which can be depicted by a pyramidal structure. For example, as presented in Fig. 1.1 this pyramid may have at its bottom layer potentially a huge amount of raw data which is prohibitive to process with human involvement. Moving up to the middle layer, aggregated, more compressed and more representative, possibly semantically meaningful groupings (which we call "*data clouds*" and differ somewhat form clusters) represented by *prototypes* and statistical parameters extracted from the data and not injected by the user. At the top layer, this pyramid can further link these *data clouds* into a relatively small (manageable) amount of highly compressed and meaningful rules or relations that can be used by the users to make predictions or classifications.
- To be computationally (and algorithmically) efficient – not to require excessive computational hardware, e.g. graphic processing unit (GPU), high performance computing (HPC), but to be able to run on normal computers, laptops or even portable boards such as Raspberry PI [2], Beaglebone [3], Movidius [4], etc.
- Last but not least, of course, to have higher or comparable performance than any other existing algorithm for the same type of problems in terms of classification accuracy, prediction error and other established metrics of performance.

Such "wish list" is extremely tough and an obvious challenge. In this book, to satisfy one possible answer to it, our conviction is that the future development of

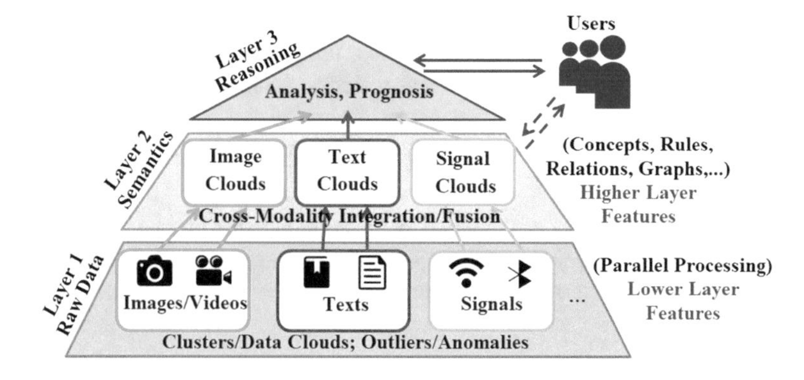

Fig. 1.1 The pyramid of processing the raw data into human-intelligible high level information

machine learning and computational intelligence research will address these challenges and lead to systems with the above mentioned characteristics.

1.2 From Handcrafting to Autonomy

One practical problem is related to the difficulties of the requirement to label all the training data—very typical for the so-called supervised machine learning. Imagine labeling manually hundreds of billions of images on Internet or even the relatively smaller amount of many millions of images from a security related application, etc. The best state-of-the-art at present, the so-called *deep learning* neural networks (DLNN) or another popular approach, support vector machines (SVM), require huge amount of labeled training data, usually operate in offline mode and require many decisions to be made by the user, including selecting thresholds, algorithmic parameters and other *ad hoc* choices. Contrary to the widely used claim that *deep learning* is solving successfully the issue of "*handcrafting*", which is typical for traditional machine learning and, especially, in image processing (the need to select features, many thresholds, parameters, etc.) [5–8], it only replaces one type of "*handcrafting*" with another. It is true that traditional *deep learning* does not require specific features to be extracted from images and its coefficients of the last fully connected layer can be used as such features [9]. We will use this as one of the alternatives in this book. However, it is also true that the structure of mainstream DLNNs such as VGG-VD (visual geometry group-very deep) models [9], AlexNet [10], GoogLeNet [11], etc., involves a lot of ad hoc and opaque decisions about the number of layers, size of windows, etc. Therefore, to claim that the problem of *handcrafting* has been solved by the use of the state-of-the-art *deep learning* is simply not correct. In reality, in traditional machine learning, including in the *deep learning* there are many more decisions, parameters, thresholds, etc. made by humans. For example, the split of the data into training and validation, the assumptions about the data generation model (e.g. most often Gaussian), the number of clusters or Gaussians in a mixture of Gaussians, etc.

Following the "wish list" described in the previous section, the proposed in this book approach will aim to reduce or fully eliminate the subjective influence of the human by using only data-derived statistical parameters such as means, standard deviation and alike (sum squared products, for example) and no user- or problem-specific thresholds, parameters or decisions. The only inputs the proposed algorithms need are the raw data and an indication of what is the expected outcome—a class label (classification), predicted value (prediction), detecting possible anomalies (anomaly detection, fault detection), autonomous data partitioning into *data clouds*. It will then autonomously perform a number of specific steps that will be described in more detail in the next chapters to produce (in most cases, in *real time* and online) this expected output. It may also require some very limited (e.g. concerning as little as 5–10% of all the data) labels or correct predicted values. It will guarantee theoretically the local optimality, convergence (in the mean) and stability. In this way, the role of the human will be very limited. It will allow high level analysis of a very small amount of rules or relations, selection of the type of the problem, providing the raw data and in some problems a very limited amount of correct answers (ground truth). However, even this small amount of labeled data/true outputs may come from automated systems and not manually from the human user. This means that the level of autonomy of the proposed approach is very high. This offers very important advantages of high throughput, productivity and in some cases may be critically important to avoid putting the human in danger. Furthermore, this approach can be used to automatically extract human-intelligible, meaningful model of the data, get insight, and understand better a practical problem from observations (thus, we call this *empirical* approach).

1.3 From Assumptions-Centered to Data-Centered Approach

The approach that we propose is centered at the experimental data and, thus, called "*empirical*". The traditional approach in the vast majority of machine learning and pattern recognition problems as well as in anomaly detection starts form a set of explicit and implicit assumptions, which concern:

- The features or model inputs—very often these are measured physical variables such as temperatures, pressures, etc. but, especially, in image processing or natural language processing (NLP) these can be of quite different nature. For example, in image processing these are often local (pixel-based) or global (spectral) quantifiers [12]. They can also be based on results of so-called segmentation and have more human interpretable nature such as size, shape of physical objects represented on the image. These are, however, often quite cumbersome to get automatically; in NLP features are often related to keywords. These assumptions are usually made explicitly;
- Data generation model (usually, they assume Gaussian distribution or a mixture of Gaussians); this assumption is often made explicitly, but is often not confirmed by the actual data distribution, see Fig. 1.2;
- Type of distance metric that will be used. In some classes of problems these are dictated by the nature of the problem. For example, in NLP they use more often cosine-related dissimilarity or divergence measures rather than Euclidean or Mahalanobis type of distance metrics; on the contrary, when dealing with signals, physical variables it is more natural to use Euclidean (due to its simplicity) or Mahalanobis type of distance metric (due to its ability to represent the shape of the data clusters much better). In image processing they more often use also Euclidean or Mahalanobis type distance metrics. This assumption is often made implicitly.
- Assumption of the specific type of model/classifier/predictor/anomaly detection method to be used. For example, why for a specific data set or stream one decides to use SVM, DLNNs, decision tree or any other classifier; same for the

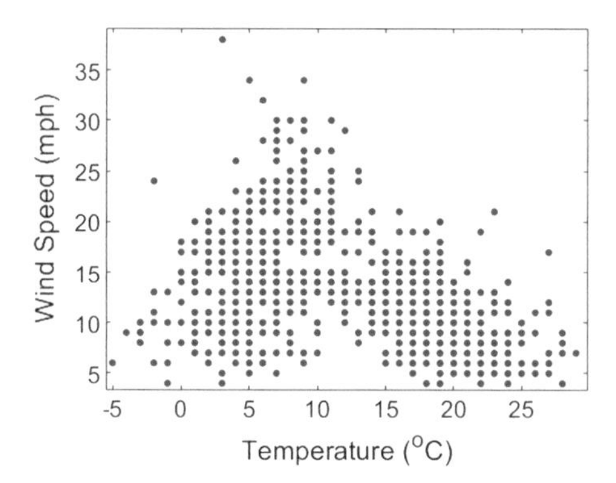

Fig. 1.2 Real climate data distribution (temperature and wind speed) measured in Manchester, UK for the period 2010–2015 (downloadable from [13])

predictive models, anomaly detection, etc. This assumption is often made implicitly and with no firm or clear arguments; very often there is also a bias due to the (lack of) subjective knowledge, publicity or temporal "popularity". However, based on different assumptions, one may get entirely different results, i.e. it is very well known from the history how the firm belief in the geocentric theory led to wrong conclusions if compared with the true, heliocentric theory as illustrated in Fig. 1.3.

- Usually the algorithms that ate being used require a number of user- or problem-dependent parameters, thresholds. For example, in k-means clustering approach [14] one has to select the number "*k*"; In many other clustering approaches one has to select a threshold which does have a direct effect on the end result. User-dependent choice means that one user may select one value while another may select a different one for the same data set. Similarly, problem-dependent means that for one problem the value of this parameter or threshold may differ from the value for another problem. Having algorithmic parameters or hidden thresholds which are data-specific and not user- or problem-specific means that they are defined as, for example, 2σ where σ denotes the standard deviation. Another example may be the maximum distance between all prototypes. It is obvious that these parameters or hidden thresholds are not subjective, user-defined and they are entirely dictated by the actual data; that is, "*empirical*". Having very few such parameters or thresholds does not preclude the algorithm to be autonomous and act without direct human intervention.

In summary, the proposed in this book new (*empirical*) approach to machine learning starts from the data and only then moves to extract from it the data generation model, data-defined statistics and parameters while the traditional approach starts form assumptions and uses the actual data only to check the validity of these hypotheses, see Fig. 1.4.

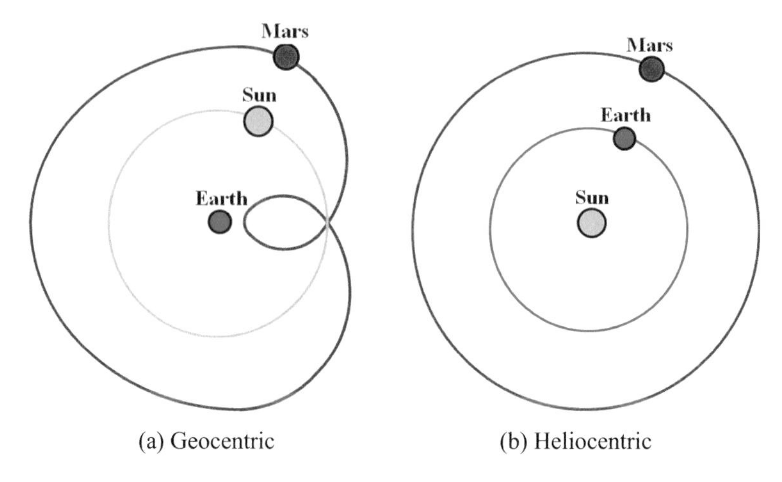

(a) Geocentric (b) Heliocentric

Fig. 1.3 Geocentric concept versus heliocentric concept

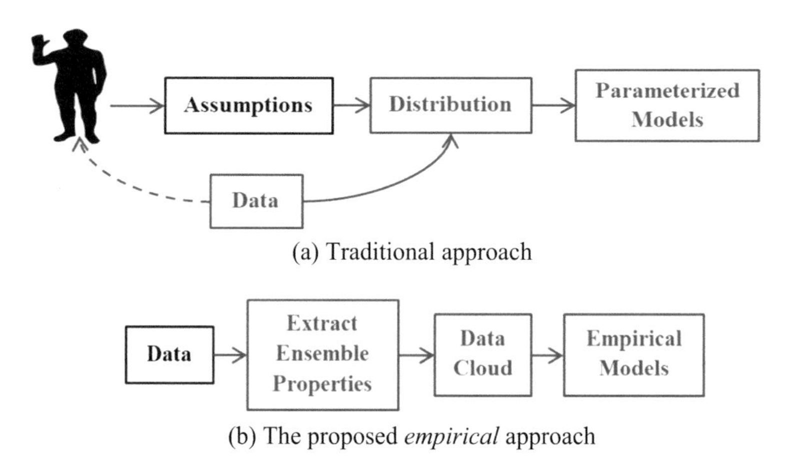

(a) Traditional approach

(b) The proposed *empirical* approach

Fig. 1.4 The proposed *empirical* approach versus the traditional approach (start from data or from assumptions)

1.4 From Offline to Online and Dynamically Evolving

As it was pointed out in Sect. 1.1, the reality we live in requires computers to be able to learn while operating and not to have a separate one-off training phase followed by an exploitation phase. In the past (and still even now) many machine learning applications do work in such offline mode when the training and design phase is one off and precedes the actual use. The challenge is to offer methods and algorithms that allow not only online learning (training while being operational), but also to be able to learn continuously, lifelong [15]. Further, the model (classifier, predictor) is usually assumed to be with a fixed structure. However, the data pattern is often non-stationary and adapting only the parameters of the model (which is what the traditional adaptive control approach offers) is not enough. It has been demonstrated that a better performance can be achieved by dynamically evolving the internal structure of the model [16, 17].

The terms "online" and "real-time" are often considered as synonymous. However, there is a subtle difference (Fig. 1.5). *Real-time* means that the frequency of update of the model is higher than the sampling frequency. For very slow processes such as wastewater treatment plants, for example sampling frequency may be several hours. In such case, any algorithm that provides an update within this time frame can be considered to work in real-time. However, it may be offline and iterative. On the contrary, there are examples of online algorithms which are, however, not real-time because the sampling frequency is very high, e.g. in high frequency trading (HFT), speech or video processing, etc.

Models designed in offline mode can have a very good performance for validation data that have similar data distribution as the training data. However, their performance dramatically worsens in cases of so-called data *shifts* and/or *drifts* [18]. In such cases, models designed offline require a recalibration or a complete re-design which may be costly not only form the point of view of model design costs, but also of the possible related industry stoppages costs [19]. In real problems, this is often the case to have significant changes due to, for example, raw material quality variations (in oil refining), illumination and lighting conditions in image processing, etc.

Another very important characteristic of an algorithm and method is whether it can work in the so-called "*one pass*" mode. This means that a data sample once taken and processed by the algorithm is then thrown away and is not stored in the

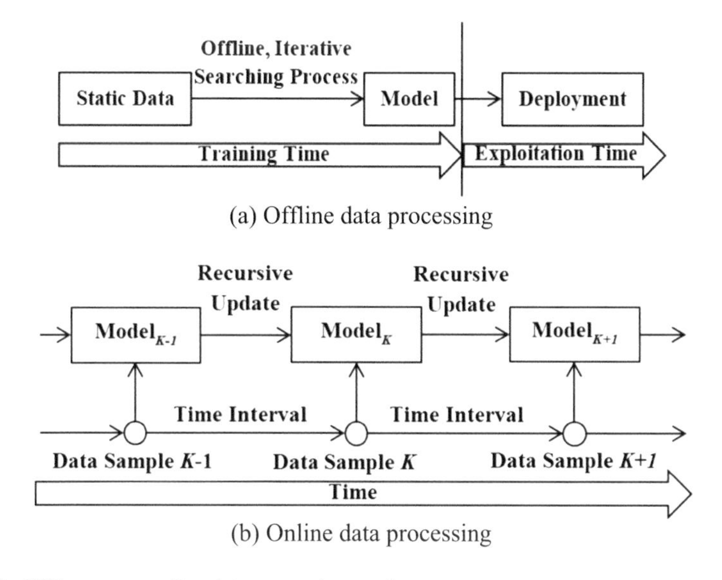

(a) Offline data processing

(b) Online data processing

Fig. 1.5 Offline versus online data processing modes

memory and not used anymore. This is a very demanding form of processing the data, but very efficient one. It requires the algorithm to rely only on the current data sample and possibly some statistical aggregations, model parameters (such as cluster centers, prototypes, etc.), but not to use a (possibly, sliding) window of data samples or so-called "data chunks".

This is very efficient mode of operation, because it is lean in terms of using memory of the machine, computational resources (search, storage) and can work lifelong. The difference is very significant. For example, in video processing where the "one pass" algorithms only process the current image frame and do not store this or past frames in the memory. Instead, it stores in the memory a small amount of aggregations, prototypes and model parameters [20].

The ability an algorithm to be "one pass" is more critical for data streams and not so important for datasets (static data). It strongly depends on it being non-iterative and recursive [16]. The reason is simple, if an algorithm involves an iterative search procedure there is no guarantee that it will get the result before the next data sample arrives. Having a recursive algorithm guarantees the ability to have "one pass" algorithm, because recursive algorithms update their parameters at the current time step using only the values of these parameters at the previous time step and some simple arithmetic operations such as summation, multiplication by numerical coefficients. Practically all "one pass" algorithms have so-called order-dependency property. This means that the data stream if re-shuffled (the data being provided in a different order), there is no guarantee that the result will be the same.

The methods and algorithms proposed in this book are in their majority *one-pass*, *online* and *dynamically evolving* unless specified otherwise.

1.5 Interpretability, Transparency, Anthropomorphic Characteristics

Traditional machine learning models have internal structures, which are largely pre-determined by the assumptions made at the start of the process as described in Sect. 1.1. For example, they strongly depend on the assumed *a priori* data generation model.

The interpretability of the traditional statistical machine learning models is typically lower than that of the so-called "first principles" models which themselves are often based on differential equations representing mass, energy balance or approximations of the laws of physics. However, the level of their interpretability is typically higher than that of some computational intelligence models such as artificial neural networks (ANNs). Their level of interpretability and transparency is somewhat close to the level of transparency and interpretability of the fuzzy rule-based (FRB) systems. Indeed, there are some parallels and dualism between FRB [21] and mixture Gaussian probabilistic or Bayesian models [22].

Currently, the most popular and successful classifiers, for example, SVM [23] and DLNNs [24, 25] like convolutional neural networks (CNNs) [10] or recurrent neural networks (RNNs) [26], etc., have very low level of interpretability. Their internal structure is not informative or interpretable to human users. It does not allow an easy analysis of the alternatives and is not directly related to the problem at hand. For the DLNNs, this problem is especially acute. The DLNNs lack interpretability and transparence. The tens or hundreds of millions of weights they use have values that cannot be linked to the problem. They do not have a clear semantics of the inner model structure, and also do not provide clear reasons for making a certain architectural decision. Even the number of their hidden layers is decided *ad hoc* by the designer. Although, they work well on certain problems, the popular DLNNs are perfect examples of a cumbersome "black box".

Anthropomorphic characteristics of a machine learning method mean "to learn like humans do" [27]. For example, the vast majority of the existing nowadays machine learning methods require a huge amount of training data to be able to work. However, people can recognize an object that they have seen only ones. Take as an example, The famous statue in Rio de Janeiro, Brazil called Christ, the Redeemer; there is a similar one on the Vasco da Gamma bridge in Lisbon, Portugal, though much smaller. We humans once saw one of them can easily appreciate the other one being similar to the first one. There is no need for tens, hundreds of thousands of training examples, lengthy training procedures and computer accelerating devices. Similarly, any Corrida stadium in Spain or Ancient Amphitheatre in Greece, Bulgaria, Italy or elsewhere are easily considered as similar (as they are) without the need to have a large amount of training images and a lengthy computational procedure. These new images are, obviously, not exactly the same as the original ones, but quite similar and we will be able to associate and make the respective reasoning, classification or anomaly detection based on a single or few examples. In summary, the ability to build a model from a single or very few training data samples, to start "from scratch", is a human-like (anthropomorphic) characteristic.

The ability to explain a decision or to have a transparent description of the internal process of making a decision is also an anthropomorphic feature. For example, when people recognize a particular image they are able to articulate why they have made this particular decision. For example, people can say, "This person looks very much like X because of the hair style, eye brows, ears, the nose, etc.". Having prototypes and segments of the image that can be associated with the prototype and, thus, help explain the reason for a certain decision is an anthropomorphic (human-like) manner. The current DLNNs (but also SVM and the other state-of-the-art approaches) cannot offer such transparency.

Furthermore, another important anthropomorphic characteristic is the ability to learn continuously, which the vast majority of the state-of-the-art approaches fail to offer. The only exception is the reinforcement learning method. However, it cannot guarantee convergence.

Humans are able to dynamically evolve their internal representation of the real world (their model of the environment)—some do this better than others, but all are able to learn new information and also to update the previously learned one. The ability to be able to do the same is another anthropomorphic characteristic, which the vast majority of the currently existing machine learning methods and algorithms do not have. Such capability is very important nowadays.

It has been reported that humans organize the data internally (in the brain) in a hierarchical, multilayered manner. It has also been reported that during sleep people perform an intelligent “data compression” and update the previously learned knowledge. This is quite similar to the pyramid we described in Sect. 1.1 and depicted in Fig. 1.1. Having such ability is another anthropomorphic characteristic which the vast majority of the currently existing methods do not possess.

Last, but not least, human brain is remarkably efficient from the point of view of energy consumption (it consumes around 12 W) but coordinates a large number of extremely complex tasks all day long. It does not have the energy or computational resources of HPC or GPUs. The key in having anthropomorphic machine learning algorithms and methods is to learn in a lean, computationally efficient manner. The vast majority of current machine learning algorithms are computationally expensive and cumbersome. The reality we live in requires a shift towards recursive, computationally efficient and lean algorithms.

The proposed and described in this book approach and related algorithms have all of the above anthropomorphic characteristics and this makes them a very interesting subject of further study and application.

1.6 The Principles and Stages of the Proposed Approach

The proposed *empirical* approach in this book has the following stages (Fig. 1.6):

Stage 0 Data pre-processing;
In some problems (e.g. image processing or NLP), it may not be obvious which features to select, but there are many existing methods and accumulated experience we can use, e.g. GIST [5], histogram of oriented gradients (HOG) [6], etc.
The most popular myths related to the data pre-processing can be formulated as

(i) No handcrafting—often used in DLNNs;
(ii) Unsupervised learning does not involve humans.

We will discuss this further in the next chapters.

Stage 1 The proposed approach to self-update the system structure based on *data density* and *typicality*, respectively, is of constructivist (bottom-up rather than the traditional top to bottom) type;

Stage 2 Local models around prototypes/peaks of *typicality/data density*;

Stage 3 (optional) Optimize (simple) locally parameterized models, e.g. linear;

Stage 4 Fuse the local outputs to form the overall decision (being classification, prediction, or else).

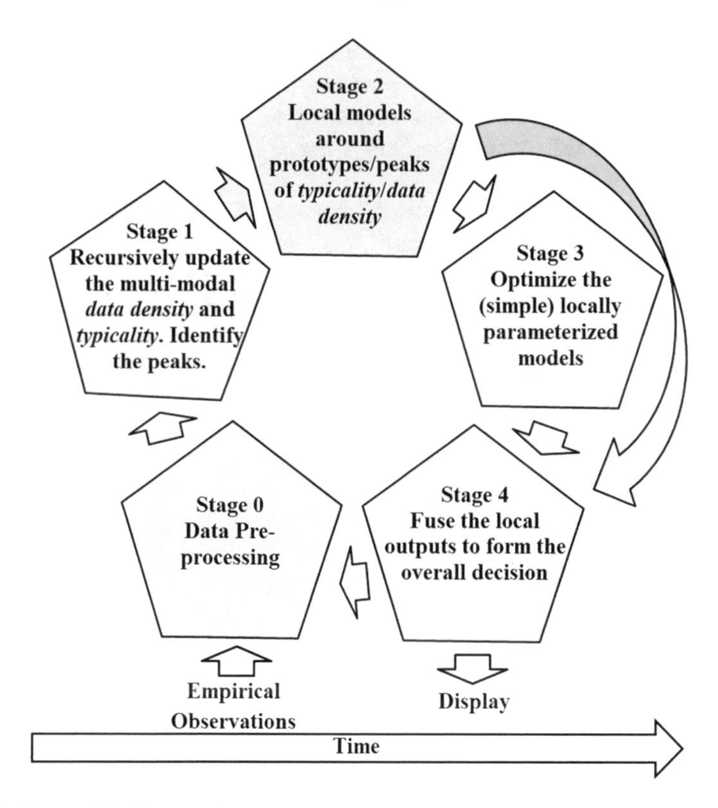

Fig. 1.6 The *empirical* approach

The principles of the new approach (why it is different from the existing ones) include:

(1) **Its high level of autonomy**—most of the algorithms do not use **any**[1,2,3] user- or problem-specific parameters; they only extract from the data statistics such as means, standard deviations, etc.

[1]*ALMMo-1* is using a single user-controlled parameter (Ω_0) which, however, **has very little influence** on the result and represents the standard for recursive least squares (RLS) algorithm initialization of the covariance matrix [28, 29]. Its value can easily be fixed to, for example, $\Omega_0 = 10$. This algorithm can **optionally** also use another two user-defined parameters (η_o and φ_0) which control the quality of the generated model.

[2]*ALMMo-0* and the DRB classifiers use a single user-controlled parameter (r_o) which, however, **has very little influence** on the result and represents the initial radius of the area of influence of the new *data cloud*. Its value can easily be fixed to, $r_o = \sqrt{2(1 - \cos(30^o))}$. Moreover, it is only required if the *ALMMo-0* and the DRB classifiers work online.

[3]SS_DRB classifier only requires two such parameters (Ω_1 and Ω_2) but they carry clear meaning and suggested value ranges are provided.

(2) **Its generic nature**—can work with any distance metric, but is most efficient when recursive Euclidean, Mahalanobis distances are used. The proposed method can work for other distance metrics, e.g. Hamming, cosine dissimilarity measures, etc. Of course, once the distance metric is selected, the result is influenced by this choice.
(3) **It is fast and computationally efficient**—non-iterative, not based on *prior* assumptions and search, "one pass".
(4) **It is anthropomorphic, human-like**—can learn from a single, handful or few examples, continue to learn and improve all the time, prototype-based, constructivist, not assumption- and hypothesis-based.
(5) **It can guarantee the local optimality; in addition, stability has been proven for the first-order learning system**.
(6) **It has hierarchically layered structure**.
(7) **Most of its algorithms are highly parallelizable**.

1.7 Book Structure

The remainder of this book has been divided into three parts, namely Parts I, II and III.

Part I consists of Chaps. 1, 2 and 3, and it provides the theoretical background of this book. Chapter 2 briefly recalls the basic concept and general principles of the probability theory and statistics. Chapter 3 presents the two main concepts of the computational intelligence, namely, FRB systems and ANNs.

The theoretical fundamentals of the proposed *empirical* approach are presented in Part II. In particular, Chap. 4 systematically describes the general concept and principles and non-parametric quantities as the fundamentals of the proposed *empirical* approach. Chapter 5 presents a new type of fuzzy sets, namely the *empirical* fuzzy sets, and, correspondingly, the *empirical* fuzzy systems. Chapter 6 introduces the *empirical* approach to anomaly detection. The *empirical* approach to data partitioning is presented in Chap. 7. The algorithmic details of the autonomous learning multi-model (*ALMMo*) systems of zero-order and first-order are given in Chap. 8. Chapter 9, as the last chapter of Part II, presents a new type of deep rule-based (DRB) classifier with a multilayer architecture, and further demonstrates its abilities of conducting semi-supervised learning and active learning.

The implementation and applications of the proposed approaches are presented in Part III. The algorithm summaries, flowchart and numerical examples based on benchmark problems of the autonomous anomaly detection algorithm, autonomous data partitioning algorithms, *ALMMo* systems, the DRB and the semi-supervised DRB (SS_DRB) classifiers are presented in Chaps. 10, 11, 12, 13 and 14, respectively.

This book is concluded by Chap. 15.

An interesting and useful feature of this book is the source code (in MATLAB) which we provide to be downloaded with extensive comments and example problems. Another useful feature is the set of lecture notes that can form a standalone postgraduate course or be used in advanced machine learning undergraduate courses.

Further, we finish each chapter with a set of questions for reinforcement as well as the so-called "take away items"—bullet points summarizing the most important ideas the reader can take from the specific chapter. All this is made to facilitate using this research monograph also as a pedagogical tool on advanced machine learning.

References

1. G.E. Moore, Cramming more components onto integrated circuits. Proc. IEEE **86**(1), 82–85 (1998)
2. https://www.raspberrypi.org/
3. https://beagleboard.org
4. https://www.movidius.com/
5. A. Oliva, A. Torralba, Modeling the shape of the scene: a holistic representation of the spatial envelope. Int. J. Comput. Vis. **42**(3), 145–175 (2001)
6. N. Dalal, B. Triggs, in *Histograms of Oriented Gradients for Human Detection*. IEEE Computer Society Conference on Computer Vision and Pattern Recognition, pp. 886–893 (2005)
7. K. Graumanand, T. Darrell, in *The Pyramid Match Kernel: Discriminative Classification with Sets of Image Features*. International Conference on Computer Vision, pp. 1458–1465 (2005)
8. G.-S. Xia, J. Hu, F. Hu, B. Shi, X. Bai, Y. Zhong, L. Zhang, AID: a benchmark dataset for performance evaluation of aerial scene classification. IEEE Trans. Geosci. Remote Sens. **55** (7), 3965–3981 (2017)
9. K. Simonyan, A. Zisserman, in *Very Deep Convolutional Networks for Large-Scale Image Recognition*. International Conference on Learning Representations, pp. 1–14 (2015)
10. A. Krizhevsky, I. Sutskever, G.E. Hinton, in *ImageNet Classification with Deep Convolutional Neural Networks*. Advances in Neural Information Processing Systems, pp. 1097–1105 (2012)
11. C. Szegedy, W. Liu, Y. Jia, P. Sermanet, S. Reed, D. Anguelov, D. Erhan, V. Vanhoucke, A. Rabinovich, C. Hill, A. Arbor, in *Going Deeper with Convolutions*. IEEE Conference on Computer Vision and Pattern Recognition, pp. 1–9 (2015)
12. A.K. Shackelford, C.H. Davis, A combined fuzzy pixel-based and object-based approach for classification of high-resolution multispectral data over urban areas. IEEE Trans. Geosci. Remote Sens. **41**(10), 2354–2363 (2003)
13. http://www.worldweatheronline.com
14. J.B. MacQueen, in *Some Methods for Classification and Analysis of Multivariate Observations*. 5th Berkeley Symposium on Mathematical Statistics and Probability, vol. 1, no. 233, pp. 281–297 (1967)
15. https://www.darpa.mil/news-events/lifelong-learning-machines-proposers-day
16. P. Angelov, *Autonomous Learning Systems: From Data Streams to Knowledge in Real Time* (Wiley, 2012)
17. P.P. Angelov, D.P. Filev, An approach to online identification of Takagi-Sugeno fuzzy models. IEEE Trans. Syst. Man Cybern.—Part B Cybern. **34**(1), 484–498 (2004)
18. E. Lughofer, P. Angelov, Handling drifts and shifts in on-line data streams with evolving fuzzy systems. Appl. Soft Comput. **11**(2), 2057–2068 (2011)
19. J. Macías-Hernández, P. Angelov, Applications of evolving intelligent systems to oil and gas industry. Evol. Intell. Syst. Methodol. Appl., 401–421 (2010)

20. R. Ramezani, P. Angelov, X. Zhou, in *A Fast Approach to Novelty Detection in Video Streams Using Recursive Density Estimation*. International IEEE Conference Intelligent Systems, pp. 14-2–14-7 (2008)
21. P. Angelov, R. Yager, A new type of simplified fuzzy rule-based system. Int. J. Gen. Syst. **41** (2), 163–185 (2011)
22. A. Corduneanu, C.M. Bishop, in *Variational Bayesian Model Selection for Mixture Distributions*. Proceedings of Eighth International Conference on Artificial Intelligence and Statistics, pp. 27–34 (2001)
23. N. Cristianini, J. Shawe-Taylor, *An Introduction to Support Vector Machines and Other Kernel-based Learning Methods* (Cambridge University Press, Cambridge, 2000)
24. Y. LeCun, Y. Bengio, G. Hinton, Deep learning. Nat. Methods **13**(1), 35 (2015)
25. I. Goodfellow, Y. Bengio, A. Courville, *Deep Learning* (MIT Press, Crambridge, MA, 2016)
26. S.C. Prasad, P. Prasad, Deep Recurrent Neural Networks for Time Series Prediction, vol. 95070, pp. 1–54 (2014)
27. P.P. Angelov, X. Gu, Towards anthropomorphic machine learning. IEEE Comput. (2018)
28. S.L. Chiu, Fuzzy model identification based on cluster estimation. J. Intell. Fuzzy Syst. **2**(3), 267–278 (1994)
29. T. Takagi, M. Sugeno, Fuzzy identification of systems and its applications to modeling and control. IEEE Trans. Syst. Man. Cybern. **15**(1), 116–132 (1985)

Part I
Theoretical Background

Chapter 2
Brief Introduction to Statistical Machine Learning

2.1 Brief Introduction to Probability Theory

Probability theory emerged as a scientific discipline about three centuries ago as an attempt to study random events and variables. It describes random variables and experiments with consequences that are not pre-determined [1], but uncertain. There are different types of uncertainty [2] and probability theory deals with uncertainty of random nature. In this section, the basic concepts of probability theory are recalled, because it provides a solid theoretical framework for the quantification and manipulation of this type of uncertainties and forms one of the central pillars for pattern recognition and data analysis [3, 4]. Probability theory serves as the mathematical foundation for statistics [3] and is essential to various areas of knowledge including engineering, economics, biology, sciences and social sciences [5].

2.1.1 *Randomness and Determinism*

Randomness and determinism are the fundamental concepts that are the two extremes on the spectrum of uncertainty. Determinism, which is the basis of the so-called "exact sciences" (mathematics, physics, chemistry, engineering) is a doctrine according to which all events and outcomes are entirely determined by the causes regarded as external to the process/system under consideration. An example is the free fall of a ball or any other physical object with a certain mass, m (of course, this describes the main phenomena in an idealistic case when there is vacuum and no air density, wind, etc.). The force of gravity will determine in a unique way the velocity, acceleration at any moment in time for a given m.

The other extreme can be represented by the so-called "fair games" often used for gambling. For example, tossing a coin, throwing dices, etc. The outcome of such

P. P. Angelov and X. Gu, *Empirical Approach to Machine Learning*, Studies in Computational Intelligence 800,
https://doi.org/10.1007/978-3-030-02384-3_2

processes is random and often used as example at the start of studying probability theory in schools and universities. Indeed, each individual outcome of such an "experiment" is random and independent of the order of outcomes and other "experiments". For example, imagine throwing dices and the answers to the two questions "What is the probability to have 3 following 1" and "What is the probability to have 6 following 1" are the same, namely, "$\frac{1}{6}$". The value $\frac{1}{6}$ is a result of applying the simplest form of so-called "frequentist" interpretation of probability as a ratio of outcomes of a specific sort in regards to all possible outcomes [3, 4]:

$$P(x_i) = \lim_{K \to \infty} \frac{K_i}{K} \tag{2.1}$$

where x_i denotes a particular event that the outcome is a specific number, e.g. "1" or "3" or "6"; K_i denotes the number of trails where x_i occurred; K is the total number of trails; $P(x_i)$ is the probability of the event x_i to take place.

That is the case, because outcomes of throwing a dice are orthogonal (uncorrelated). This is true for "fair games".

However, the vast majority of processes in the real life which are of interest to people, such as financial time series, climate, economic variables, human behavior, etc. are positioned in the spectrum of uncertainty between the clear determinism and clear randomness [6]. Indeed, let's take a simple example. The probability to have 11 °C following two days with a temperature, respectively 10 and 12 °C is much higher than to have, for example, 18 °C. One can provide a similar example for the stock market, etc. The explanation is simple: while clearly, not being deterministic processes, the climate, stock market, etc. are also not strictly random. To avoid this confusion, they consider so-called "probability distribution" models [3, 4]. The most popular of probability distribution is the so-called Gaussian (named after the German mathematician Friedrich Gauss, 1777–1855) also called "normal distribution", which is parametrized in terms of the mean, μ and standard deviation of the data (x), namely, σ (Fig. 2.1):

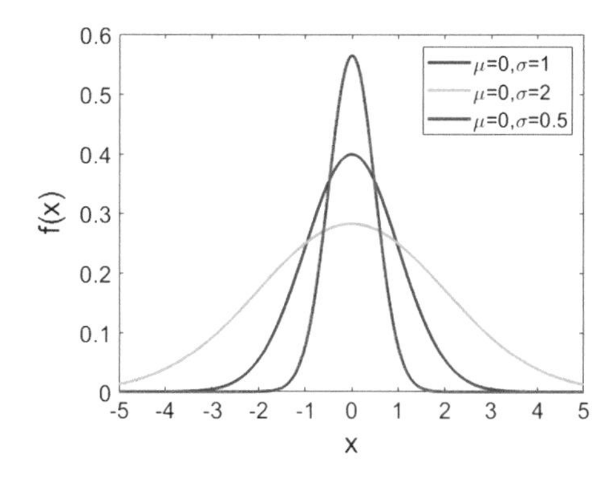

Fig. 2.1 Examples of Gaussian functions

$$f(x;\ \mu,\sigma) = \frac{1}{\sqrt{2\pi\sigma^2}} e^{-\frac{(x-\mu)^2}{2\sigma^2}} \tag{2.2}$$

The power of the Gaussian/normal distribution stems from the so-called "Central Limit Theorem (CLT)" [3, 4]. According to the CLT, for independent and identically distributed variables, $x_1,\ x_2,\ x_3,\ \ldots\ x_n$ with $E(x_1) = 0$ and $E\left(x_1^2\right) = 1$, the distribution of the normalized variable average $y = \frac{1}{\sqrt{K}}\sum_{i=1}^{K} x_i$ tends towards the normal distribution when $K \to \infty$ even if the original variables themselves are not normally distributed.

CLT is of paramount importance to the probability theory and statistics because it implies that probabilistic and statistical approaches for normal distributions are applicable to many problems involving other types of distributions providing infinite amount of observations.

However, in reality, we never have infinite amount of observations and they are not orthogonal (as demonstrated above). As a result, the mathematical functions smooth and convenient to work with, like Gaussian, Cauchy, Laplace distribution, remain idealistic assumptions that deviate significantly from the real experimental data observations. However, real distributions of processes with practical significance for people are more complex, multimodal (with multiple local peaks and troughs and look like real mountains). We will return to this issue in Chap. 4 (more specifically, Sect. 4.7).

2.1.2 Probability Mass and Density Functions

The probability theory [1, 3, 4] starts with the concept of the *discrete random variable*. Such variables take only (countably) finite values in the data space. A typical example is the outcome of a *fair game*, e.g. tossing a coin, throwing dices, etc. Indeed, the outcome of tossing a coin is binary with one of the two possible outcomes; the outcome of throwing a dice is one out of six possible numbers (1, 2, 3, 4, 5 or 6) and these outcomes randomly appear.

More complex, multivariate discrete random variables can also exist. They are described by a vector of outcomes. For example, if we throw several (for the sake of illustration, let us consider two) dices at the same time we can encode their outcome by: $x = \{3; 6\}$. This obviously means that one of the dices has a value of 3 and the other one, respectively, 6.

Discrete random variables are being described by the so-called *probability mass function* (PMF) [1, 5]. For a discrete random variable x with the cardinality $\{x\} = \{x_1, x_2, x_3, \ldots\}$ (finite or countable finite), the PMF of x, $P(x)$, is a function such that:

$$P(x_i) = \Pr(x = x_i)\ \ for\ i = 1, 2, 3, \ldots \tag{2.3}$$

As one can see from Eq. (2.3), PMF is a function describing the probability of a random variable x to have a specific value, x_k. It is defined within a certain range, which can be reformulated in a general form as:

$$P(x) = \begin{cases} \Pr(x) & x \in \{x\} \\ 0 & otherwise \end{cases} \tag{2.4}$$

PMFs have the following properties [1]:

$$0 \le P(x) \le 1 \tag{2.5}$$

$$\sum_{x \in \{x\}} P(x) = \sum_{x \in \{x\}} \Pr(x) = 1 \tag{2.6}$$

For

$$\{x\}_o \in \{x\}, \quad P\left(x \in \{x\}_o\right) = \sum_{x \in \{x\}_o} \Pr(x) \tag{2.7}$$

Another important quantifier of the probability of random variables is the cumulative distribution function (CDF) evaluated at x_o and defined as [7]:

$$F(x_o) = \Pr(x \le x_o) = \sum_{y \in \{x\}, y \le x_o} \Pr(x) \tag{2.8}$$

The CDF of the discrete random variable x accumulates the values of the probability for each discrete point (precisely the values that the random variable x may take, $x \in \{x_1, x_2, x_3, \ldots\}$). In this way, the value of the CDF is constant for the intervals between such values and jumps/increments to a higher value instantaneously. Therefore, the CDF of a discrete random variable is a discontinuous function.

Now we are ready to introduce/recall the most important quantifier of a continuous random variable, namely its *probability density function* (PDF), $f(x)$. It is defined as a function whose value at any given point in its cardinality quantifies the relative likelihood that the value of the random variable is equal to that sample, meanwhile, the absolute likelihood for a continuous random variable have any other particular value is 0 [1, 3–5].

For a continuous random variable, x the probability for it to fall into the value range of $[x_1, x_2]$ is calculated as [1, 7]:

$$\Pr(x_1 < x < x_2) = \int_{x=x_1}^{x_2} P(x)dx \tag{2.9}$$

where $P(x)$ stands for the PDF of x. And the CDF of x calculated in regards to x_o is defined as [1, 7]:

$$F(x_o) = \Pr(x \leq x_o) = \int_{x=-\infty}^{x_o} P(x)dx \tag{2.10}$$

from which one can see that, the CDF of a continuous random variable is a continuous function.

In fact, the PDF is used to specify the probability of the random variable falling within a particular range of values, as opposed to taking on any single value (through the integral of this variable's PDF over that range for the continuous case and summation of this variable's PDF over that range for the discrete case).

PDFs have the following properties which are similar to the properties of the PMFs, namely:

$$0 \leq P(x) \tag{2.11}$$

$$\int_{x=-\infty}^{+\infty} P(x)dx = 1 \tag{2.12}$$

One of the commonly used PDFs is the Gaussian function. Its univariate form has been given in Eq. (2.2); its multivariate form is expressed as:

$$f(\boldsymbol{x};\, \boldsymbol{\mu}, \boldsymbol{\Sigma}) = \frac{1}{(2\pi)^{\frac{N}{2}}|\boldsymbol{\Sigma}|^{\frac{1}{2}}} e^{-(\boldsymbol{x}-\boldsymbol{\mu})^T\boldsymbol{\Sigma}^{-1}(\boldsymbol{x}-\boldsymbol{\mu})} \tag{2.13}$$

where $\boldsymbol{x} = [x_1, x_2, \ldots, x_N]^T \in \mathbf{R}^N$; $\mathbf{R}^N$ is a real metric space with the dimensionality of N; $\boldsymbol{\mu}$ and $\boldsymbol{\Sigma}$ are the mean and the covariance matrix, respectively; $|\boldsymbol{\Sigma}|$ is the determinant of $\boldsymbol{\Sigma}$.

Another widely used type of PDF is the Cauchy type. The univariate Cauchy type PDF is formulated as:

$$f(x;\, \mu, \sigma) = \frac{1}{\pi\sigma\left(1 + \left(\frac{x-\mu}{\sigma}\right)^2\right)} \tag{2.14}$$

And its multivariate form is given as:

$$f(\boldsymbol{x};\, \boldsymbol{\mu}, \boldsymbol{\Sigma}) = \frac{\Gamma\left(\frac{1+N}{2}\right)}{\Gamma\left(\frac{1}{2}\right)\pi^{\frac{N}{2}}|\boldsymbol{\Sigma}|^{\frac{1}{2}}\left[1 + (\boldsymbol{x}-\boldsymbol{\mu})^T\boldsymbol{\Sigma}^{-1}(\boldsymbol{x}-\boldsymbol{\mu})\right]^{\frac{1+N}{2}}} \tag{2.15}$$

One can also represent the PMF of the discrete random variable x with a generalized PDF by using Dirac delta function $\delta(x)$, which is frequently used in the field of signal processing [8]:

$$P(x) = \sum_{i=1}^{K} \Pr(x_i)\delta(x - x_i) \tag{2.16}$$

where $\Pr(x_i)$ is the probability associated with these values $x_i \in \{x\}_K = \{x_1, x_2, \ldots, x_K\}$.

2.1.3 *Probability Moments*

In statistics, moments are the specific quantitative measures that can be used for characterizing the PDF. The hth moment around value j can be mathematically formulated as follows [9]:

$$m(j, h) = \frac{1}{K}\sum_{i=1}^{K} (x_i - j)^h \tag{2.17}$$

where $\{x\}_K = \{x_1, x_2, \ldots, x_K\}$ is a set of data consisting of K discrete observations.

In practice, the most commonly used (and at the same time, the most important) moments in the probability theory and statistics are the first moment around 0 expressed as follows:

$$m(0, 1) = \frac{1}{K}\sum_{i=1}^{K} x_i \tag{2.18}$$

and the second moment around $m(0, 1)$, which is formulated as:

$$m(m(0, 1), 2) = \frac{1}{K}\sum_{i=1}^{K} (x_i - m(0, 1))^2 \tag{2.19}$$

Higher moments are usually used in physics, and in this book, we will not cover them.

The two notations $m(0, 1)$ and $m(m(0, 1), 2)$, namely Eqs. (2.18) and (2.19), are, in fact, the mean (μ_K) and variance (σ_K^2) of $\{x\}_K$, respectively:

$$\mu_K = m(0, 1) \tag{2.20a}$$

$$\sigma_K^2 = m(m(0, 1), 2) \tag{2.20b}$$

Mean, μ_K defines the average of the possible values of a random variable and variance, σ_K^2 is the expectation of the squared deviation of a random variable from its mean. With the first and second moments, the Gaussian type PDF of the random variable can be directly built, see Eq. (2.2).

However, the density distribution of real data is usually complex, multimodal and varies with time. In practice, only several "standard" types of PDF are considered thanks to their convenient mathematical formulation. The most widely used ones include the Gaussian and Cauchy types [Eqs. (2.2) and (2.14)].

One of the main problems of the traditional probability theory and statistics is that the data distribution in reality often can differ significantly from the assumed model of data generation. Actual data is usually discrete (or discretized), and the real distribution is often multimodal (many local peaks), not smooth and nonstationary. In addition, the a priori distribution is unknown. If the prior data generation hypothesis is confirmed, results are usually good. However, the alternative, which is more often the case, leads to significant discrepancies.

2.1.4 *Density Estimation Methods*

Density estimation aims to recover the underlying PDF from the observed data. The commonly used density estimation methods include histogram, kernel density estimation (KDE), Gaussian mixture model (GMM), etc.

2.1.4.1 Histogram

Histogram is the simplest form of density estimation. A histogram is a visualization of the distribution of the data where bins are defined, and the number of data samples within each bin is counted. It gives a rough sense of the density of the underlying distribution of the data.

However, the major problems with histograms are the choice of the size of the bin, which can lead to different representations and cause confusion, and the large (potentially, huge) number of data points required linked to the so-called "curse of dimensionality". In the illustrative example presented in Fig. 2.2, two histograms of 20 normally distributed data samples are given. They use two different scales (16 and 8 bins, respectively), and the height of each bar is the value of the normalized density, which is equal to the relative number of observations (number of observations in the bin/total number of observations), and the sum of the bar heights is less than or equal to 1. As one can see, using different number of bins (namely, the bin size), the histograms may vary a lot.

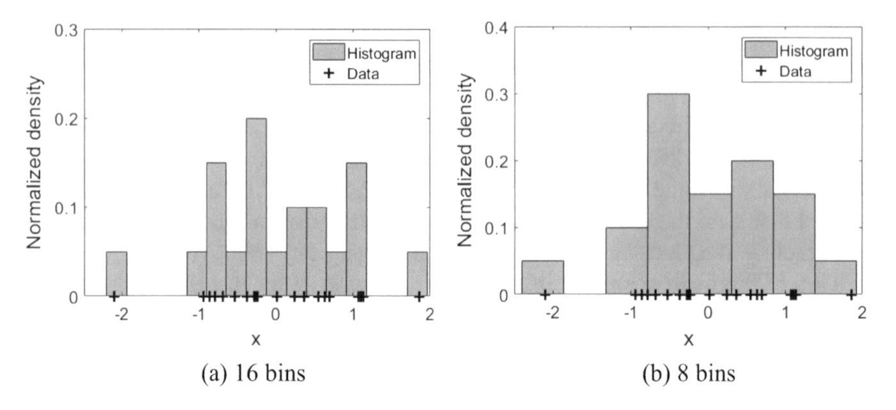

(a) 16 bins (b) 8 bins

Fig. 2.2 Histograms of the normally distributed data

2.1.4.2 Kernel Density Estimation

Kernel density estimation (KDE) is a genetic approach for estimating the PDF of a random variable. It derives a multimodal distribution from a set of simpler, kernel representations that are valid locally [6] and helps reduce the complexity of the problem.

A one-dimensional kernel is a positive smoothing function $\kappa(x)$ with the following constraints:

$$\int \kappa(x,h)dx = 1 \tag{2.21a}$$

$$\int x\kappa(x,h)dx = 0 \tag{2.21b}$$

$$\int x^2\kappa(x,h)dx > 0 \tag{2.21c}$$

Some commonly used kernels are as follows:

(1) Gaussian kernel

$$\kappa(x) = \frac{1}{\sqrt{2\pi}}\exp\left(-\frac{x^2}{2}\right) \tag{2.22}$$

(2) Tophat kernel

$$\kappa(x) = \begin{cases} \frac{1}{2} & -1 \le x \le 1 \\ 0 & else \end{cases} \tag{2.23}$$

(3) Epanechnikov kernel

$$\kappa(x) = \begin{cases} \frac{3}{4}(1-x^2) & -1 \le x \le 1 \\ 0 & else \end{cases} \quad (2.24)$$

(4) Exponential kernel

$$\kappa(x) = \frac{1}{2}\exp(-|x|) \quad (2.25)$$

where $|x|$ denotes the absolute value of x.

(5) Linear kernel

$$\kappa(x) = \begin{cases} 1-|x| & -1 \le x \le 1 \\ 0 & else \end{cases} \quad (2.26)$$

(6) Cauchy kernel

$$\kappa(x) = \frac{1}{\pi(1+x^2)} \quad (2.27)$$

The curves of the six kernels are given in Fig. 2.3.

Assume a particular dataset defined as $\{\boldsymbol{x}\}_K = \{\boldsymbol{x}_1, \boldsymbol{x}_2, \ldots, \boldsymbol{x}_K\} \in \mathbf{R}^N$ ($\boldsymbol{x}_i = \left[x_{i,1}, x_{i,2}, \ldots, x_{i,N}\right]^T$, $i = 1, 2, \ldots, K$), where subscripts denote data samples (for a set).

Given a kernel $\kappa(x)$ and a positive number h, which is the bandwidth, the KDE is defined to be [6]:

$$P(\boldsymbol{y}) = \frac{1}{Kh^N}\sum_{i=1}^{K}\kappa\left(\frac{\boldsymbol{y}-\boldsymbol{x}_i}{h}\right) \quad (2.28)$$

From the above one can find out that KDE has a single parameter, namely the bandwidth, h, which can largely influence the result, see Fig. 2.4, where the same data as in Fig. 2.2 and a Gaussian kernel are used. It can be quite difficult to choose the most suitable bandwidth without any prior knowledge about the problem, and thus, it is a major weakness of the KDE method. Another problem of the KDE is its computational complexity, because of which, it is limited to offline applications.

2.1.4.3 Gaussian Mixture Model

Gaussian mixture is a probabilistic model composed of a number of multivariate normal density components. It assumes that all the data samples are generated from a mixture of Gaussian distributions with unknown parameters. Gaussian mixture

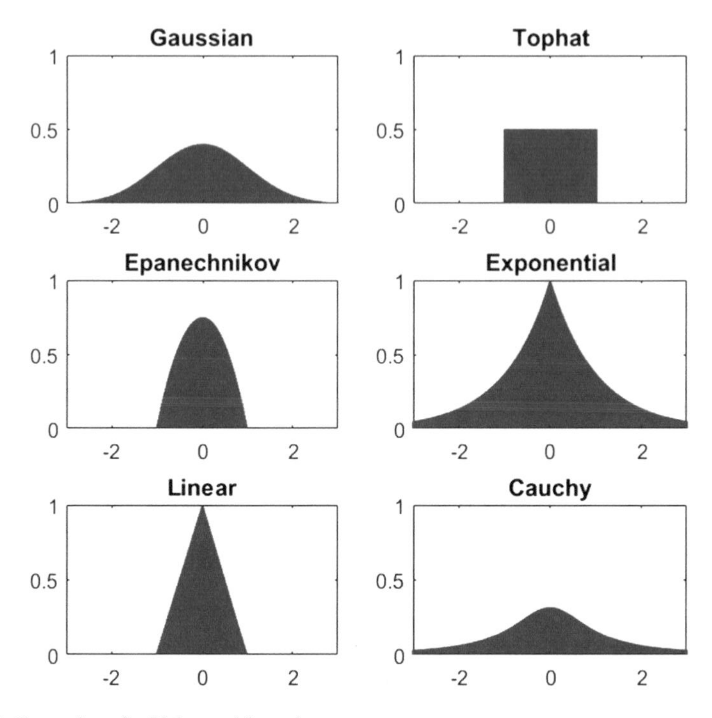

Fig. 2.3 Examples of widely used kernels

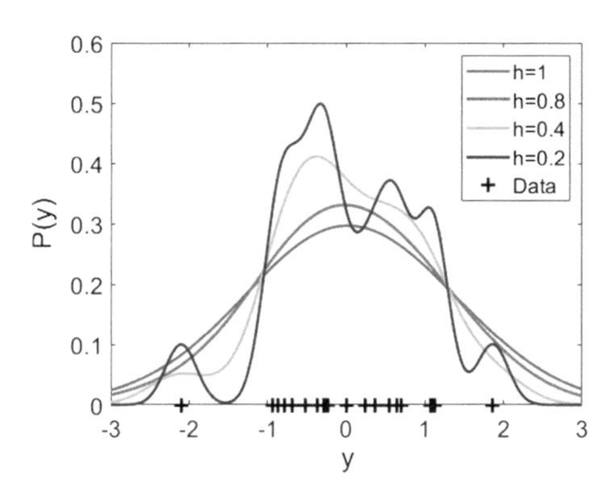

Fig. 2.4 KDE using Gaussian kernel with different bandwidths

model (GMM) has been used for density estimation, clustering, feature extraction, etc.

A GMM is parameterized by the mixture component weights, the component means and variances/covariance. The one-dimensional GMM is formulated as follows [9, 10]:

$$p(x) = \sum_{i=1}^{M} w_i f(x; \mu_i, \sigma_i) \tag{2.29a}$$

$$f(x; \mu_i, \sigma_i) = \frac{1}{\sqrt{2\pi}\sigma_i} e^{-\frac{(x-\mu_i)^2}{\sigma_i^2}} \tag{2.29b}$$

$$\sum_{i=1}^{M} w_i = 1 \tag{2.29c}$$

where M is the number of models; w_i, μ_i and σ_i are the respective weight, mean and variance of the ith model $(i = 1, 2, \ldots, M)$.

The multi-dimensional GMM is formulated as follows [9, 10]:

$$p(\boldsymbol{x}) = \sum_{i=1}^{M} w_i f(\boldsymbol{x}; \boldsymbol{\mu}_i, \boldsymbol{\Sigma}_i) \tag{2.30a}$$

$$f(\boldsymbol{x}; \boldsymbol{\mu}_i, \boldsymbol{\Sigma}_i) = \frac{1}{\sqrt{(2\pi)^N |\boldsymbol{\Sigma}_i|^{\frac{1}{2}}}} e^{-(\boldsymbol{x}-\boldsymbol{\mu}_i)^T \boldsymbol{\Sigma}_i^{-1} (\boldsymbol{x}-\boldsymbol{\mu}_i)} \tag{2.30b}$$

$$\sum_{i=1}^{M} w_i = 1 \tag{2.30c}$$

where $\boldsymbol{\Sigma}_i$ is the covariance matrix of the ith model.

The expectation maximization (EM) algorithm [11] is the most commonly used technique to iteratively estimate the parameters of the GMM. The EM algorithm for GMM consists of two steps. The first step is the Expectation step, which calculates the expectation of the component assignments for each data samples given the model parameters. The second step is the Maximization step, which is using expectation to update the model parameters. More details regarding the EM algorithm can be found in [11].

One of the major drawbacks of the GMM is that this approach assumes that all the data samples are distributed normally, which is often not the case in reality, and, therefore, the estimated density distribution may differs a lot from the real distribution. Another major drawback is that the number of component models, M is required to be predefined, which is not possible without prior knowledge of the problem. Otherwise, it requires the user to guess the value of M and make the best trade-off between the performance and the number of components. The very low computational efficiency is also a problem that GMM suffers from.

2.1.5 Bayesian Approach and Other Branches of Probability Theory

The Bayesian approach to analysis of probabilities stems from the fundamental work by Thomas Bayes (1701–1761) "An essay towards solving a problem in the doctrine of chances" [12]—a cornerstone publication by the Irish reverend who was fascinated by the gambling and interested in insurance.

Bayesian approach evaluates the probability of a hypothesis by specifying a prior probability and then updating to a *posterior* probability based on new evidence. The main result can be summarized by the so-called Bayesian theorem which links the prior and *posterior* probabilities.

Bayesian theorem derives the *posterior* probability $P(Y = y|X = x)$ as a consequence of two antecedents, a likelihood function $P(X = x|Y = y)$ (PMF or PDF) for the evidence $X = x$ given the hypothesis $Y = y$, and a prior probability for hypothesis, $P(Y = y)$ reflecting the probability of $Y = y$ before the evidence $X = x$ is observed:

$$P(Y = y|X = x) = \frac{P(X = x|Y = y)P(Y = y)}{P(X = x)} \tag{2.31}$$

where $P(X = x)$ is the marginal likelihood, which is the probability of evidence $X = x$ considering all the hypotheses:

$$P(X = x) = \int P(X = x|Y = y)P(Y = y)dy \tag{2.32}$$

Let us consider a simple illustrative example. Suppose that in a school exam, 80% of the students who have prepared for it will pass (regardless of the quality of the preparation), and only 10% of the students who did not spend time for preparation will pass. Suppose that there are 10% of students who did not prepare. What is the probability that a randomly selected student passed exam but did not prepare for it?

To calculate the *posterior* probability, the probability for passing the exam (which is the marginal likelihood) is calculated firstly as follows:

$$\begin{aligned} &P(exam\,passed) \\ &\quad = P(exam\,passed|student\,prepared)P(student\,prepared) \\ &\qquad + P(exam\,passed|student\,unprepared)P(student\,unprepared) \end{aligned} \tag{2.33a}$$

and in this example, there is:

$$\begin{aligned} P(exam\,passed) &= 80\% \times 90\% + 10\% \times 10\% \\ &= 73\% \end{aligned} \tag{2.33b}$$

Then, the probability that a randomly selected student passed the exam without preparation is given as:

$$P(\textit{student unprepared}|\textit{exam passed}) = \frac{P(\textit{exam passed}|\textit{student unprepared})P(\textit{student unprepared})}{P(\textit{exam passed})} \tag{2.34a}$$

and in this example, there is:

$$P(\textit{student unprepared}|\textit{exam passed}) = \frac{10\% \times 10\%}{73\%} \approx 1.37\% \tag{2.34b}$$

One can notice the restrictive assumption: pass and fail are mutually exclusive and binary. In real life, the possible values of the variables are more complexly intertwined.

Bayesian probability is, probably, one of the most important branches of the probability theory, and Bayesian approach is the most widely used method to inference, but there are also other important branches including, but also not limited to:

(1) ***Empirical probability***: This type of probability is called statistical probability, which is derived from the past records by the frequentist approach (see Sect. 2.1.3).

(2) ***Conditional probability***: Conditional probability considers the probability of a dependent event. For two dependent events, $X = x$ and $Y = y$, the conditional probability of X given Y is defined as:

$$P(X = x|Y = y) = \frac{P(X = x, Y = y)}{P(Y = y)} \tag{2.35}$$

where $P(X = x, Y = y)$ is the probability of the joint event of $X = x$ and $Y = y$. If X and Y are independent from each other, then:

$$P(X = x|Y = y) = P(X = x) \tag{2.36}$$

(3) ***Marginal probability***: It gives the probability of one event irrespective of other events [the continuous version is given by Eq. (2.32)]:

$$\begin{aligned} P(X = x) &= \sum_y P(X = x, Y = y) \\ &= \sum_y P(X = x|Y = y)P(Y = y) \end{aligned} \tag{2.37}$$

The marginal probability is also called unconditional.

(4) ***Markov chain***: A Markov chain, named after the Soviet mathematician Andrey Markov (1856–1992), is a statistic model that describes a sequence of possible events transitioning from one to another according to certain probabilistic rules. For the so-called first-order Markov chain, the probability of transitioning to any particular event is dependent solely on the current event. An illustrative example of a Markov chain is visualized in Fig. 2.5.

2.1.6 Analysis

While the traditional probability theory is based on simple and self-evident principles and mathematical expressions (equations and inequalities) it also makes some strong assumptions that are often not satisfied in the real-life applications. For example,

- a random nature of the variables;
- pre-defined smooth and "convenient to use" of PDF;
- independent and identically distributed variables;
- infinite amount of observations.

PDF is the derivative of the CDF; however, differentiation can create numerical problems in both practical and theoretical aspects and is a challenge for functions that are not analytically defined or are complex. In many cases, one may also notice that the PDF of a variable $x, f(x)$, has positive values for the infeasible values of x if there are no any constraints imposed beforehand over the range of integration.

The appeal of the traditional statistical approach comes from its solid mathematical foundation and the ability to provide guarantees of performance on condition that data is plenty, and generated from the same distribution independently as being assumed in the probability theory. The actual data is usually discrete (or discretized), which in traditional probability theory and statistics are modelled as realizations of the random variables. The problem is that their distribution is not known a priori. If the prior data generation assumption is confirmed, the results can be good; otherwise, they can be quite poor.

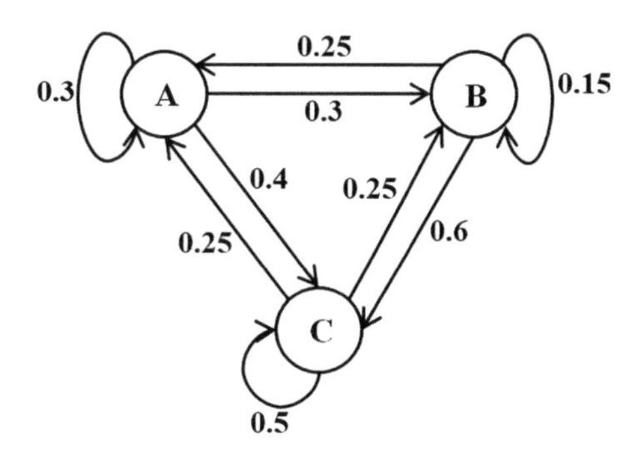

Fig. 2.5 Example of a Markov chain

Even if the hypothesis meets the reality, the difference of working with realizations and random variables is still needed to be addressed, and choosing estimators of statistical quantities necessary for data analysis is also an issue. Different estimators usually behave differently in the finite (and even in the infinite) sample case and lead to different results [13]. The most likely reason behind this is that the functional properties of the chosen estimator do not preserve all the properties embodied in the statistical quantity. The measure of the random variable in the probability law that explains the collected data is a hard problem studied in density estimation [3, 4, 14].

The traditional probability theory is used as the theoretical foundation of the statistics and statistical data analysis [15]. Probability theory and statistical learning are essential and widely used tools for quantitative analysis of the data representation of stochastic type of uncertainties. However, there are a few critical issues existing in the traditional probability theory and statistics:

(1) The assumptions of independent and identically distributed (*iid*) variables with a random nature that the traditional probability theory made fail to meet the reality for real processes of interest (such as climate, economic, social, mechanical, electronic, biological, etc.).
(2) The solid mathematical foundation of the traditional statistical approach and its ability to provide guarantees of performance come from the condition that data is plenty, and generated from the same distribution independently. However, the actual data is usually discrete (or discretized) and their distribution is not known a priori.
(3) Different estimators usually behave differently in the finite (and even in the infinite) sample case and lead to different results [13]. Choosing estimators of statistical quantities necessary for data analysis is an issue because the chosen estimator does not preserve all the properties embodied in the statistical quantity.
(4) The difference of working with realizations and random variables needs to be addressed. The measure of the random variable in the probability law that explains the collected data is a hard problem studied in density estimation [3, 4, 14].

2.2 Introduction to Statistical Machine Learning and Pattern Recognition

Statistical approaches to machine learning and pattern recognition are based on the probability theory discussed in the previous subsection. We will now briefly introduce the basic stages of machine learning from data, which will also be used further in this book in our proposed *empirical* approach to machine learning. For a more comprehensive review on statistical learning the interested readers are directed, for example, to [14].

2.2.1 Data Pre-processing

The first stage after acquiring the data is called data pre-processing [16, 17]. It is an important step that precedes the actual learning from data and concerns the preparation for it. Data gathered from different sensors and observations can have missing or infeasible values. Data, sometimes, can be highly correlated and difficult to distinguish. High dimensionality is also a tough problem for common techniques because highly dimensional data is very hard to process and visualize. Therefore, pre-processing techniques are often used to reduce the dimensionality and/or to deal with missing and anomalous data.

If anomalies exist in the observed data, i.e. outliers, novelties, noise, and exceptions, machine learning algorithms will produce misleading results. For that reason, anomaly detection is an important part of the pre-processing. On the other hand, anomaly detection is an important statistical analysis method on its own which is at the heart of the applications to fault detection in technical systems, intruder detection in security systems, novelty and objects detection in image processing, insurance analysis and many other problems. Because of its value as a stand-alone method we will describe anomaly detection in more detail in Chap. 6.

In this subsection, we will describe the following commonly used data pre-processing techniques:

- normalization and standardization;
- feature/input selection, extraction and orthogonalization.

The similarity and distance measures will be covered as well.

First of all, let us consider particular data set/stream within the real metric space $\mathbf{R}^N$, defined as $\{\boldsymbol{x}\}_K = \{\boldsymbol{x}_1, \boldsymbol{x}_2, \ldots, \boldsymbol{x}_K\} \in \mathbf{R}^N$; $\boldsymbol{x}_i = \left[x_{i,1}, x_{i,2}, \ldots, x_{i,N}\right]^T$; $i = 1, 2, \ldots, K$, where subscripts denote individual data samples (in the case of time series—the time instant). Static multi-variate dataset, $\{\boldsymbol{x}\}_K$ can be presented in a convenient matrix form as follows:

$$\mathbf{X}_K = [\boldsymbol{x}_1, \boldsymbol{x}_2, \ldots, \boldsymbol{x}_K] = \begin{bmatrix} x_{1,1} & x_{2,1} & \cdots & x_{K,1} \\ x_{1,2} & x_{2,2} & \cdots & x_{K,2} \\ \vdots & \vdots & \ddots & \vdots \\ x_{1,N} & x_{2,N} & \cdots & x_{K,N} \end{bmatrix} \tag{2.38}$$

2.2.1.1 Similarity and Distance Measures

Similarity and distance measures play a critical role in machine learning and pattern recognition. Classification, clustering, regression, and many other problems employ various types of similarity and distance to measure the separation between data samples [18].

An important definition is that of a (full) distance metric. From the strict mathematical point of view a full distance metric must satisfy the following conditions [19]:

(1) Non-negativity:

$$d(\boldsymbol{x},\boldsymbol{y}) \geq 0 \tag{2.39a}$$

(2) Identity of indiscernible:

$$d(\boldsymbol{x},\boldsymbol{y}) = 0 \; \textit{iff} \; \boldsymbol{x} = \boldsymbol{y} \tag{2.39b}$$

(3) Symmetry:

$$d(\boldsymbol{x},\boldsymbol{y}) = d(\boldsymbol{y},\boldsymbol{x}) \tag{2.39c}$$

(4) Triangle inequality:

$$d(\boldsymbol{x},z) + d(\boldsymbol{y},z) \geq d(\boldsymbol{y},\boldsymbol{x}) \tag{2.39d}$$

where $\boldsymbol{x},\boldsymbol{y} \in \mathbf{R}^N$ are two data samples in the data space.

In data analysis, another important measure, but weaker than the distance metric, is the similarity. It is a real value function that quantifies the similarity between two data samples. Although there is no unique definition for it, the similarity measure usually takes on large values for similar data samples and small values for dissimilar data samples. For a similarity measure, it only has to follow the property of symmetry [Eq. (2.39c)]. One of the most widely used similarities is the cosine dissimilarity [20–24], which will be briefly described later in this subsection.

Divergence is a function in statistics that measures the directed (asymmetric) difference of two probability distributions [25]. The divergence is also a weaker measure than the distance metric, and it has the following two properties [25–27]:

(1) Non-negativity:

$$\text{div}(P_1(x) \| P_2(x)) \geq 0 \tag{2.40a}$$

(2) Identity of indiscernible:

$$\text{div}(P_1(x) \| P_2(x)) = 0 \; \textit{iff} \; P_1(x) = P_2(x) \tag{2.40b}$$

where $P_1(x)$ and $P_2(x)$ are the two probability distributions. However, divergence does not have to satisfy the properties of symmetry [Eq. (2.39c)] and triangle inequality [Eq. (2.39d)].

The most well-known example of divergence in statistics and pattern recognition is the Kullback–Leibler (KL) divergence [25–27]. If both $P_1(x)$ and $P_2(x)$ are PMFs, namely, x is a discrete variable, the KL divergence between them, $\text{div}_{KL}(P_1(x) \| P_2(x))$, is defined as [27]:

$$\mathrm{div}_{KL}(P_1(x)\|P_2(x)) = \sum_x P_1(x)\log\left(\frac{P_1(x)}{P_2(x)}\right) \tag{2.41}$$

In contrast, if both $P_1(x)$ and $P_2(x)$ are PDFs, namely, x is a continuous variable, $\mathrm{div}_{KL}(P_1(x)\|P_2(x))$ is formulated as [27]:

$$\mathrm{div}_{KL}(P_1(x)\|P_2(x)) = \int P_1(x)\log\left(\frac{P_1(x)}{P_2(x)}\right)dx \tag{2.42}$$

In this subsection, we will briefly recall the most commonly used (dis)similarity and distance measures.

A. ***Euclidean distance***

The Euclidean distance, named after the ancient Greek mathematician Euclid (about 325 BC—about 265 BC), between two data samples, $\boldsymbol{x}_i, \boldsymbol{x}_j \in \{\boldsymbol{x}\}_K$ is calculated based on the following equation:

$$d(\boldsymbol{x}_i, \boldsymbol{x}_j) = \|\boldsymbol{x}_i - \boldsymbol{x}_j\| = \sqrt{\sum_{l=1}^{N}(x_{i,l} - x_{j,l})^2} \tag{2.43}$$

The Euclidean distance is the most commonly used type of distance metric. In most cases, the distance used in daily life refers to Euclidean distance, see Fig. 2.6a for a simple illustrative example in a *2D* space (Euclidean distance between two data samples $\boldsymbol{x}_i = [-2, 1]^T$ and $\boldsymbol{x}_j = [2, -1]^T$).

Its main advantage is the simplicity and computational efficiency. Its main disadvantage is the fact that it gives equal weight/importance to each of the N dimensions. The Euclidean distance is a full metric. Indeed, it is easy to check that for the Euclidean distance, all four conditions are satisfied [18, 19].

B. ***City block distance***

City block (also called Manhattan) type distance between $\boldsymbol{x}_i$ and $\boldsymbol{x}_j$ is expressed as follows:

$$d(\boldsymbol{x}_i, \boldsymbol{x}_j) = \sum_{l=1}^{N}|x_{i,l} - x_{j,l}| \tag{2.44}$$

where $|\cdot|$ denotes the absolute value. From the above equation one can see that the city block distance between any two points is the sum of the absolute differences of their Cartesian coordinates. Following the same example as in Fig. 2.6a, the city block distance between points $\boldsymbol{x}_i$ and $\boldsymbol{x}_j$ is given in Fig. 2.6b. The comparison in *1D* between the Euclidean and city block distances for the same two points is given in Fig. 2.6 as well. One can see from the figure that the city block and Euclidean distances between the two points ($\boldsymbol{x}_i$ and $\boldsymbol{x}_j$) are not the same. City block distance is always larger than the Euclidean distance because of the triangle inequality.

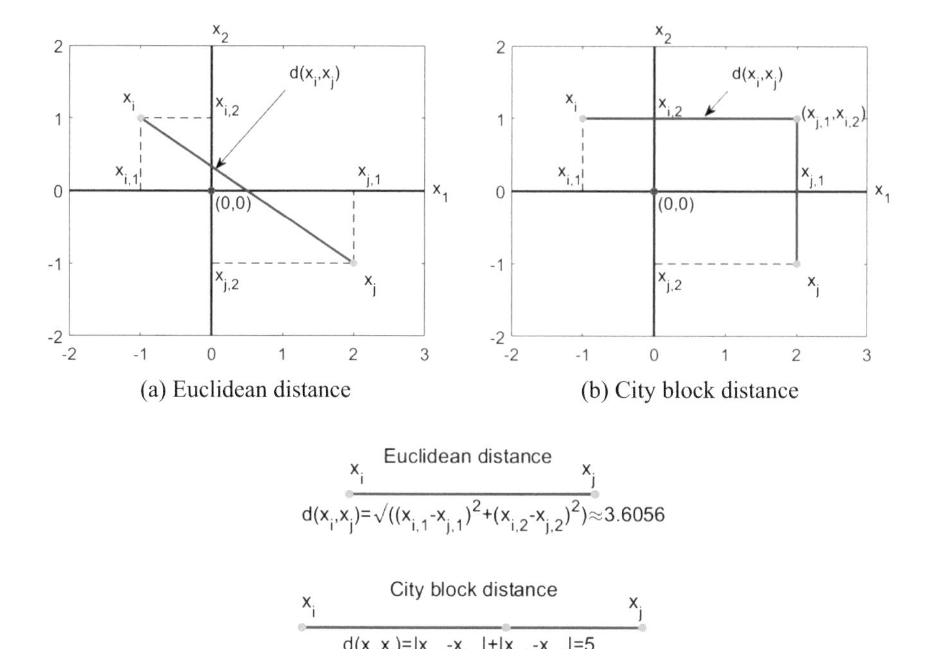

(c) Comparison between the Euclidean and city block distances in *1D*

Fig. 2.6 Illustrative examples of the Euclidean and city block distances

C. ***Minkowski distance***

The above two types of distance metrics can be generalized with the so-called "Minkowski type distance" named after the German mathematician Hermann Minkowski (1864–1909), which has the following form:

$$d(\boldsymbol{x}_i, \boldsymbol{x}_j) = \left(\sum_{l=1}^{N} \left| x_{i,l} - x_{j,l} \right|^h \right)^{\frac{1}{h}} \tag{2.45}$$

where $h \geq 1$. Minkowski type distance is also called the "L_h metric" and it is easy to see that it does provide a whole set of distance metrics with city block distance being Minkowski with $h = 1$ or L_1, Euclidean distance being Minkowski distance with $h = 2$ or L_2, etc., though the Minkowski type distances for other values of h do not have specific names.

D. ***Mahalanobis distance***

Mahalanbois type of distance metric does take into account the standard deviations of $\boldsymbol{x}_i$ from $\boldsymbol{x}_j$ in a vector form through the covariance matrix $\boldsymbol{\Sigma}_K$. The Mahalanbois distance between two data samples, $\boldsymbol{x}_i$ and $\boldsymbol{x}_j$, is calculated by:

$$d\left(\boldsymbol{x}_i,\boldsymbol{x}_j\right) = \sqrt{\left(\boldsymbol{x}_i - \boldsymbol{x}_j\right)^T \boldsymbol{\Sigma}_K^{-1} \left(\boldsymbol{x}_i - \boldsymbol{x}_j\right)} \tag{2.46}$$

where $\boldsymbol{\mu}_K = \frac{1}{K}\sum_{i=1}^{K} \boldsymbol{x}_i$; $\boldsymbol{\Sigma}_K$ is the covariance matrix of $\{\boldsymbol{x}\}_K$ obtained by:

$$\boldsymbol{\Sigma}_K = \frac{1}{K}\sum_{i=1}^{K} (\boldsymbol{x}_i - \boldsymbol{\mu}_K)(\boldsymbol{x}_i - \boldsymbol{\mu}_K)^T \tag{2.47}$$

Mahalanobis type distance can be seen as an extension of the Euclidean distance. Indeed, for the special case when $\boldsymbol{\Sigma}_K = \mathbf{I}$, Mahalanobis type distance becomes Euclidean. Through the covariance matrix the Mahalanobis type distance, in fact, automatically assigns different weight to each variable (dimension of the data vector). The values and the rank of the covariance matrix can also serve as indicators of the orthogonality (independence) of these variables: if the data are fully orthogonal (independent) the covariance matrix will be equal to the identity matrix:

$$\boldsymbol{\Sigma}_K = \mathbf{I} = \begin{bmatrix} 1 & 0 & \cdots & 0 \\ 0 & 1 & \cdots & 0 \\ \vdots & \vdots & \ddots & \vdots \\ 0 & 0 & \cdots & 1 \end{bmatrix} \tag{2.48}$$

One can also easily see that the Euclidean type of distance, in fact, assumes complete and ideal orthogonality and independence of the data which in reality is usually not the case (as discussed earlier, it is true for *fair games,* but not for real processes of interest).

For a better understanding, let us continue the example given in Fig. 2.6 and calculate the Mahalanobis distance between $\boldsymbol{x}_i$ and $\boldsymbol{x}_j$. However, because the covariance matrix of two data samples is invertible, there is no point in calculating the Mahalanobis distance between them. Therefore, eight randomly generated data samples are added to the data space, denoted by $\boldsymbol{x}_1, \boldsymbol{x}_2, \ldots, \boldsymbol{x}_8$, and they form a data matrix together with $\boldsymbol{x}_i$ and $\boldsymbol{x}_j$, denoted by $\mathbf{X}_{10} = \left[\boldsymbol{x}_i, \boldsymbol{x}_j, \boldsymbol{x}_1, \ldots, \boldsymbol{x}_8\right]$. Then, to calculate the Mahalanobis distance between $\boldsymbol{x}_i$ and $\boldsymbol{x}_j$, we need to, firstly, centralize the dataset to its mean. Centralization is an operation that is like standardization (will be described in detail in the next subsection), but without dividing by the standard deviation:

$$\mathbf{X}_{10} \leftarrow \mathbf{X}_{10} - \left[\overbrace{\boldsymbol{\mu}_{10}, \boldsymbol{\mu}_{10}, \ldots, \boldsymbol{\mu}_{10}}^{10}\right] \tag{2.49}$$

Secondly, the eigenvalues and eigenvectors of the covariance matrix of $\mathbf{X}_{10}$ are obtained:

$$\boldsymbol{\Sigma}_{10}[\boldsymbol{q}_1, \boldsymbol{q}_2] = [\boldsymbol{q}_1, \boldsymbol{q}_2]\begin{bmatrix} \lambda_1 & 0 \\ 0 & \lambda_2 \end{bmatrix} \tag{2.50}$$

where $\boldsymbol{\Sigma}_{10}$ is the covariance matrix; λ_1 and λ_2 denote the two eigenvalues and $\boldsymbol{q}_1, \boldsymbol{q}_2$ are the respective eigenvectors.

An eigenvector, $\boldsymbol{q}$ is a specific non-zero vector that is not rotated by a linear transformation using a transformation matrix, $\boldsymbol{\Sigma}$ applied on it. The corresponding eigenvalue, λ is the amount by which the eigenvector is stretched, namely,

$$\boldsymbol{\Sigma q} = \lambda \boldsymbol{q} \tag{2.51}$$

Thirdly, $\mathbf{X}_{10}$ is rotated using the eigenvectors after centralization:

$$\mathbf{X}_{10} \leftarrow [\boldsymbol{w}_1, \boldsymbol{w}_2] \cdot \mathbf{X}_{10} \tag{2.52}$$

Finally, after centralization and rotation, $\mathbf{X}_{10}$ is standardized using the eigenvalues:

$$\mathbf{X}_{10} \leftarrow \begin{bmatrix} \frac{1}{\sqrt{\lambda_1}} & 0 \\ 0 & \frac{1}{\sqrt{\lambda_2}} \end{bmatrix} \cdot \mathbf{X}_{10} \tag{2.53}$$

After the above process, the Euclidean distance between $\boldsymbol{x}_i$ and $\boldsymbol{x}_j$ is the Mahalanobis distance [28]. This process is also depicted in Fig. 2.7.

E. ***Cosine (dis)similarity***

Cosine similarity measures the cosine of the angle between two non-zero vectors of the inner product space. The illustrative example regarding the angle between $\boldsymbol{x}_i$ and $\boldsymbol{x}_j$ in Fig. 2.6 is given in Fig. 2.8.

The cosine similarity between $\boldsymbol{x}_i$ and $\boldsymbol{x}_j$ is formulated as [29]:

$$d(\boldsymbol{x}_i, \boldsymbol{x}_j) = \cos\left(\theta_{\boldsymbol{x}}^{i,j}\right) \tag{2.54}$$

where $\theta_{\boldsymbol{x}}^{i,j}$ denotes the angle between $\boldsymbol{x}_i$ and $\boldsymbol{x}_j$ in Euclidean space.

This can also be represented by the inner product of the two vectors:

$$d(\boldsymbol{x}_i, \boldsymbol{x}_j) = \frac{\langle \boldsymbol{x}_i, \boldsymbol{x}_j \rangle}{\|\boldsymbol{x}_i\| \|\boldsymbol{x}_j\|} \tag{2.55}$$

where $\langle \boldsymbol{x}_i, \boldsymbol{x}_j \rangle = \sum_{l=1}^{N} x_{i,l} x_{j,l}$ and $\|\boldsymbol{x}_i\| = \sqrt{\langle \boldsymbol{x}_i, \boldsymbol{x}_i \rangle}$.

We can further modify this expression as follows [30]:

$$d(\boldsymbol{x}_i, \boldsymbol{x}_j) = 1 - \frac{1}{2}\left\| \frac{\boldsymbol{x}_i}{\|\boldsymbol{x}_i\|} - \frac{\boldsymbol{x}_j}{\|\boldsymbol{x}_j\|} \right\|^2 \tag{2.56}$$

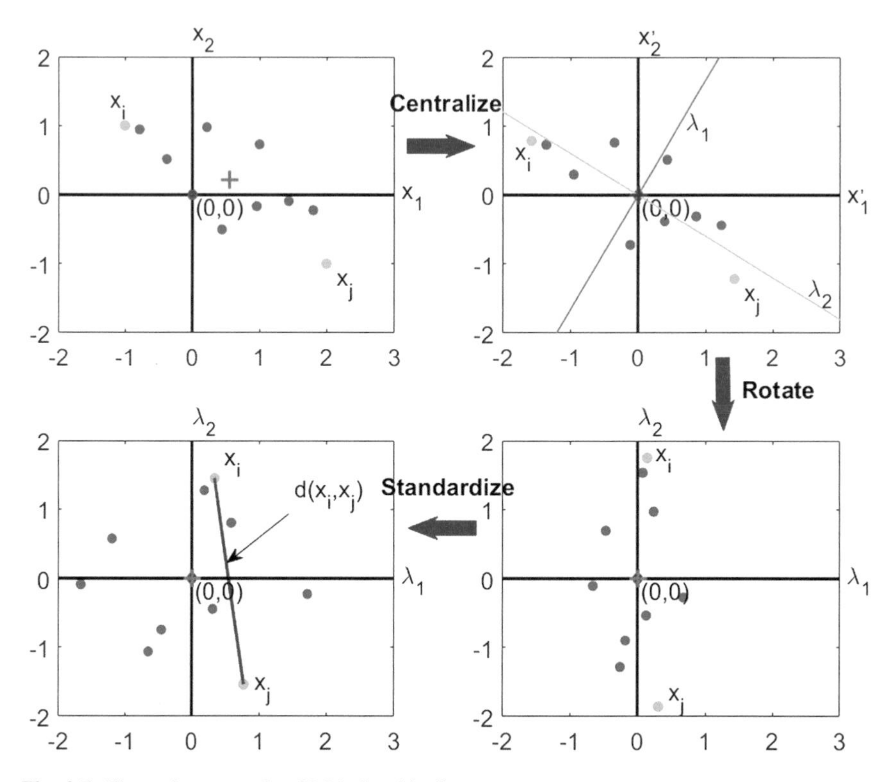

Fig. 2.7 Illustrative example of Mahalanobis distance

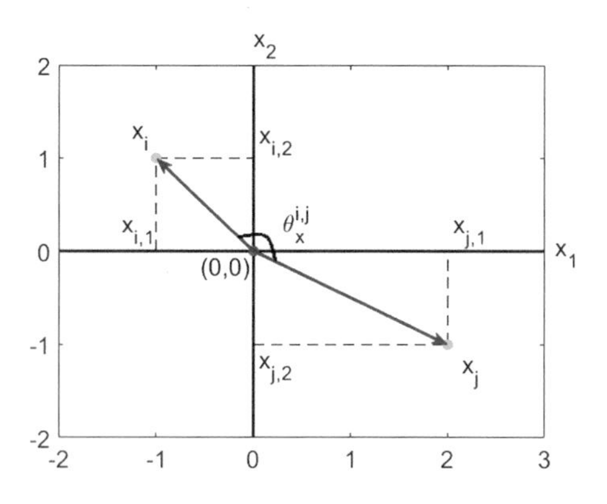

Fig. 2.8 Illustrative example of the angle between $\boldsymbol{x}_i$ and $\boldsymbol{x}_j$

The cosine similarity can be simplified in the form of Eq. (2.56) because of the following transformation [30]:

$$\frac{\sum_{l=1}^{N} x_{i,l}x_{j,l}}{\|\boldsymbol{x}_i\|\|\boldsymbol{x}_j\|} = \frac{\sum_{l=1}^{N} x_{i,l}x_{j,l}}{\|\boldsymbol{x}_i\|\|\boldsymbol{x}_j\|} - \frac{\sum_{l=1}^{N} x_{j,l}^2}{2\|\boldsymbol{x}_j\|^2} - \frac{\sum_{l=1}^{N} x_{i,l}^2}{2\|\boldsymbol{x}_i\|^2} + 1$$
$$= 1 - \frac{1}{2}\sum_{l=1}^{N}\left(\frac{x_{i,l}}{\|\boldsymbol{x}_i\|} - \frac{x_{j,l}}{\|\boldsymbol{x}_j\|}\right)^2 \tag{2.57}$$

Cosine dissimilarity is also often formulated as:

$$d(\boldsymbol{x}_i, \boldsymbol{x}_j) = 1 - \cos\left(\theta_{\boldsymbol{x}}^{i,j}\right) = \frac{1}{2}\left\|\frac{\boldsymbol{x}_i}{\|\boldsymbol{x}_i\|} - \frac{\boldsymbol{x}_j}{\|\boldsymbol{x}_j\|}\right\|^2 \tag{2.58}$$

F. ***Analysis***

The Euclidean distance is the most widely used and versatile distance metric due to its simplicity and computational efficiency. The shapes of clusters formed using the Euclidean and Mahalanobis distances (hyper-spherical and hyper-ellipsoidal, respectively) are not flexible and often fail to cover well the actual data. Furthermore, Mahalanobis type distance requires the covariance matrix of the data to be calculated and inverted which itself is a computationally very expensive operation and sometimes it leads to so-called *singularity* [31]. For some problems, such as natural language processing (NLP), for example, Minkowski type distances give contradictory to the common logic results, and cosine dissimilarity is usually used [32, 33]. There are several reasons for this. One of them is related to the fact that the number of features in NLP problems (usually the keywords used in a text) is often high (hundreds or thousands). Another reason is that in NLP problems quite often the dimensionality of each individual data item is different and the representation is sparse (with many zeros). For example, one may be interested to compare the (dis)similarity between two texts or textual documents whereby in one of them there are N_1 keywords and in the other one—N_2 keywords. Last, but not least, logically, if we (as humans) compare two texts or textual documents and in one of them there is 65 times a certain keyword of interest (say "*Trump*") and in another text, we have 70 times the same keyword we can conclude that both texts are about the current USA President. If, however, we compare two textual documents and in one of them we have 5 times the same keyword and in the other one we have no mention of it (0 times), we can conclude that the first document/text is talking about the current United States President while the second one is **not**. Even if we have 6 and 1 times mentioning instead of 5 and 0, respectively, this will still have different meaning in terms of human perception: one of them is talking about him while the second one is barely mentioning the name once. If, however, we use Euclidean distance between these two texts/documents the result in all three cases (70 and 65; 5 and 0; 6 and 1) will be the same and equal to 5. Such a contradiction is not to be ignored. For all these reasons in NLP they use cosine (dis)similarity even if it is not a full distance metric [18].

In any case, it has to be stressed that the most suitable choice of distance measure is usually problem- and domain-specific.

2.2.1.2 Normalization and Standardization

Normalization and standardization are examples of a data pre-processing technique that is practically compulsory at the start of the process (except the cases when the variables in all dimensions have comparable nominal value ranges). In practice, using either normalization, standardization or both (standardization first followed by normalization) combining with outlier detection is necessary [6]. These pre-processing operations re-scale the values of features (input variables) into a more familiar range, and make different features comparable in terms of their values and ranges [6]. Thus, the two techniques can reduce the influence of differences in the magnitudes of the input variables. It is important to remove outliers before applying normalization, because otherwise they will highly influence the result. The *normalization*, which is also called *feature scaling*, converts the nominal numerical values on different features (dimensions of the input data vector) to the interval $[0, 1]$ or sometimes to $[-1, 1]$:

$$x'_{j,i} = \frac{x_{j,i} - x_{\min,i}}{x_{\max,i} - x_{\min,i}}, \quad x'_{j,i} \in [0, 1], \quad \boldsymbol{x}_j \in \{x\}_K \tag{2.59a}$$

$$x'_{j,i} = \frac{2\left(x_{j,i} - x_{\min,i}\right)}{x_{\max,i} - x_{\min,i}} - 1, \quad x'_{j,i} \in [-1, 1], \quad \boldsymbol{x}_j \in \{x\}_K \tag{2.59b}$$

where $x_{\min,i}$ and $x_{\max,i}$ are the minimum and maximum values of the ith feature of $\{x\}_K$.

In this way, the data space is transformed into a (hyper-)cube with size $[0, 1]^N$ (or $[-1, 1]^N$), see Fig. 2.9.

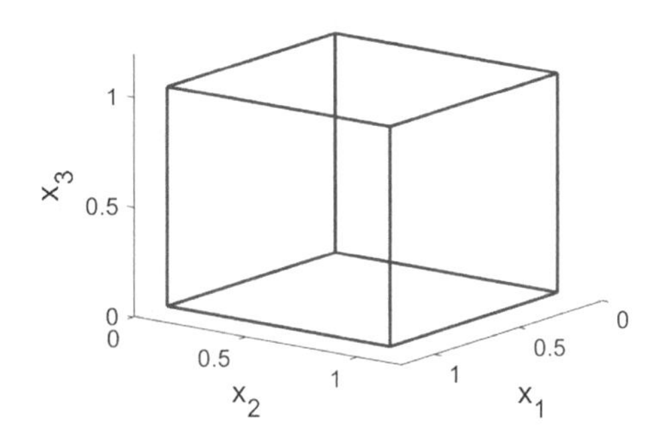

Fig. 2.9 Normalization results in compressing the data space into a hypercube with size $[0, 1]^N$, $N = 3$

Standardization is an alternative form for re-scaling the values of the features into a more convenient range, and making them comparable in terms of their values and ranges. It transforms the data space into a hypercube with size $[-3,3]^N$ for the vast majority of the data (this is a consequence of the Chebyshev theorem and will be further discussed again in more detail). It is using the statistical parameters of the data per feature (input variable) [6]:

$$x'_{j,i} = \frac{x_{j,i} - \mu_{K,i}}{\sigma_{K,i}}, \quad \boldsymbol{x}_j \in \{x\}_K \tag{2.60}$$

where $\mu_{K,i}$ and $\sigma_{K,i}$ denote the mean and standard deviation of the ith feature of $\{x\}_K$, respectively.

Standardization results in compressing the vast majority of data ($> \frac{8}{9}$ in any case, rising up to >99.7 if data follows a Gaussian distribution) into a (hyper-) cube with size $[-3,3]^N$. Outliers are left out of the (hyper-) cube. The hypercube after the standardization is illustrated in Fig. 2.10.

Analysis

Normalization requires the range (and, respectively, the minimum and maximum values) per feature/input to be known and fixed in advance. Thus, normalization is often limited to offline applications or assumption of the ranges of the variables or a recalculation of the ranges [6]. In contrast, standardization is very convenient for streaming data processing as both, the mean and variance can be updated online. Compared with normalization, the ranges of different standardized features may not be the same because different features may follow different distributions and, thus, have different statistical characteristics, most notably, the standard deviation. Nonetheless, it is theoretically proven with the Chebyshev inequality [34] that *for an arbitrary distribution* the probability that the standardized data will lie outside the range $[-3;3]$ is less than $1/9$ and for the Gaussian distribution it is easy to demonstrate that this percentage is less than 0.3%.

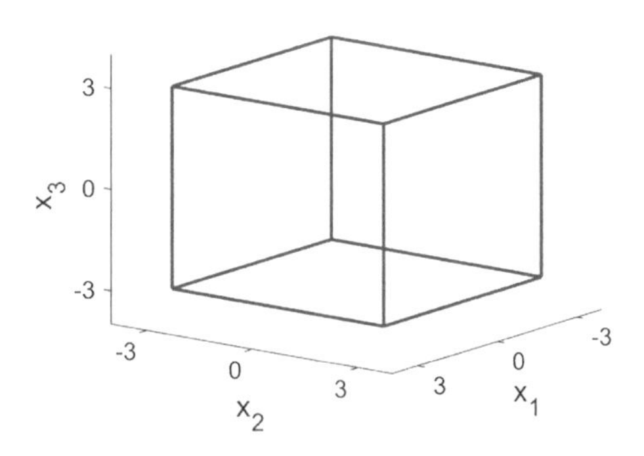

Fig. 2.10 Standardization results in compressing the data space into a hypercube with size $[-3,3]^N$, $N = 3$

The recommended sequence of applying the pre-processing techniques is as follows:

Step 1. Apply standardization to the raw data;

Step 2. Analyze the candidate outliers (points that are outside the $[-3,3]^N$). If the natural distribution of the data is Gaussian then the amount of such suspected to be outliers data points will be no more than 0.3% of all data samples, but in any case according to the Chebyshev inequality the amount of such data points will not be more than $\frac{1}{9}$ (approximately 11%) of all data points. In reality, their number can be much smaller and one may use domain knowledge, expert judgment or other techniques or simply rely on statistical characteristics and apply a threshold.

Step 3. Once outliers are detected and removed from the raw data, one can apply normalization to the already standardized data. Now, we know for certain that for all the remaining data, the range of all input variables/features is $[-3,3]^N$. This means that the range in each dimension $(1,2,\ldots,N)$ is equal to 6. Therefore, such an approach is suitable for online applications as well.

2.2.1.3 Feature Extraction

In many machine learning problems, the input variables (or features as they are called in pattern recognition literature) are clear and their choice is obvious. For example, in technical systems these are often temperatures, pressures, weight, flow rates, etc. Their choice often requires domain and expert knowledge. In some problems, e.g. image processing, the features are pixel color or intensity properties or high level properties of the image, such as size, shape, roundness, etc. [35].

2.2.1.4 Feature Selection

High dimensional, complex datasets usually contain a number of redundant and/or irrelevant features with trivial information. Processing all the features of this kind at the same time requires a large amount of memory and computational resources without gaining much information. Learning algorithms may result in overfitting due to the low generalization [36]. In addition, this may result in the so-called "curse of dimensionality". Therefore, it is of great importance to remove such features from the data.

Feature/input selection techniques are commonly used in the fields of machine learning and statistics for selecting out a subset of the most effective features from the original features for further processing, i.e. building robust learning models for clustering, classification and prediction. There are four major benefits for conducting feature/input selection [37]:

(1) improve the interpretability of the learning model;
(2) enhance the generalization ability of the learning model;
(3) speed up the learning process;
(4) avoid the "curse of dimensionality".

Furthermore, in some specific problems, e.g. spectroscopy studies in biomedical research this can lead to identification of a relatively small number of so-called "biomarkers" [38]. This is related to the fact that relatively narrow bands of the spectra correspond to specific proteins and, respectively, amino acids, and this information can be very useful to isolate specific cures or medications to treat some conditions [39, 40].

Another important aspect of inputs and feature selection is the ability to do this automatically and online (while the classifier or predictor) is in use [6].

2.2.1.5 Feature Orthogonalization

Orthogonalization is an effective approach for eliminating the correlation between vectors. As it was stated in the previous subsection, many features/inputs in high dimensional datasets are often (highly) correlated, and processing these kinds of datasets without feature reduction may result in problems, such as overfitting, a waste of storage and computation resources.

Principle component analysis (PCA) [14] is a widely used in machine learning and pattern recognition technique for reducing both, the complexity and independence of input features, and, thus, achieving orthogonalization as well as feature reduction/selection. The idea of PCA is to project the original set of inputs/features into a new feature space resulting in a new set of inputs/features that are orthogonal and independent. This new set is called principal components (PC). Thus, the redundancy and interdependence between the original inputs/features are eliminated or reduced. The newly generated set of features/inputs is linear combinations of the original ones and thus, preserves the variance in the original set. As the most of the variance is captured by the first few PCs, the rest of the PCs can be ignored and the dimensionality of the feature/input space can be significantly reduced [6].

The PCA is an offline approach that is based on singular-value decomposition (SVD) [4, 41]. The first step of the PCA is to centralize the data

$$\mathbf{X}'_K = \mathbf{X}_K - \left[\overbrace{\boldsymbol{\mu}_K, \boldsymbol{\mu}_K, \ldots, \boldsymbol{\mu}_K}^{K}\right] = [\boldsymbol{x}_1, \boldsymbol{x}_2, \ldots, \boldsymbol{x}_K] - \left[\overbrace{\boldsymbol{\mu}_K, \boldsymbol{\mu}_K, \ldots, \boldsymbol{\mu}_K}^{K}\right] \tag{2.61}$$

Then, SVD is applied using $\mathbf{X}'_K$ as follows:

$$\mathbf{X}'_K = \mathbf{QSE}^T \tag{2.62}$$

where $\mathbf{Q}$ is a $N \times N$ matrix of the eigenvectors of $\mathbf{X}'_K(\mathbf{X}'_K)^T$, which performs a rotation of the original axes of the inputs; $\mathbf{E}$ is a $K \times K$ matrix of the eigenvectors of $(\mathbf{X}'_K)^T\mathbf{X}'_K$, which performs another rotation; $\mathbf{S}$ is the rectangular diagonal matrix with non-negative real numbers on the main diagonal.

The PCs of $\{\boldsymbol{x}\}_K$ are extracted using the eigenvectors in $\mathbf{Q}$ corresponding to the first few largest eigenvalues (assuming the first n) in the following form:

$$\mathbf{X}''_K = [\boldsymbol{q}_1, \boldsymbol{q}_2, \ldots, \boldsymbol{q}_n]^T \cdot \mathbf{X}'_K \tag{2.63}$$

where $\mathbf{X}''_K$ is the $n \times K$ dimensional PC matrix of $\{\boldsymbol{x}\}_K$; $\boldsymbol{q}_i$ is the $N \times 1$ dimensional eigenvector corresponding to the ith largest eigenvalue.

Other problems to be considered as part of pre-processing include:

(1) dealing with missing data [42];
(2) dealing with categorical/non-numerical/qualitative data [43].

In this book, we focus on the other aspects and the interested readers are directed, for example, to [6, 16, 17] for more details.

2.2.2 Unsupervised Learning (Clustering)

Clustering is a form of data partitioning and is considered as an unsupervised machine learning technique for finding out the underlying groups and pattern within the data. Clustering algorithms might have a wide variety of goals, but the ultimate one is to group or partition a collection of data into subsets or "clusters" such that data samples within the same cluster are more closely related to each other than to data samples from other clusters [14]. Clustering technique is also intrinsically related to complex system structure identification [6], where the technique is used for identifying the multimodal distributions and forming local model architecture.

There is a wide variety of clustering algorithms and techniques which can be grouped into [44]:

(1) hierarchical clustering algorithms;
(2) centroid-based clustering algorithms;
(3) density-based clustering algorithms;
(4) distribution-based clustering algorithms;
(5) fuzzy clustering algorithms.

Since it is impossible to cover all of the existing clustering algorithm, in this subsection, the most typical and representative ones are briefly reviewed. For more details, the interested readers are referred, for example, to [44, 45].

2.2.2.1 Hierarchical Clustering Algorithms

Hierarchical clustering algorithms [46] produce dendrograms representing nested groupings of patterns controlled by the distances or linkages between them at different granularity levels. The clustering result is achieved by cutting the dendrogram at the desired distance value using a threshold. The dendrogram generated using the Fisher iris dataset [47] (downloadable from [48]) is given in Fig. 2.11 as an illustrative example.

Currently, there are two major types (based on the bottom-up or top-down approaches, respectively) of hierarchical clustering algorithms, namely:

(1) agglomerative clustering approaches [49, 50], which initially treat each data sample as a cluster on its own, and then, merge them successively until the desired clustering structure is obtained.
(2) divisive clustering approaches [51, 52], which achieve the clustering result via a contrary direction by starting with treating all the data samples as a single cluster and successively dividing the cluster into sub-clusters until the desired clustering structure is obtained.

The most well-known hierarchical clustering algorithms include the BIRCH [53] and a more recently introduced one named *affinity propagation* [54].

2.2.2.2 Centroid-Based Clustering Algorithms

Centroid-based clustering algorithms begin with an initial partitioning and relocate instances by moving them from one cluster to another. The methods require an exhaustive enumeration process of all possible partitions and use certain greedy heuristics for iterative optimization.

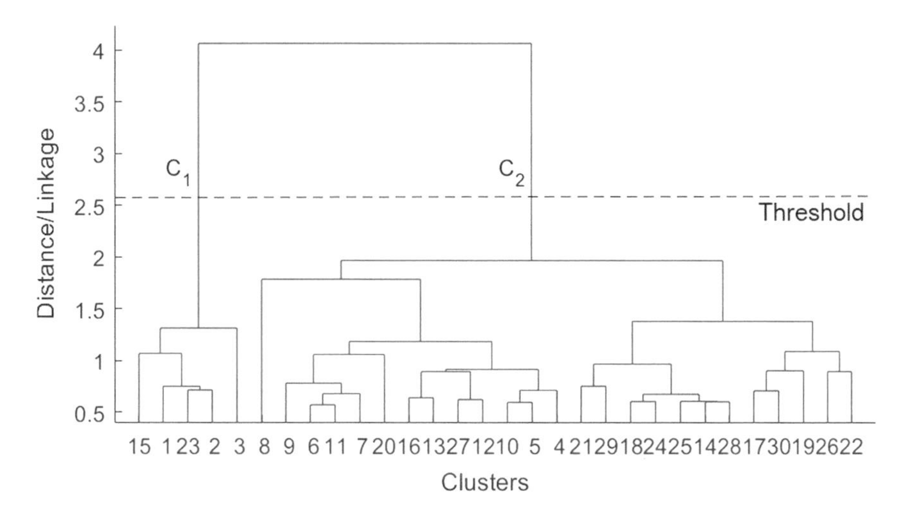

Fig. 2.11 An example of a dendrogram based on Fisher Iris dataset

The basic idea of the centroid-based clustering algorithms is to find a clustering structure that minimizes a certain error criterion that measures the distance of each data sample to its representative value (centroid), and the process is called error minimization. The most well-known criterion is the sum of squared error minimization:

$$J = \sum_{i=1}^{M} \sum_{\boldsymbol{x} \in \mathbf{C}_i} \left\| \boldsymbol{x} - \boldsymbol{c}_i \right\|^2 \rightarrow \min \tag{2.64}$$

where $\boldsymbol{c}_i$ is the center of the cluster $\mathbf{C}_i$; $i = 1, 2, \ldots, M$; M is the number of centers.

The most commonly used centroid-based algorithm is the k-means [55]. The k-means algorithm starts by randomly initializing k cluster centers, and then, the algorithm iteratively assigns data samples to the closest centers and updates the centers until the predefined termination condition is satisfied [44].

2.2.2.3 Density-Based Clustering Algorithms

Density-based clustering approaches assume that clusters exist in areas of higher density of the data space. Each cluster is characterized by a local mode or maximum of the density function.

One of the most popular density based clustering method is the DBSCAN (density-based spatial clustering of applications with noise) algorithm [56]. The main idea of DBSCAN is to group data samples that are very close together in the data space, and mark data samples that lie alone in low-density as outliers. The algorithm begins with an arbitrary data sample that has not been visited before by extracting the neighborhood and checking to see whether this area contains a sufficient number of data samples. If so, a cluster is started; otherwise, it is labelled as noise. If a data sample is found to be in a dense part of a cluster, its neighboring area is also a part of this cluster, and, thus, all the data samples located in that area are added to the cluster. The process continues until the cluster is completely built. After this, an unvisited data sample is retrieved and processed to form another cluster or be identified as noise.

Mean-shift [57, 58] is another widely used density-based algorithm built upon the concept of KDE (see Sect. 2.1.4.2). Mean-shift algorithm implements the KDE idea by iteratively shifting each data sample to the densest area in its vicinity until all the data samples converge to local maxima of the density. As the mean-shift algorithm is limited to offline applications, the evolving local means (ELM) algorithm inheriting the basic concept of the mean-shift was introduced for online scenarios [59].

eClustering algorithm [60] is the most popular online density-based clustering approach for streaming data processing, which can self-evolve its structure and update its parameters in a dynamic way. It is able to successfully handle the drifts and shifts of the data pattern in the data streams [61]. eClustering algorithm opens the door for the evolving clustering approaches and a number of modifications have been introduced later [59, 62, 63]. It is also one of the theoretical bases of the *empirical* machine learning techniques described in this book in the later chapters.

2.2.2.4 Distribution-Based Clustering Algorithms

Distribution-based clustering algorithms assume that the data samples of each cluster are generated from a specific probability distribution, and the overall data distribution is *assumed* to be a mixture of several distributions. Thus, algorithms of this type are closely related to the statistics.

The most prominent distribution-based algorithm is known as Gaussian mixture models (GMM) [64–67]. GMM algorithm assumes the generation model of data to be a mixture of Gaussian distributions. It *randomly* initializes a number of Gaussian distributions and *iteratively* optimizes the parameters to fit the model.

2.2.2.5 Fuzzy Clustering Algorithms

Traditional clustering algorithms generate partitioning, in which each data sample belongs to one and only one cluster, and thus, the clusters are disjointed. Fuzzy clustering, however, allows a partial and dual or multiple memberships which leads to the concept of fuzzy clusters [68, 69]. This means a data sample can belong to different clusters at the same time.

The most representative fuzzy clustering approach is the well-known fuzzy c-means algorithm [68, 69]. Fuzzy c-means (FCM) algorithm is based on the minimization of the following equation:

$$J = \sum_{j=1}^{K}\sum_{i=1}^{M} \lambda_{j,i}^{n}\left\|\boldsymbol{x}_j - \boldsymbol{c}_i\right\|^2 \rightarrow \min \tag{2.65}$$

where $\lambda_{j,i}$ is the degree of membership of $\boldsymbol{x}_j$ in the ith cluster; n is fuzzy partition matrix component for controlling the degree of fuzzy overlap. The values of the degrees of membership, $\lambda_{j,i}$ are between 0 and 1.

The general procedure of the FCM algorithm is quite similar to the k-means approach. While in FCM algorithm, each cluster is a fuzzy set. Larger membership degrees suggest higher confidence in the assignment and, vice versa. A non-fuzzy clustering result can also be achieved by applying a threshold over the membership degrees.

2.2.2.6 Analysis

Clustering is considered to be the typical example of unsupervised machine learning technique. However, current clustering algorithms require various types of parameters to be predefined by the users. Taking the representative clustering algorithms mentioned in this subsection as examples, the hierarchical clustering algorithms usually require the threshold to be set for cutting the dendrogram [46, 53]; the k-means algorithm requires the number of clusters, k to be predefined [55]; DBSCAN requires two parameters: the maximum radius of the neighborhood and the minimum number of points required to form a dense region [56]; Mean-shift algorithm requires the kernel function as well as the threshold to be chosen in advance [57, 58]; ELM algorithm requires the initial radius of the clusters to be pre-defined [59]; eClustering algorithm requires the initial radius of the clusters to be pre-defined [70]; GMM requires the number of models to be given beforehand [64–67]; FCM requires the membership function type and number of clusters to be decided [68, 69].

These algorithmic parameters significantly influence the performance and efficiency of the unsupervised clustering algorithms. Pre-setting the suitable values of these algorithmic parameters is always a challenging task and requires a certain amount of prior knowledge. In real situations, however, prior knowledge is very limited and not enough for users to decide the best inputs for these algorithms in advance. Without carefully chosen parameters, the efficiency as well as the effectiveness of the clustering algorithms may be quite poor.

Moreover, many clustering algorithms impose pre-assumed data generation model. For example, DBSCAN, Mean-shift and GMM algorithms assume that the data is drawn from a normal/Gaussian distributions [56–58, 64–67]. However, the prior assumptions usually are too strong requirement to be satisfied in real-world applications. If the prior assumptions are satisfied, clustering results can be very good; otherwise, the quality of the clustering results is very poor.

2.2.3 *Supervised Learning*

2.2.3.1 Classification

Classification is the task of assigning a class label to an input data sample. The class label indicates one of a given set of classes, thus, it usually is an integer value. In contrast to clustering, classification is usually considered as a supervised machine learning technique [71]. We should stress that some forms of classification can be performed in a semi-supervised or even in a fully unsupervised formed as well as in an evolving form of machine learning [72–75]. We will come back to this in Part II of the book. In this section, the following most widely used and representative classification approaches are reviewed, which include:

(1) Naïve Bayes classifier;
(2) k-nearest neighbour (kNN) classifier;
(3) Support vector machine (SVM) classifier;
(4) Offline fuzzy rule-based (FRB) classifier;
(5) eClass classifier.

For more details, the interested readers are referred, for example, to [3, 6, 71].

A. ***Naïve Bayes Classifier***

Naive Bayes classifier is one of the most widely studied and used classification algorithms deeply rooted in the traditional probability theory and statistics [36]. The naïve Bayes classifier is based on the conditional PDFs [Eq. (2.35)] derived from the training data samples:

$$P(\boldsymbol{x}|\mathbf{C}_i) = \prod_{j=1}^{N} P\left(x_j|\mathbf{C}_i\right) \tag{2.66}$$

which is obtained based on the prior assumption that different features of the data are statistically independent.

The computational efficiency of the naïve Bayes classifier is quite high, and it only requires a small number of training data. However, the drawbacks of the naïve Bayes classifier are also obvious:

(1) Its prior assumption, although simple, is often not held in real cases;
(2) The choice of the PDFs requires prior knowledge and can influence the efficiency and accuracy of the classifier if it is not properly set;
(3) Its model is over simplified, which makes it insufficient in dealing with complex problems.

B. ***kNN Classifier***

Nearest neighbor rule [76] is the simplest non-parametric procedure for deciding the label of an unlabeled data sample. kNN classifier [76–78] is the most representative algorithm employing the nearest neighbor rule directly. The algorithm is also among the simplest of all machine learning algorithms. The label of a particular data sample is decided by the labels of its k nearest neighbors based on a voting mechanism. The kNN algorithm mainly conducts computation during the classification stage.

kNN is also one of the most widely used classifiers, however, the main drawbacks of the algorithm are:

(1) The best choice of k is data-dependent, which means that it requires prior knowledge to make this decision;
(2) kNN classifier is also sensitive to the structure of the data. Its performance severely deteriorates with the noisy, irrelevant features or unbalanced feature scales.

C. ***SVM Classifier***

SVM [79] is one of the most popular classification algorithms. In essence, SVM is an algorithm for maximizing the class separation margin with respect to a given collection of data [80]. There are four very important basic concepts within the SVM classifier [80].

(1) hyperplane, which separates the data samples of two different classes [36, 80, 81].
(2) maximum margin, which is the zone around the hyperplane with the maximum distance from it to the nearest support vector on each side [36, 80, 81]. The maximum-margin hyperplane is the key to the success of the SVM [81];
(3) soft margin, which allows the SVM to deal with errors and overfitting by allowing few anomalies to fall on the wrong side of the separating hyperplane without affecting the final result [80]. Essentially, it provides some tolerance;
(4) kernel function, which itself is a mathematical expression. It is applied to get the maximum margin. The kernel function provides a way around the nonlinear separability problems by changing the dimensionality of the data and it projects the data from a low dimensional space to a higher, infinite dimensional space [80].

Some common kernels include [82]:

- Polynomial:

$$\kappa(\boldsymbol{x}_i, \boldsymbol{x}_j) = \left(\boldsymbol{x}_i^T \boldsymbol{x}_j + 1\right)^n; \quad i, j = 1, 2, \ldots, K \tag{2.67}$$

- Gaussian radial basis function:

$$\kappa(\boldsymbol{x}_i, \boldsymbol{x}_j) = e^{-\frac{\left\|x_i - x_j\right\|^2}{2\sigma^2}}; \quad i, j = 1, 2, \ldots, K \tag{2.68}$$

There is also a TEDA (*typicality* and *eccentricity*-based data analytics) kernel recently introduced into the SVM, which is based on the *eccentricity* (see Sect. 4.3 for more details) and is directly learned from the *empirically* observed data [83]. Moreover, the TEDA kernel function can be further updated in an online manner.

Although, the SVM is one of the most widely used classifiers and is able to exhibit very good performance in various classification problems, there are some major drawbacks [36]:

(1) It requires prior knowledge for choosing the kernel function. If the kernel function is not correctly chosen, its performance is not guaranteed;
(2) Its computational efficiency decreases quickly in large-scale problems;
(3) It requires a full retraining if more new training samples are available;
(4) It is less efficient in handling multi-class classification problems;
(5) It requires few parameters to be pre-defined, especially in regards to the soft margin.

D. ***Offline FRB Classifiers***

A FRB classifier is a system consisting of a number of fuzzy rules for classification purpose on the basis of the fuzzy sets theory and fuzzy logic introduced by Lotfi A. Zadeh in 1965 [84]. Typically, there are two types of FRB classifiers, namely, the zero-order type and the first-order type. Zero-order classifiers use the fuzzy rules of Zadeh-Mamdani type [85], and the first-order ones employs the fuzzy rules of

Takagi-Sugeno type [60, 70, 86]. More on the topic of fuzzy sets will be discussed in Sect. 3.1.

Traditional FRB classifiers were initially designed for expert-based systems, their applications are seriously restricted. The design of a particular FRB classifier requires a large amount of prior knowledge of the problem as well as heavy involvement of human expertise. Furthermore, once implemented, they are limited to offline scenarios [87, 88]. Data-driven FRB classifiers were also developed [89, 90].

More detailed descriptions of the fuzzy set theory and FRB systems are given in Sect. 3.1.

E. ***eClass Classifier***

One of the first applications of the so-called evolving FRB (which will be also discussed in Chap. 3 and are at the basis of the approach proposed in this book) is the dynamically self-evolving classifiers [91].

The most widely used and successful evolving fuzzy classifier is eClass [73], which is a classification approach for streaming data processing. It is able to self-evolve the classifier structure and update its meta-parameters on a sample-by-sample basis. eClass classifier [73] has two versions, the first one is eClass0, which uses Zadeh-Mamdani type fuzzy rules [Eq. (3.1)] and can be even trained in an unsupervised manner, and the other one is eClass1, which employs the Takagi-Sugeno type fuzzy rules [Eq. (3.2)]. There are a lot of evolving classifiers introduced on the basis of the eClass classifier [73] including: simpl_eClass [92], autoClass [93], TEDAClass [94], FLEXFIS-Class [95], PANFIS (parsimonious network based on fuzzy inference system) [96], GENFIS (generic evolving neuro-fuzzy inference system) [97], etc.

2.2.3.2 Regression and Prediction

Regression is a statistical process for estimating the relationships among variables and it is commonly used for prediction and forecasting in various areas including engineering [98, 99], biology [14], economy and finance [100, 101]. The most widely used regression algorithm is the linear regression [14]. Linear regression is a simple, linear, offline algorithm which has been studied rigorously, and used extensively in practical applications in the pre-computer area of statistics [102]. However, even now, it is still the predominant *empirical* tool in economics [14]. Another widely used algorithm is the adaptive-network-based fuzzy inference system (ANFIS), which was introduced in 1993 [103] as a kind of artificial neural network (ANN) based on Takagi-Sugeno fuzzy inference system.

Nowadays, due to the fact that we are faced not only with large datasets, but also with huge data streams, the traditional simple offline algorithms are not sufficient to meet the need. The more advanced evolving intelligent systems are also being developed and widely applied for the purpose of prediction [61]. The two most representative algorithms to learn evolving intelligent system are the evolving Takagi-Sugeno (eTS) fuzzy model [70, 104–106] and the dynamic evolving neural-fuzzy inference system (DENFIS) [107]. There are also many other well-known approaches including, flexible fuzzy inference systems (FLEXFIS) [108], sequential adaptive fuzzy inference system (SAFIS) [109, 110], correntropy-based evolving fuzzy neural system (CENFS) [111], interval-based evolving granular system (IBeGS) [112], etc.

In this section, the following four well-known algorithms are briefly reviewed.

(1) Linear regression;
(2) ANFIS;
(3) eTS;
(4) DENFIS.

For more details, the interested readers are referred, for example, to [3, 4, 6, 14].

A. ***Linear Regression***

This model assumes that the regression function is linear in terms of inputs. The linear model is one of the most important tools in the area of statistics. Linear models are quite simple, but can provide an adequate and interpretable description of the relationship between the inputs and outputs [14].

Using $\boldsymbol{x} = [x_1, x_2, \ldots, x_N]^T$ as the input vector, the linear model to predict the output y is given as:

$$y = a_0 + \sum_{j=1}^{N} a_j x_j = \bar{\boldsymbol{x}}^T \boldsymbol{a} \tag{2.69}$$

where $\bar{\boldsymbol{x}} = [1, x_1, x_2, \ldots, x_N]^T$ and $\boldsymbol{a} = [a_0, a_1, a_2, \ldots, a_N]^T$.

The most popular approach to identification of parameters $\boldsymbol{a}$ is the least squares (LS) method [14]:

$$\boldsymbol{a} = \left(\bar{\boldsymbol{x}}^T \bar{\boldsymbol{x}}\right)^{-1} \bar{\boldsymbol{x}} y \tag{2.70}$$

There are also many different modifications of the linear regression algorithm. One of the most representative such modifications is the sliding window linear regression, which has been widely used in the finance and economics [113].

Although, the linear regression model is still one of the most popular regression algorithms, its major drawbacks are also obvious:

(1) Linear model oversimplifies the problems;
(2) Linear model is insufficient in dealing with complex and large-scale problems;
(3) Its training process is offline.

B. ***ANFIS***

ANFIS is a data learning technique that uses fuzzy logic to transform given inputs into a desired output through a simple neural network (NN) with a predefined structure [103]. Since ANFIS integrates both NNs and fuzzy logic principles, it has potential to capture the benefits of both in a single framework and has the learning capability to approximate nonlinear functions. Since, for both, ANN and FRB systems, it has been proven that they are universal approximators (see [114] and

[115]), ANFIS is also a universal approximator. The details of the ANFIS will be descripted in Sect. 3.3.

ANFIS was developed in the era when the datasets were predominantly static and not complicated. Therefore, ANFIS system is insufficient for the contemporary real applications nowadays due to the following drawbacks:

(1) The structure of the fuzzy inference system needs to be predefined, which requires prior knowledge and a large number of *ad hoc* decisions.
(2) Its structure is not self-evolving and its parameters cannot be updated online.

C. ***eTS***

The eTS method was firstly introduced in [105, 116] and ultimately in [70, 106]. Nowadays, together with its numerous extensions and modifications, it is the most widely used approach for learning of evolving intelligent system. The learning mechanism of the eTS system is computationally very efficient because it is fully recursive. Unlike most of the alternative techniques, it can uniquely start "from scratch". The two phases include:

(1) Data space partitioning, which is achieved by the eClustering algorithm [60] (also see Sect. 2.2.2.4), and based on the partitioning to form and update the fuzzy rule-based models structure;
(2) Learning parameters of the consequent part of the fuzzy rules.

In eTS system, there are outputs and the aim is to find such (perhaps overlapping) clustering of the joint input-output data space that fragments the input-output relationship into locally valid simpler (possibly linear) dependences [61]. The consequent parameters of the eTS system are learned by the fuzzily weighted recursive least squares (FWRLS) [70]. The detailed formulation of the FWRLS algorithm can be found in Sect. 8.3.3.2.

Due to its generic nature, the eTS system has been widely applied to different problems including, but not limited to clustering, time series prediction, control.

D. ***DENFIS***

DENFIS is another popular approach for learning of evolving intelligent system [107], which emerged in parallel and independently of eTS. DENFIS is able to generate a neural fuzzy model through an efficient adaptive offline learning process, and can conduct accurate dynamic time series prediction. The online version requires an initial offline seeding training. Its online learning is achieved by the evolving clustering method (ECM), which, essentially, can be viewed as a greedy clustering algorithm using a threshold with a mechanism of incrementally gaining new clusters [107]. Its offline learning process is also very similar to the k-means algorithm, which requires the number of clusters to be predetermined [107].

DENFIS [107], both online and offline versions, uses Takagi-Sugeno type inference engine (described in more detail in the next chapter) [60, 86]. At each time moment, the output of DENFIS is calculated based on the most activated fuzzy rules, which are dynamically chosen from the fuzzy rule base.

Despite of being widely used, the major drawbacks of the DENFIS algorithm are:

(1) It requires prior assumptions and predefined parameters, i.e. number of initial rules, parameters of the membership function;
(2) As an online algorithm, it requires offline training and cannot start "from scratch".

2.2.4 *Brief Introduction to Image Processing*

2.2.4.1 Image Transformation Techniques

Image transformation is one of the most important areas of image processing, and plays a critical role in many machine learning algorithms as a pre-processing technique. The commonly used image transformation techniques presented in this book include:

- normalization,
- elastic distortion, and
- affine distortion, including:
  - scaling;
  - rotation, and
  - segmentation.

The normalization and affine distortion are pre-processing transformations which are generally applicable to various image processing problems, i.e. remote sensing [117], object recognition [118], etc. Affine distortion can be done by applying affine displacement fields to images, computing for every pixel a new target location with respect to the original one. Affine distortions are very effective to improve the generalization and decrease the overfitting [119–121]. In contrast, the elastic distortion is applicable mostly to the handwritten digits and/or letters recognition problems, i.e. MNIST (Modified National Institute of Standards and Technology) database [122].

A. ***Normalization***

The normalization is a common technique in image processing that changes the value range of the pixels within the image by mapping into a more familiar or normal range. This operation can be used to readjust the degree of illumination of the images as well.

B. ***Scaling***

Image scaling refers to the resampling and resizing of a digital image [123, 124]. There are two types of image scaling:

(1) image contraction, and
(2) image expansion.

Image scaling is achieved by using an interpolation function. There is a number of different interpolation methods for image resizing reported in the literature [123–126], e.g. nearest neighbor, bilinear and bicubic interpolation methods. The most commonly used one is the bicubic interpolation method [125, 126], which considers the nearest 16 pixels (4×4) in the neighborhood of certain pixel and calculates the output pixel value as their weighted average. Since the 16 neighboring pixels are at various similarity levels from the output pixel, closer pixels are given a higher weighting in the calculation.

C. ***Rotation***

Image rotation is another common affine distortion technique performed by rotating an image at certain angle around the center point [127]. Usually, the nearest neighbor interpolation in terms of pixel values is used after the rotation and the values of pixels that are outside the rotated images are set to 0 (black). An illustration of image rotation performed on an image from UCMerced remote sensing image set [117] (downloadable from [128]) is given in Fig. 2.12.

D. ***Segmentation***

Segmentation is the process of partitioning an image into smaller pieces to extract local information or discard the less informative part of the image [21]. The main purpose of employing the segmentation are:

(1) improving the generalization ability, and
(2) increasing the efficiency of the feature descriptors in harvesting information from the image.

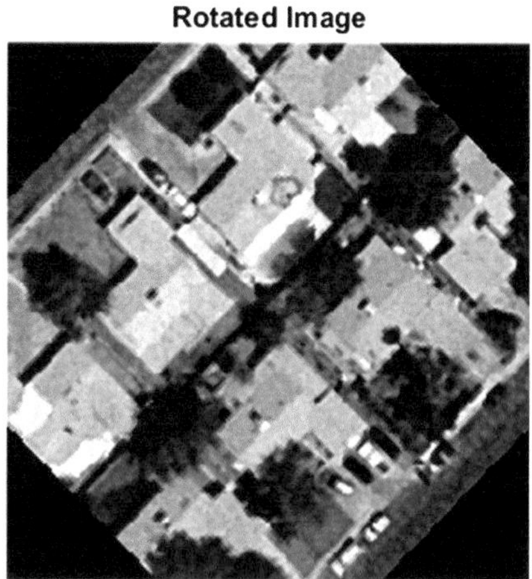

Fig. 2.12 An illustration of image rotation

E. ***Elastic Distortion***

Elastic distortion (see Fig. 2.13) is a more complex and controversial technique to expand the dataset and improve the generalization [119–121]. In Fig. 2.13, we apply elastic distortion on an image from UCMerced remote sensing image set [117] for illustration.

It is done by, firstly, generating random displacement fields and then convolving the displacement fields with a Gaussian function of standard deviation σ (in pixels) [119]. This type of image deformation has been widely used in the state-of-art deep convolutional neural networks (DCNN) for handwritten recognition [120, 121] and largely improved the recognition accuracy.

However, this kind of distortion exhibits a significant randomness that puts in question the achieved results' repeatability and requires a cross-validation that further obstructs the online applications and the reliability of the results. In addition, it adds user-specific parameters that can be chosen differently. For a particular image, each time the elastic distortion is performed, an entirely new image is being generated.

Another problem the elastic distortion creates is that there is no evidence or experiment supporting that the elastic distortion can be applied to other types of image recognition problems. In fact, elastic distortion destroys the original images. We did not use elastic distortion in our research and will not use it further in this book.

2.2.4.2 Image Feature Extraction Techniques

Image feature extraction is the key to solve computer vision problems such as object recognition, content-based image classification and image retrieval [129]. The extracted features have to be informative, non-redundant, and, most

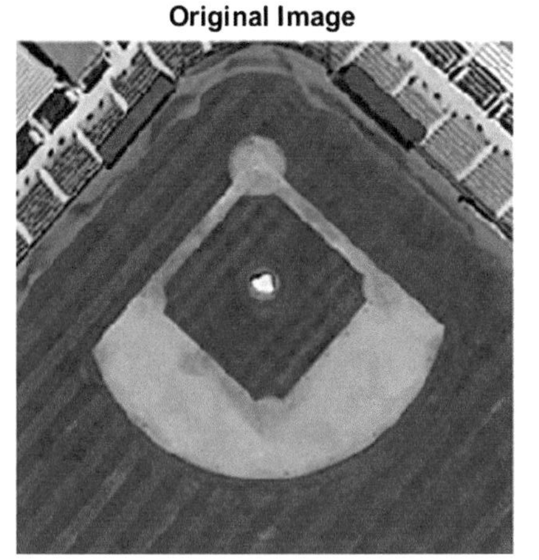

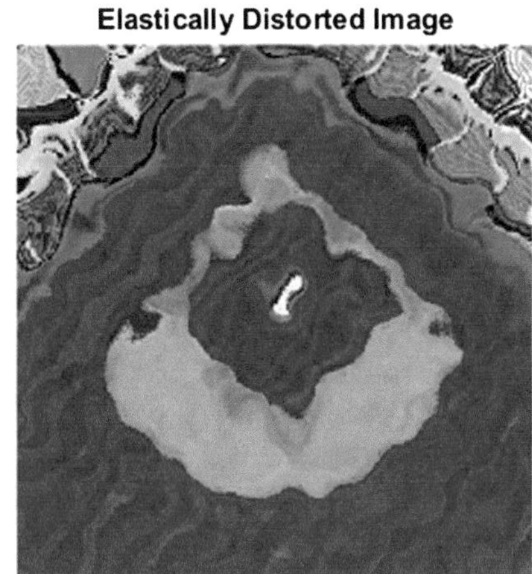

Fig. 2.13 An illustration of image elastic distortion

importantly, to be able to facilitate the subsequent learning and generalization. Based on their descriptive abilities, image feature descriptors can be divided into the following categories [130]:

(1) low-level feature descriptors;
(2) medium-level feature descriptors, and
(3) high-level feature descriptors.

Different feature descriptors have different advantages and are applicable in different scenarios. Selecting the most suitable image transformation technique(s) and feature descriptor(s) for a particular problem usually requires prior knowledge.

In this subsection, a brief introduction covering the commonly used image transformation techniques and feature descriptors is presented.

A. ***Low-level Image Feature Descriptors***

Low-level image features work very well in the problems where low-level visual features, e.g., spectral, texture, and structure, play the dominant role. Low-level descriptors extract feature vectors directly from the images in a straightforward manner involving neither external learning process nor global statistical analysis. The commonly used low-level descriptors include:

(1) GIST [131];
(2) histogram of oriented gradient (HOG) [132];
(3) scale invariant feature transform (SIFT) [133];
(4) color histogram (CH) [134] and
(5) Haar-like [135].

GIST descriptor [131]

GIST feature descriptor gives an impoverished and coarse version of the principal contours and textures of the image [131]. It provides multi-scale rotation invariant features globally. The GIST feature descriptor as proposed in [131] is implemented by convolving the grey-level image with 4 scales, 8 orientations Gabor filters, producing 32 feature maps. Then, the feature maps are divided into 16 regions by a 4×4 spatial grid, and the mean vectors of each grid of the 32 feature maps are concatenated resulting in a $16 \times 4 \times 8 = 512$ dimensional GIST feature vector.

Histogram of oriented gradient (HOG) descriptor [132]

HOG feature descriptor [132, 136] has been proven to be very successful in various computer vision tasks such as object detection, texture analysis and image classification. The key idea behind the HOG feature descriptor is that local object appearance and shape within an image can be described by the distribution of intensity gradients or edge directions. The HOG feature vector of a true-color/grey-level image is obtained as the concatenation of the cell histograms of gradient directions of pixel intensities of the block regions of this image.

Scale invariant feature transform (SIFT) descriptor [133]

SIFT feature descriptor consists of four main components:

(1) scale-space detection;
(2) key point localization;
(3) orientation assignment, and
(4) key point descriptor [133].

The best results achieved by SIFT features based on experiments are computed with a 4×4 array of histograms with 8 orientation bins in each. Thus, the final SIFT feature vector has a dimension of 128 ($4 \times 4 \times 8$) [137].

Color histogram (CH) descriptor [134]

CH feature descriptor extracts a number of pixels in the image that have colors in a fixed list of color ranges that span the color space. CH feature vectors are flexible constructs that can be built from images in various color spaces, whether RGB, or any other color space of any dimension [137].

Haar-like descriptor [135]

The Haar-like feature descriptor considers adjacent rectangular regions at a specific location in a detection window. The feature vector of an image is obtained by summing up the pixel intensities in each region and calculating the difference between these sums.

B. ***Medium-level Image Feature Descriptors***

Mid-level feature descriptors are built upon a global analysis of low-level local feature descriptors with the possible involvement of external knowledge [130]. The most popular medium-level image feature descriptors include:

(1) bag of visual words (BOVW) [117],
(2) spatial pyramid matching (SPM) [138], and
(3) local-constrained linear coding (LCLC) [139].

Bag of visual words (BOVW) [117]

BOVW model is widely used in natural language processing (NLP) for document classification by representing the documents with sparse vectors consisting of the occurrence frequencies of the words [140]. In the computer vision, the BOVW model [117, 141] treats the image features as "words" and represents the image as a sparse vector of the occurrence frequencies of local image features. To achieve this, the BOVW model conducts the following three steps:

(1) feature extraction,
(2) feature description, and
(3) codebook generation [117].

The feature extraction is usually done by grid segmentation or using some key-points detection methods, i.e. the speeded up robust feature (SURF) detector [142]. The feature description is achieved by using low-level feature descriptors, i.e. the ones mentioned in the previous subsection, to extract local features from the segments (arounds the key-points if some detection methods are used). The codebook composed by a number of visual words is often produced by clustering local image features using unsupervised learning algorithm, i.e. the k-means algorithm [55].

Spatial pyramid matching (SPM) [138]

SPM extends BOVW model to spatial pyramids by partitioning the image into increasingly fine sub-regions and concatenating the weighted local image features in each sub-region at different scales [138]. The SPM image representation is then combined with a kernel-based pyramid matching scheme [143] that efficiently computes approximate global geometric correspondence between sets of features in two images. The spatial pyramid representation has strong discriminative power with the global spatial information and, thus, allows the SPM model to outperform the BOVM model on many challenging image categorization tasks [144].

Local-constrained linear coding (LCLC) [139]

LCLC replaces the vector quantization coding method in the traditional SPM model with an effective coding scheme [139]. LCLC model uses the locality constraints to project each local feature descriptor into its local coordinate system, and integrates the projected coordinates with max pooling to generate the final representation with the same size of the dictionary [130, 139].

C. ***High-level Image Feature Descriptors***

Deep convolutional neural networks (DCNNs) involve a number of hidden layers determined *ad hoc* to hierarchically learn high-level representations of the images. Compared with low-level feature descriptors, high-level deep learning descriptors work better on classifying images with high-diversity and nonhomogeneous spatial distributions because they can learn more abstract and discriminative semantic features. The commonly used pre-trained DCNNs as feature descriptors include:

(1) VGG-VD (visual geometry group-very deep) models [147];
(2) CAFFE (convolutional architecture for fast feature embedding) model [146];
(3) GoogLeNet model [147];
(4) AlexNet model [148].

The pre-trained DCNNs are used directly as feature descriptors in many researches and are setting the state-of-the-art performance in the area of image classification [149–151]. The common practice of using pre-trained DCNNs as feature descriptors is to use the activations from the first fully connected layer as the feature vector of the image [130]. However, fine-tuning can be applied to the pre-trained DCNN for further boosting the performance [152].

2.3 Conclusions

In this chapter we made a relatively brief overview of the theory of probability, statistical and machine learning. There is a lot of literature on these topics and some excellent books, e.g. [14] but in order to have a self-sustained "one stop shop" type of text we introduce and discuss briefly the main ideas and the most popular and widely used methods in this area.

We start with discussing the randomness and determinism and stress the fact that most of the problems of interest as real applications (e.g. weather/climate, financial, human behavior, etc.) are not clearly deterministic or random. We then move to the probability mass and distribution, probability density and moments, density estimation. Then we move towards the Bayesian and other branches of the probability theory. We then analyze critically these basic and well-known topics stressing the relation to the *empirical* approach we propose in this book.

We then move to the topic of statistical machine learning and pattern recognition. We start this review with data pre-processing, including the distance metrics (Euclidean, city block, the more general Minkowski type, the more flexible Mahalanobis type) and the cosine dissimilarity measure. We then analyze various aspects of the issue of proximity measures.

Then, we move towards the normalization and standardization, feature selection and orthogonalization. We continue further with the unsupervised type of learning including hierarchical, centroid-based, density-based, distribution-based and fuzzy methods.

After these, we move to supervised machine learning methods. This includes classification methods: naive Bayesian, kNN, SVM classifiers, offline FRB and evolving FRB classifiers. We then describe briefly the regression type predictive systems (we consider several methods of this group, namely, linear regression, ANFIS, DENFIS, eTS.

Finally, we summarize the topic of image processing starting with image transformation techniques (including normalization, scaling, rotation, segmentation and elastic distortion). We then describe very briefly image feature transformation techniques (low level, such as GIST, HOG, CH and Haar-like; medium level, such as bag of visual words, spatial pyramid matching, local constrained near coding; high level features such as various pre-trained DCNNs, e.g. VGG-VD, GoogleNet, etc., working as encoders only).

2.4 Questions

(1) What are the restrictive assumptions on which the probability theory is based upon. Give examples when they do not hold in practice.
(2) Are the processes like climate, financial and economics, medicine, human behavior strictly random or deterministic and why?

(3) Provide examples of different distance metrics and compare them.
(4) Is cosine dissimilarity a full distance metric and why?
(5) Provide examples of unsupervised learning.
(6) Provide examples of supervised learning.
(7) Why image processing is more difficult problem in comparison to dealing with physical variables like temperature, pressure, mass, energy, etc.?
(8) What is the goal of data pre-processing? Is it compulsory (unavoidable) step of machine learning?

2.5 Take Away Items

- The well-established theory of probability and statistical learning apply to random variables. The so-called fair games (throwing dices, coins, etc.) are good examples of random processes. Deterministic models are another widely used alternative apparatus in physics and mathematics. The majority of the processes we are interested to describe, predict, analyse, classify or cluster (e.g. climate, financial data, economics, human behaviour, etc.) are, however, neither strictly deterministic nor random.
- Because of the importance and closeness of the elements of statistical learning to the newly proposed method, in this chapter we describe them starting with distance and proximity measures, data pre-processing, unsupervised and supervised learning.
- Image processing is briefly described separately because the data in the form of images and video are nowadays widely available and form a part of many problems unlike in the past. We finish this chapter with some examples of application of mainstream DCNNS to image processing.

References

1. G. Grimmett, D. Welsh, *Probability: an Introduction* (Oxford University Press, 2014)
2. P. Angelov, S. Sotirov (eds.), *Imprecision and Uncertainty in Information Representation and Processing* (Springer, Cham, 2015)
3. C.M. Bishop, *Pattern Recognition and Machine Learning* (Springer, New York, 2006)
4. R.O. Duda, P.E. Hart, D.G. Stork, *Pattern Classification*, 2nd edn. (Chichester, West Sussex, UK,: Wiley-Interscience, 2000)
5. M.S. de Alencar, R.T. de Alencar, *Probability Theory* (Momentum Press, New York, 2016)
6. P. Angelov, *Autonomous Learning Systems: From Data Streams to Knowledge in Real Time* (Wiley, Ltd., 2012)
7. J. Nicholson, *The Concise Oxford Dictionary of Mathematics*, 5th edn. (Oxford University Press, 2014)
8. S. Haykin, *Communication Systems* (Wiley, 2008)
9. W.H. Press, S.A. Teukolsky, W.T. Vetterling, B.P. Flannery, *Numerical Recipes: The Art of Scientific Computing*, 3rd edn. (Cambridge university press, 2007)

10. J.-M. Marin, K. Mengersen, C.P. Robert, Bayesian modelling and inference on mixtures of distributions, in *Handbook of statistics* (2005), pp. 459–507
11. T.K. Moon, The expectation-maximization algorithm. IEEE Signal Process. Mag. **13**(6), 47–60 (1996)
12. T. Bayes, An essay towards solving a problem in the doctrine of chances. Philos. Trans. R. Soc. **53**, 370 (1763)
13. J. Principe, *Information Theoretic Learning: Renyi's Entropy and Kernel Perspectives* (Springer, 2010)
14. T. Hastie, R. Tibshirani, J. Friedman, *The Elements of Statistical Learning: Data Mining, Inference, and Prediction* (Springer, Burlin, 2009)
15. V. Vapnik, R. Izmailov, Statistical inference problems and their rigorous solutions. Stat. Learn. Data Sci. **9047**, 33–71 (2015)
16. S.B. Kotsiantis, D. Kanellopoulos, P.E. Pintelas, Data preprocessing for supervised learning. Int. J. Comput. Sci. **1**(2), 111–117 (2006)
17. M. Kuhn, K. Johnson, Data pre-processing, in *Applied Predictive Modeling* (Springer, New York, NY, 2013) pp. 27–59
18. X. Gu, P.P. Angelov, D. Kangin, J.C. Principe, A new type of distance metric and its use for clustering. Evol. Syst. **8**(3), 167–178 (2017)
19. B. McCune, J.B. Grace, D.L. Urban, *Analysis of Ecological Communities* (2002)
20. F.A. Allah, W.I. Grosky, D. Aboutajdine, Document clustering based on diffusion maps and a comparison of the k-means performances in various spaces, in *IEEE Symposium on Computers and Communications*, 2008, pp. 579–584
21. N. Dehak, R. Dehak, J. Glass, D. Reynolds, P. Kenny, "Cosine Similarity Scoring without Score Normalization Techniques," in *Proceedings of Odyssey 2010—The Speaker and Language Recognition Workshop (Odyssey 2010)*, 2010, pp. 71–75
22. N. Dehak, P. Kenny, R. Dehak, P. Dumouchel, P. Ouellet, Front end factor analysis for speaker verification. IEEE Trans. Audio. Speech. Lang. Process. **19**(4), 788–798 (2011)
23. V. Setlur, M.C. Stone, A linguistic approach to categorical color assignment for data visualization. IEEE Trans. Vis. Comput. Graph. **22**(1), 698–707 (2016)
24. M. Senoussaoui, P. Kenny, P. Dumouchel, T. Stafylakis, Efficient iterative mean shift based cosine dissimilarity for multi-recording speaker clustering, in *IEEE International Conference on Acoustics, Speech and Signal Processing (ICASSP)*, 2013 pp. 7712–7715
25. J. Zhang, Divergence function, duality, and convex analysis. Neural Comput. **16**(1), 159–195 (2004)
26. S. Eguchi, A differential geometric approach to statistical inference on the basis of contrast functionals. Hiroshima Math. J. **15**(2), 341–391 (1985)
27. J.R. Hershey, P.A. Olsen, Approximating the Kullback Leibler divergence between Gaussian mixture models, in *IEEE International Conference on Acoustics, Speech and Signal Processing*, 2007, pp. 317–320
28. R.G. Brereton, The mahalanobis distance and its relationship to principal component scores. J. Chemom. **29**(3), 143–145 (2015)
29. R.R. Korfhage, J. Zhang, A distance and angle similarity measure method. J. Am. Soc. Inf. Sci. **50**(9), 772–778 (1999)
30. X. Gu, P. Angelov, D. Kangin, J. Principe, Self-organised direction aware data partitioning algorithm. Inf. Sci. (Ny) **423**, 80–95 (2018)
31. R.A. Horn, C.R. Johnson, *Matrix Analysis* (Cambridge University Press, 1990)
32. C.C. Aggarwal, A. Hinneburg, D.A. Keim, On the surprising behavior of distance metrics in high dimensional space, in *International Conference on Database Theory*, 2001, pp. 420–434
33. K. Beyer, J. Goldstein, R. Ramakrishnan, U. Shaft, When is 'nearest neighbors' meaningful?, in *International Conference on Database Theoryheory*, 1999, pp. 217–235
34. J.G. Saw, M.C.K. Yang, T.S.E.C. Mo, Chebyshev inequality with estimated mean and variance. Am. Stat. **38**(2), 130–132 (1984)

35. G. Kumar, P.K. Bhatia, A detailed review of feature extraction in image processing systems, in *IEEE International Conference on Advanced Computing and Communication Technologies*, 2014, pp. 5–12
36. S.T.K. Koutroumbas, *Pattern Recognition*, 4th edn. (Elsevier, New York, 2009)
37. I. Guyon, A. Elisseeff, An introduction to variable and feature selection. J. Mach. Learn. Res. **3**(3), 1157–1182 (2003)
38. J. Trevisan, P.P. Angelov, A.D. Scott, P.L. Carmichael, F.L. Martin, IRootLab: a free and open-source MATLAB toolbox for vibrational biospectroscopy data analysis. Bioinformatics **29**(8), 1095–1097 (2013)
39. X. Zhang, M.A. Young, O. Lyandres, R.P. Van Duyne, Rapid detection of an anthrax biomarker by surface-enhanced Raman spectroscopy. J. Am. Chem. Soc. **127**(12), 4484–4489 (2005)
40. P.C. Sundgren, V. Nagesh, A. Elias, C. Tsien, L. Junck, D.M.G. Hassan, T.S. Lawrence, T. L. Chenevert, L. Rogers, P. McKeever, Y. Cao, Metabolic alterations: a biomarker for radiation induced normal brain injury-an MR spectroscopy study. J. Magn. Reson. Imaging **29**(2), 291–297 (2009)
41. G.H. Golub, C. Reinsch, Singular value decomposition and least squares solutions. Numer. Math. **14**(5), 403–420 (1970)
42. J. Scheffer, Dealing with missing data. Res. Lett. Inf. Math. Sci. **3**, 153–160 (2002)
43. A. Agresti, *Categorical Data Analysis* (Wiley, 2003)
44. O. Maimon, L. Rokach, *Data Mining and Knowledge Discovery Handbook* (Springer, Boston, MA, 2005)
45. C.C. Aggarwal, C.K. Reddy (eds.), *Data Clustering: Algorithms and Applications* (CRC Press, 2013)
46. S.C. Johnson, Hierarchical clustering schemes. Psychometrika **32**(3), 241–254 (1967)
47. R.A. Fisher, The use of multiple measurements in taxonomic problems. Ann. Eugen. **7**(2), 179–188 (1936)
48. http://archive.ics.uci.edu/ml/datasets/Iris
49. G. Karypis, E.-H. Han, V. Kumar, Chameleon: hierarchical clustering using dynamic modeling. Comput. (Long. Beach. Calif) **32**(8), 68–75 (1999)
50. W.H.E. Day, H. Edelsbrunner, Efficient algorithms for agglomerative hierarchical clustering methods. J. Classif. **1**, 7–24 (1984)
51. A. Gucnoche, P. Hansen, B. Jaumard, Efficient algorithms for divisive hierarchical clustering with the diameter criterion. J. Classif. **8**, 5–30 (1991)
52. T. Xiong, S. Wang, A. Mayers, E. Monga, DHCC: divisive hierarchical clustering of categorical data. Data Min. Knowl. Discov. **24**, 103–135 (2012)
53. T. Zhang, R. Ramakrishnan, M. Livny, BIRCH: a new data clustering algorithm and its applications. Data Min. Knowl. Discov. **1**(2), 141–182 (1997)
54. B.J. Frey, D. Dueck, Clustering by passing messages between data points, Science (80-.) **315** (5814), pp. 972–976 (2007)
55. J.B. MacQueen, Some methods for classification and analysis of multivariate observations, in *5th Berkeley Symposium on Mathematical Statistics and Probability*, vol. 1, no. 233, (1967) pp. 281–297
56. M. Ester, H.P. Kriegel, J. Sander, X. Xu, A density-based algorithm for discovering clusters in large spatial databases with noise, in *International Conference on Knowledge Discovery and Data Mining*, vol. 96 (1996) pp. 226–231
57. D. Comaniciu, P. Meer, Mean shift: a robust approach toward feature space analysis. IEEE Trans. Pattern Anal. Mach. Intell. **24**(5), 603–619 (2002)
58. K.L. Wu, M.S. Yang, Mean shift-based clustering. Pattern Recognit. **40**(11), 3035–3052 (2007)
59. R. Dutta Baruah, P. Angelov, Evolving local means method for clustering of streaming data, in *IEEE International Conference on Fuzzy Systems*, 2012, pp. 10–15
60. P. Angelov, An approach for fuzzy rule-base adaptation using on-line clustering. Int. J. Approx. Reason. **35**(3), 275–289 (2004)

61. P.P. Angelov, D.P. Filev, N.K. Kasabov, *Evolving Intelligent Systems: Methodology and Applications* (2010)
62. R. Hyde, P. Angelov, A fully autonomous data density based clustering technique, in *IEEE Symposium on Evolving and Autonomous Learning Systems*, 2014, pp. 116–123
63. R. Hyde, P. Angelov, A.R. MacKenzie, Fully online clustering of evolving data streams into arbitrarily shaped clusters. Inf. Sci. (Ny) **382–383**, 96–114 (2017)
64. A. Corduneanu, C.M. Bishop, Variational Bayesian model selection for mixture distributions, in *Proceedings of the Eighth International Joint Conference on Artificial statistics*, 2001, pp. 27–34
65. C.A. McGrory, D.M. Titterington, Variational approximations in Bayesian model selection for finite mixture distributions. Comput. Stat. Data Anal. **51**(11), 5352–5367 (2007)
66. D.M. Blei, M.I. Jordan, Variational methods for the Dirichlet process, in *Proceedings of the Twenty-First International Conference on Machine Learning*, 2004, p. 12
67. D.M. Blei, M.I. Jordan, Variational inference for Dirichlet process mixtures. Bayesian Anal. **1**(1A), 121–144 (2006)
68. J.C. Dunn, A fuzzy relative of the ISODATA process and its use in detecting compact well-separated clusters. J. Cybern. **3**(3) (1973)
69. J.C. Dunn, Well-separated clusters and optimal fuzzy partitions. J. Cybern. **4**(1), 95–104 (1974)
70. P.P. Angelov, D.P. Filev, An approach to online identification of Takagi-Sugeno fuzzy models. IEEE Trans. Syst. Man, Cybern. Part B Cybern. **34**(1), 484–498 (2004)
71. M.N. Murty, V.S. Devi, *Introduction to Pattern Recognition and Machine Learning* (World Scientific, 2015)
72. P. Angelov, X. Zhou, D. Filev, E. Lughofer, Architectures for evolving fuzzy rule-based classifiers, in *IEEE International Conference on Systems, Man and Cybernetics*, 2007, pp. 2050–2055
73. P. Angelov, X. Zhou, Evolving fuzzy-rule based classifiers from data streams. IEEE Trans. Fuzzy Syst. **16**(6), 1462–1474 (2008)
74. P. Angelov, Fuzzily connected multimodel systems evolving autonomously from data streams. IEEE Trans. Syst. Man, Cybern. Part B Cybern. **41**(4), 898–910 (2011)
75. X. Gu, P.P. Angelov, Semi-supervised deep rule-based approach for image classification. Appl. Soft Comput. **68**, 53–68 (2018)
76. T. Cover, P. Hart, Nearest neighbor pattern classification. IEEE Trans. Inf. Theory **13**(1), 21–27 (1967)
77. P. Cunningham, S.J. Delany, K-nearest neighbour classifiers. Mult. Classif. Syst. **34**, 1–17 (2007)
78. K. Fukunage, P.M. Narendra, A branch and bound algorithm for computing k-nearest neighbors. IEEE Trans. Comput. **C-24**(7), 750–753 (1975)
79. N. Cristianini, J. Shawe-Taylor, *An Introduction to Support Vector Machines and Other Kernel-Based Learning Methods* (Cambridge University Press, Cambridge, 2000)
80. W.S. Noble, What is a support vector machine? Nat. Biotechnol. **24**(12), 1565–1567 (2006)
81. V. Vapnik, A. Lerner, Pattern recognition using generalized portrait method. Autom. Remote Control **24**(6), 774–780 (1963)
82. C.J.C. Burges, A tutorial on support vector machines for pattern recognition. Data Min. Knowl. Discov. **2**(2), 121–167 (1998)
83. D. Kangin, P. Angelov, Recursive SVM based on TEDA, in *International Symposium on Statistical Learning and Data Sciences*, 2015, pp. 156–168
84. L.A. Zadeh, Fuzzy sets. Inf. Control **8**(3), 338–353 (1965)
85. E.H. Mamdani, S. Assilian, An experiment in linguistic synthesis with a fuzzy logic controller. Int. J. Man Mach. Stud. **7**(1), 1–13 (1975)
86. T. Takagi, M. Sugeno, Fuzzy identification of systems and its applications to modeling and control. IEEE Trans. Syst. Man. Cybern. **15**(1), 116–132 (1985)
87. H. Ishibuchi, K. Nozaki, H. Tanaka, Distributed representation of fuzzy rules and its application to pattern classification. Fuzzy Sets Syst. **52**(1), 21–32 (1992)

88. H. Ishibuchi, K. Nozaki, N. Yamamoto, H. Tanaka, Selecting fuzzy if-then rules for classification problems using genetic algorithms. IEEE Trans. Fuzzy Syst. **3**(3), 260–270 (1995)
89. L. Kuncheva, *Combining Pattern Classifiers: Methods and Algorithms* (Wiley, Hoboken, New Jersey, 2004)
90. H. Ishibuchi, T. Nakashima, M. Nii, *Classification and Modeling with Linguistic Information Granules: Advanced Approaches to Linguistic Data Mining* (Springer Science & Business Media, 2006)
91. C. Xydeas, P. Angelov, S.Y. Chiao, M. Reoullas, Advances in classification of EEG signals via evolving fuzzy classifiers and dependant multiple HMMs. Comput. Biol. Med. **36**(10), 1064–1083 (2006)
92. R.D. Baruah, P.P. Angelov, J. Andreu, Simpl _ eClass : simplified potential-free evolving fuzzy rule-based classifiers, in *IEEE International Conference on Systems, Man, and Cybernetics (SMC)*, 2011, pp. 2249–2254
93. P. Angelov, D. Kangin, D. Kolev, Symbol recognition with a new autonomously evolving classifier AutoClass, in *IEEE Conference on Evolving and Adaptive Intelligent Systems (EAIS)*, 2014, pp. 1–7
94. D. Kangin, P. Angelov, J.A. Iglesias, Autonomously evolving classifier TEDAClass. Inf. Sci. (Ny) **366**, 1–11 (2016)
95. P. Angelov, E. Lughofer, X. Zhou, Evolving fuzzy classifiers using different model architectures. Fuzzy Sets Syst. **159**(23), 3160–3182 (2008)
96. M. Pratama, S.G. Anavatti, P.P. Angelov, E. Lughofer, PANFIS: a novel incremental learning machine. IEEE Trans. Neural Networks Learn. Syst. **25**(1), 55–68 (2014)
97. M. Pratama, S.G. Anavatti, E. Lughofer, Genefis: toward an effective localist network. IEEE Trans. Fuzzy Syst. **22**(3), 547–562 (2014)
98. T. Isobe, E.D. Feigelson, M.G. Akritas, G.J. Babu, Linear regression in astronomy. Astrophys. J. **364**, 104–113 (1990)
99. R.E. Precup, H.I. Filip, M.B. Rədac, E.M. Petriu, S. Preitl, C.A. Dragoş, Online identification of evolving Takagi-Sugeno-Kang fuzzy models for crane systems. Appl. Soft Comput. J. **24**, 1155–1163 (2014)
100. V. Bianco, O. Manca, S. Nardini, Electricity consumption forecasting in Italy using linear regression models. Energy **34**(9), 1413–1421 (2009)
101. X. Gu, P.P. Angelov, A.M. Ali, W.A. Gruver, G. Gaydadjiev, Online evolving fuzzy rule-based prediction model for high frequency trading financial data stream, in *IEEE Conference on Evolving and Adaptive Intelligent Systems (EAIS)*, 2016, pp. 169–175
102. X. Yan, X. Su, *Linear Regression Analysis: Theory and Computing* (World Scientific, 2009)
103. J.S.R. Jang, ANFIS: adaptive-network-based fuzzy inference system. IEEE Trans. Syst. Man Cybern. **23**(3), 665–685 (1993)
104. P. Angelov, R. Buswell, Identification of evolving fuzzy rule-based models. IEEE Trans. Fuzzy Syst. **10**(5), 667–677 (2002)
105. P. Angelov, R. Buswell, Evolving rule-based models: a tool for intelligent adaption, in *IFSA World Congress and 20th NAFIPS International Conference*, 2001, pp. 1062–1067
106. P. Angelov, D. Filev, On-line design of takagi-sugeno models, in *International Fuzzy Systems Association World Congress* (Springer, Berlin, Heidelberg, 2003), pp. 576–584
107. N.K. Kasabov, Q. Song, DENFIS: dynamic evolving neural-fuzzy inference system and its application for time-series prediction. IEEE Trans. Fuzzy Syst. **10**(2), 144–154 (2002)
108. E.D. Lughofer, FLEXFIS: a robust incremental learning approach for evolving Takagi-Sugeno fuzzy models. IEEE Trans. Fuzzy Syst. **16**(6), 1393–1410 (2008)
109. H.J. Rong, N. Sundararajan, G. Bin Huang, P. Saratchandran, Sequential adaptive fuzzy inference system (SAFIS) for nonlinear system identification and prediction. Fuzzy Sets Syst. **157**(9), 1260–1275 (2006)
110. H.J. Rong, N. Sundararajan, G. Bin Huang, G.S. Zhao, Extended sequential adaptive fuzzy inference system for classification problems. Evol. Syst. **2**(2), 71–82 (2011)

111. R. Bao, H. Rong, P.P. Angelov, B. Chen, P.K. Wong, Correntropy-based evolving fuzzy neural system. IEEE Trans. Fuzzy Syst. (2017). https://doi.org/10.1109/TFUZZ.2017.2719619
112. D. Leite, P. Costa, F. Gomide, Interval approach for evolving granular system modeling, in *Learning in Non-stationary Environments* (New York, NY: Springer, 2012), pp. 271–300
113. W. Leigh, R. Hightower, N. Modani, Forecasting the New York stock exchange composite index with past price and interest rate on condition of volume spike. Expert Syst. Appl. **28** (1), 1–8 (2005)
114. J. Park, I.W. Sandberg, Universal approximation using radial-basis-function networks. Neural Comput. **3**(2), 246–257 (1991)
115. L.X. Wang, J.M. Mendel, Fuzzy basis functions, universal approximation, and orthogonal least-squares learning. IEEE Trans. Neural Networks **3**(5), 807–814 (1992)
116. P.P. Angelov, *Evolving Rule-Based Models: A Tool for Design of Flexible Adaptive Systems* (Springer, Berlin Heidelberg, 2002)
117. Y. Yang, S. Newsam, Bag-of-visual-words and spatial extensions for land-use classification, in *International Conference on Advances in Geographic Information Systems*, 2010, pp. 270–279
118. L. Fei-Fei, R. Fergus, P. Perona, One-shot learning of object categories. IEEE Trans. Pattern Anal. Mach. Intell. **28**(4), 594–611 (2006)
119. P.Y. Simard, D. Steinkraus, J.C. Platt, Best practices for convolutional neural networks applied to visual document analysis, in *Proceedings of Seventh International Conference on Document Analysis and Recognition*, 2003, pp. 958–963
120. D.C. Cireşan, U. Meier, L.M. Gambardella, J. Schmidhuber, Convolutional neural network committees for handwritten character classification, in *International Conference on Document Analysis and Recognition*, vol. 10, , 2011, pp. 1135–1139
121. D. Ciresan, U. Meier, J. Schmidhuber, Multi-column deep neural networks for image classification, in *Conference on Computer Vision and Pattern Recognition*, 2012, pp. 3642–3649
122. Y. LeCun, L. Bottou, Y. Bengio, P. Haffner, Gradient-based learning applied to document recognition. Proc. IEEE **86**(11), 2278–2323 (1998)
123. T.M. Lehmann, C. Gönner, K. Spitzer, Survey: interpolation methods in medical image processing. IEEE Trans. Med. Imaging **18**(11), 1049–1075 (1999)
124. P. Thevenaz, T. Blu, M. Unser, Interpolation revisited. IEEE Trans. Med. Imaging **19**(7), 739–758 (2000)
125. R. Keys, Cubic convolution interpolation for digital image processing. IEEE Trans. Acoust. **29**(6), 1153–1160 (1981)
126. J.W. Hwang, H.S. Lee, Adaptive image interpolation based on local gradient features. IEEE Signal Process. Lett. **11**(3), 359–362 (2004)
127. R.G. Casey, Moment Normalization of Handprinted Characters. IBM J. Res. Dev. **14**(5), 548–557 (1970)
128. http://weegee.vision.ucmerced.edu/datasets/landuse.html
129. S.B. Park, J.W. Lee, S.K. Kim, Content-based image classification using a neural network. Pattern Recognit. Lett. **25**(3), 287–300 (2004)
130. G.-S. Xia, J. Hu, F. Hu, B. Shi, X. Bai, Y. Zhong, L. Zhang, AID: a benchmark dataset for performance evaluation of aerial scene classification. IEEE Trans. Geosci. Remote Sens. **55** (7), 3965–3981 (2017)
131. A. Oliva, A. Torralba, Modeling the shape of the scene: A holistic representation of the spatial envelope. Int. J. Comput. Vis. **42**(3), 145–175 (2001)
132. N. Dalal, B. Triggs, Histograms of oriented gradients for human detection, in *IEEE Computer Society Conference on Computer Vision and Pattern Recognition*, 2005, pp. 886–893
133. D.G. Lowe, Distinctive image features from scale-invariant keypoints. Int. J. Comput. Vis. **60**(2), 91–110 (2004)
134. M.J. Swain, D.H. Ballard, Color indexing. Int. J. Comput. Vis. **7**(1), 11–32 (1991)

135. P. Viola, M. Jones, Rapid object detection using a boosted cascade of simple features, in *Proceedings of the 2001 IEEE Computer Society Conference on Computer Vision and Pattern Recognition, CVPR 2001*, 2001, p. I-511–I-518
136. Y. Lin, F. Lv, S. Zhu, M. Yang, T. Cour, K. Yu, L. Cao, T. Huang, Large-scale image classification: Fast feature extraction and SVM training, in *IEEE Computer Society Conference on Computer Vision and Pattern Recognition*, 2011, pp. 1689–1696
137. M.M. El-Gayar, H. Soliman, N. Meky, A comparative study of image low level feature extraction algorithms. Egypt Inform. J. **14**(2), 175–181 (2013)
138. S. Lazebnik, C. Schmid, J. Ponce, Beyond bags of features : spatial pyramid matching for recognizing natural scene categories, in *IEEE Computer Society Conference on Computer Vision and Pattern Recognition*, 2006, pp. 2169–2178
139. J. Wang, J. Yang, K. Yu, F. Lv, T. Huang, Y. Gong, Locality-constrained linear coding for image classification, in *IEEE Conference on Computer Vision and Pattern Recognition*, 2010, pp. 3360–3367
140. T. Joachims, Text categorization with support vector machines: learning with many relevant features, in *European Conference on Machine Learning*, 1998, pp. 137–142
141. X. Peng, L. Wang, X. Wang, Y. Qiao, Bag of visual words and fusion methods for action recognition: comprehensive study and good practice. Comput. Vis. Image Underst. **150**, 109–125 (2015)
142. H. Bay, T. Tuytelaars, L. Van Gool, SURF : Speeded - Up Robust Features, in *European Conference on Computer Vision*, 2006, pp. 404–417
143. K. Graumanand, T. Darrell, The pyramid match kernel: discriminative classification with sets of image features, in *International Conference on Computer Vision*, 2005, pp. 1458–1465
144. S. Lazebnik, C. Schmid, J. Ponce, Spatial pyramid matching, in *Object Categorization: Computer and Human Vision Perspectives*, 2009, pp. 1–19
145. K. Simonyan, A. Zisserman, Very deep convolutional networks for large-scale image recognition, in *International Conference on Learning Representations*, 2015, pp. 1–14
146. Y. Jia, E. Shelhamer, J. Donahue, S. Karayev, J. Long, R. Girshick, S. Guadarrama, T. Darrell, Caffe: convolutional architecture for fast feature embedding∗, in *ACM International Conference on Multimedia*, 2014, pp. 675–678
147. C. Szegedy, W. Liu, Y. Jia, P. Sermanet, S. Reed, D. Anguelov, D. Erhan, V. Vanhoucke, A. Rabinovich, C. Hill, A. Arbor, Going deeper with convolutions, in *IEEE conference on computer vision and pattern recognition*, 2015, pp. 1–9
148. A. Krizhevsky, I. Sutskever, G.E. Hinton, ImageNet classification with deep convolutional neural networks, in *Advances In Neural Information Processing Systems*, 2012, pp. 1097–1105
149. M.D. Zeiler, R. Fergus, Visualizing and understanding convolutional networks, in *European Conference on Computer Vision*, 2014, pp. 818–833
150. A.B. Sargano, X. Wang, P. Angelov, Z. Habib, Human action recognition using transfer learning with deep representations, in *IEEE International Joint Conference on Neural Networks (IJCNN)*, 2017, pp. 463–469
151. Q. Weng, Z. Mao, J. Lin, W. Guo, Land-use classification via extreme learning classifier based on deep convolutional features. IEEE Geosci. Remote Sens. Lett. **14**(5), 704–708 (2017)
152. G.J. Scott, M.R. England, W.A. Starms, R.A. Marcum, C.H. Davis, Training deep convolutional neural networks for land-cover classification of high-resolution imagery. IEEE Geosci. Remote Sens. Lett. **14**(4), 549–553 (2017)

Chapter 3
Brief Introduction to Computational Intelligence

Computational intelligence is a term that was coined two decades ago [1, 2] as an umbrella term for several nature-inspired computational methodologies which address real-world problems that traditional mathematical modelling cannot solve. It includes techniques such as fuzzy sets and systems (inspired by human reasoning), artificial neural networks (ANN; inspired by the human brain), evolutionary algorithms such as particle swarm optimization, genetic algorithms etc. (inspired by the natural evolution of population of species), dynamically evolving systems (inspired by the self-development of individual humans and animals). *Computational intelligence* differs from the traditional artificial intelligence (AI) by being much closer to machine learning and control theory (system identification, optimization). In this book we will focus on fuzzy systems, ANN and dynamically evolving systems and not on evolutionary algorithms.

3.1 Introduction to Fuzzy Sets and Systems

Fuzzy sets theory was introduced by Lotfi A. Zadeh in 1965 [3] and then extended to fuzzy variable and fuzzy rule-based (FRB) systems [4]. Since then, multiple engineering, decision making and other applications of FRB systems were developed and reported [5–19]. In this section, the concepts of fuzzy sets and FRB systems are presented briefly. For more details the interested reader is directed, for example, to [5, 15].

3.1.1 Fuzzy Set and Membership Function

Fuzzy sets differ from traditional, crisp (non-fuzzy) sets, which only offer a binary choice between the following two options:

P. P. Angelov and X. Gu, *Empirical Approach to Machine Learning*, Studies in Computational Intelligence 800,
https://doi.org/10.1007/978-3-030-02384-3_3

(1) a data sample to belong to this set, and
(2) a data sample not to belong to the set.

Instead, fuzzy sets allow a data sample to *partially* belong to a fuzzy set [3]. This partial membership is a form of flexibility to represent uncertainties that very often exist in the real life.

A fuzzy set, **A** is characterized by its membership function $\mu_{\mathbf{A}}(x)$ [3, 20] that can take any value from the interval $[0, 1]$: $\mu_{\mathbf{A}}(x) \in [0, 1]$. Contrasting this to the binary choice of traditional/crisp/non-fuzzy sets, **B**, which represent the membership function as a Boolean: $\mu_{\mathbf{B}}(x) \in \{0, 1\}$. It is obvious that fuzzy sets are generalizations of the concept of traditional/crisp/non-fuzzy sets with the two extremes as described above being special cases:

- $\mu_{\mathbf{A}}(x) = 1$ for the case (1) and
- $\mu_{\mathbf{A}}(x) = 0$ for the case (2).

The meaning of the membership function, $\mu_{\mathbf{A}}(x)$ is of a degree of *truth, fulfilment, satisfaction* of the fact that the data sample x belongs to the fuzzy set **A**. In the fuzzy sets theory several types of membership functions are being usually used, such as:

- Triangular;
- Gaussian;
- Trapezoidal;
- bell shaped, etc.

In practice, the type of membership functions that are being used is usually decided *ad hoc*, based on the experience and preferences of the human designer and are not necessarily supported well by the experimental data [21]. Parameters of the membership functions are either determined as a result of an offline optimization problem [22] or are handcrafted based on human experience and preferences.

Let us consider a simple illustrative example. Using the real climate dataset measured in Manchester, UK for the period 2010–2015 (downloadable from [23]) the Gaussian type membership functions ($f(x) = e^{-\frac{(x-p)^2}{2\sigma_o^2}}$) of the two attributes (temperature, °C) and wind speed (mph) of the dataset are given in Fig. 3.1, where the prototype p of each membership function is the mean of the data samples of the corresponding class and σ_o is set to be 3.

One may also have noticed that there is a similarity between the membership functions (especially of Gaussian type) and the probability density function (PDF) described in the previous chapter in terms of their appearance. However, there are substantial differences in several important aspects including of principle nature.

Firstly, while the shape of the membership function and PDF may be (in some cases) the same or similar, the maximum value attained by the membership function is usually equal to 1 (the so-called completeness property for the membership functions requires the maximum to be 1) while the maximum value of the PDF is usually much less than 1, but the integral or sum of all values of the PDF is exactly equal to 1. Obviously, the value of this integral or sum for the membership function is always >1.

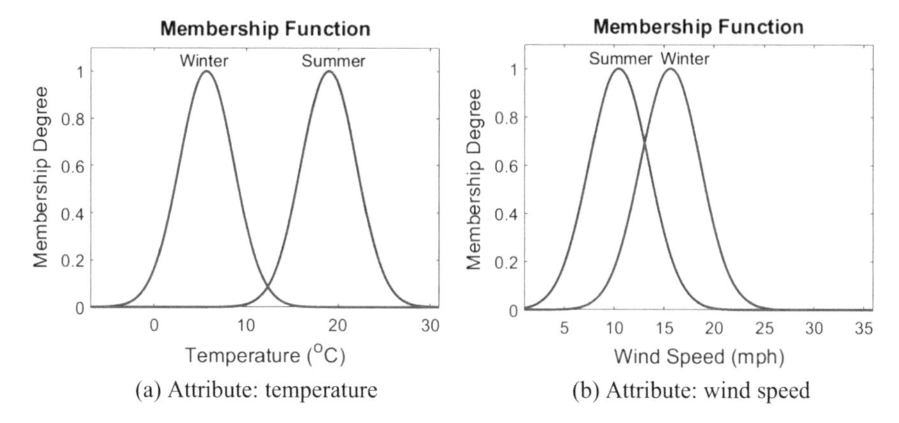

(a) Attribute: temperature (b) Attribute: wind speed

Fig. 3.1 Examples of Gaussian type membership functions with the real climate dataset

Secondly, a difference of a principle nature is that the probability defines the uncertainty based on lack of knowledge about an outcome of an event—we do not know for certain ***if*** an event (e.g. rain) will take place or not. However, once this event takes place it is fully and entirely existing, not partially, while the fuzzy sets represent uncertainty in terms of partial degree of membership (for example, *torrential* rain, *moderate* rain, *light* drizzle, etc.). A specific case may not be precisely characterized by any of these three types of rain, but to some degree *torrential* rain and to some degree a *moderate* rain or to some degree a *moderate* rain but also to some degree *light* drizzle. Other examples may be: probability a person to be smoker versus the degree of smoking (*chain* smoker, *occasional* smoker, etc.); probability that a door (or a valve) is open or closed may be, for example, 30% or 0.3 versus the degree of opening the door or the valve which may vary from 0 (closed) to 100% (fully open). It is obvious that the very nature of the uncertainty that is being described by the fuzzy sets is different from the uncertainty being described by the probabilities. Another observation is that the fuzzy sets represent the duality and partial membership (the real world is not binary, black and white as the mathematical representations we often use). In other words, a single object can be represented by multiple labels, descriptions, models or structures at the same time to a certain degree. This is a very powerful mathematical tool which has some philosophical links with Hegel's dialectics [24]. Georg Wilhelm Friedrich Hegel (1770–1831) was a German philosopher and an important figure of German idealism.

3.1.2 Different Types of Fuzzy Rule-Based Systems

Fuzzy sets can be linked into simple, but very powerful fuzzy rule-based (FRB) systems forming very intuitive and easy to interpret by humans linguistic rules of the so-called IF...THEN form. The importance of the FRB systems grew significantly following the theoretical proof (which itself stems from Kolmogorov's theorem formulated in 1959. Andrey Nikolaevich Kolmogorov is a 20th-century Soviet mathematician (1903–1987)) that they have so-called *universal approximator* property [25]. That is, fuzzy systems can describe arbitrarily well any nonlinear (complex) functions.

In simple terms, this means that any nonlinear continuous function defined on the unit hypercube can be approximated arbitrarily well (the deviation between the approximation and the real values can asymptotically tend to zero). This is extremely powerful result, even if in practical terms this also may require a huge number of rules and fuzzy sets, their parametrization and time and data to train or expert knowledge to design them.

There are three main types of FRB systems, namely:

- Zadeh-Mamdani, introduced in 1970s by L. A. Zadeh and E. Mamdani and Assilian [4, 26, 27];
- Takagi-Sugeno [6, 28, 29], introduced in 1985 by Tomohiro Takagi and Michio Sugeno;
- *AnYa* [21], introduced in 2010 by the lead author of this book in co-authorship with Prof. Ronald R. Yager, one of the main figures in the fuzzy sets theory.

Both Zadeh-Mamdani and Takagi-Sugeno types fuzzy rules share the same type of antecedent (IF) parts and differ in the consequent (THEN) parts. The *AnYa* type FRB, on the contrary, uses a new type of antecedent (IF) part. In this subsection, the three types of fuzzy rules and FRB systems are described and analyzed. In the remainder of this book, however, only the *AnYa* FRB will be used.

3.1.2.1 Zadeh-Mamdani Type FRB Systems

Let us start with the FRB system that was introduced first historically, the so-called *Zadeh-Mamdani* type. It includes a number of linguistic IF…THEN rules in which both, the antecedent/premise (IF) part as well as the consequents (THEN) part are represented by fuzzy sets [27]:

$$\boldsymbol{R}_i : \begin{array}{c} IF\left(x_1\ is\ LT_{i,1}\right) AND \left(x_2\ is\ LT_{i,2}\right) AND\ldots AND \left(x_N\ is\ LT_{i,N}\right) \\ THEN\left(y_i\ is\ LT_{i,out}\right) \end{array} \tag{3.1}$$

where $\boldsymbol{x} = \left[x_1, x_2, \ldots, x_N\right]^T$; $LT_{i,j}$ $(i = 1, 2, \ldots, M,\ j = 1, 2, \ldots, N)$ is the linguistic term of the jth fuzzy set of the ith fuzzy rule; M is the number of fuzzy rules; $LT_{i,out}$ is linguistic term of the output of the ith fuzzy rule; y_i is the output of the ith fuzzy rule.

Examples of linguistic terms can be *Low* (speed, temperature, pressure, income, etc.), *Medium*, *Young*, *Old*, *High*, etc. Linguistic terms are defined by membership functions and signify the semantic representations of respective fuzzy sets.

3.1.2.2 Takagi-Sugeno Type FRB Systems

FRB of Takagi-Sugeno type differ by their consequent (THEN) part, which is represented by a mathematical (usually, linear) function [6, 28, 29]:

$$\boldsymbol{R}_i : \begin{array}{c} IF\left(x_1\ is\ LT_{i,1}\right)AND\left(x_2\ is\ LT_{i,2}\right)AND\ldots AND\left(x_N\ is\ LT_{i,N}\right) \\ THEN\left(y_i = \bar{\boldsymbol{x}}^T\boldsymbol{a}_i\right) \end{array} \tag{3.2}$$

where $\bar{\boldsymbol{x}} = \left[1, \boldsymbol{x}^T\right]^T$; $\boldsymbol{a}_i = \left[a_{i,0}, a_{i,1}, a_{i,2}, \ldots, a_{i,N}\right]^T$ is a $(N+1)\times 1$ dimensional vector of consequent parameters of the *i*th fuzzy rule.

It is clear that the consequent part of each Takagi-Sugeno fuzzy rule is a (linear) regression model. It is, perhaps, less clear, but each such consequence is a local model in the data space. The degree of its validity is defined by the antecedent (IF) part. We will come back to this again when we discuss the design of FRB systems.

3.1.2.3 *AnYa* Type FRB Systems

While both, Zadeh-Mamdani type and Takagi-Sugeno type FRB systems have the same (fuzzy set-based) antecedent (IF) part, the *AnYa* type FRB systems has a different, prototype-based antecedent (IF) part. It does not require the membership function to be pre-defined, but only its apex/peak. This significantly simplifies the design process regardless whether it is being defined subjectively by a human expert or objectively, from the experimental data. We will return to this again in Chap. 5. The consequent part of the *AnYa* type FRB systems is the same as in the Takagi-Sugeno models (functional, rather than a fuzzy set). Although, in principle, it may be any mathematical function, practically, most convenient and widely used are the linear functions (models of this type are called first-order) as well as singletons (zero-order). Singletons are constant values and are very convenient to use in classifiers because the output of a classifier is an integer (often binary) value. An example of a zero-order *AnYa* type FRB system is provided below [21]:

$$\boldsymbol{R}_i : IF(\boldsymbol{x} \sim \boldsymbol{p}_i) \quad THEN\left(y_i \sim LT_{i,out}\right) \tag{3.3}$$

and a first order *AnYa* type fuzzy rule is expressed as [21]:

$$\boldsymbol{R}_i : IF(\boldsymbol{x} \sim \boldsymbol{p}_i) \quad THEN(y_i = \bar{\boldsymbol{x}}^T\boldsymbol{a}_i) \tag{3.4}$$

where $\boldsymbol{p}_i$ is the prototype of the *i*th fuzzy rule; "$\sim$" denotes similarity, which can also be seen as a fuzzy degree of satisfaction/membership [21, 30] or *typicality* [31] which will be discussed in more details in Chap. 5.

3.1.2.4 Design of FRB Systems

The design of FRB systems of both, Zadeh-Mamdani and Takagi-Sugeno types include the definition of the membership functions of the respective fuzzy sets. In addition, the design of the Takagi-Sugeno type FRB systems also includes the parameterization of the consequent (local, usually, linear) models.

Historically, this has been made using a number of *ad hoc* choices, including a heavy involvement of human expertise and *prior* knowledge concerning [30, 32]:

(1) The types of the membership functions, i.e. triangular or Gaussian or bell shaped, trapezoidal, etc.
(2) The linguistic terms for each rule (their number, semantics);
(3) The parameters of the membership functions.

In addition, for the Takagi-Sugeno models a parametrization method (most often an optimization procedure) is necessary to determine the parameter values for the consequent models.

In contrast, the *AnYa* type FRB systems simplify the design process significantly, reducing it to the choice of prototypes only [21, 30]. Prototypes are actual data samples/vectors that are representative, and serve as the focal points of *data clouds*. *Data clouds* consisting of nearby data samples attracted around these focal points form Voronoi tessellation in the data space [33] as it will be described further in more detail in Chaps. 5 and 6.

3.1.2.5 Analysis of Different Types of FRB Systems

A brief comparative analysis of the three types of FRB systems is given in Table 3.1 [5, 21].

Table 3.1 A comparison between Zadeh-Mamdani, Takagi-Sugeno and *AnYa* types fuzzy rules

<table>
<tr><th colspan="2">Type</th><th>Antecedent (IF) part</th><th>Consequent (THEN) part</th><th>Defuzzification</th></tr>
<tr><td colspan="2">Zadeh-Mamdani</td><td rowspan="2">Parameterized fuzzy sets</td><td>Parameterized fuzzy sets</td><td>Central of gravity averaging of individual rule's contributions</td></tr>
<tr><td colspan="2">Takagi-Sugeno</td><td>Parameterized crisp (usually, linear) function</td><td>Fuzzily weighted sum of outputs per rule</td></tr>
<tr><td rowspan="3">AnYa</td><td>Zero-order</td><td rowspan="3">Prototypes, data clouds forming Voronoi tessellation [33]</td><td>Scalar singleton for classifiers</td><td rowspan="3">Winner takes all or few winners take all or fuzzily weighted sum</td></tr>
<tr><td>First-order</td><td>Linear</td></tr>
<tr><td>Higher-order</td><td>Non-linear, e.g. Gaussian, exponential</td></tr>
</table>

3.1.3 *Fuzzy Rule-Based Classifiers, Predictors and Controllers*

3.1.3.1 Introduction

Fuzzy sets theory gained significant popularity in the 1980–1990s with its applications to different engineering problems for which the traditional (often based on first principles such as mass, energy balance, differential equations and deterministic assumptions) or the statistical methods (which themselves heavily rely on *prior* assumptions) were not able to provide a solution or these solutions were too complex or ineffective [5, 34]. For example, fuzzy controllers [35] were successfully applied to automotive [36], space [37], transport [38–40] and other industries [41, 42]. FRB systems were successfully used to address difficult to solve otherwise prediction problems in financial [43], biotechnological [44], human behavior [45, 46] problems, etc. FRB systems were also successfully used to develop classifiers [47–49], clustering [34, 50] algorithms and other machine learning, control theory and operations research problems, e.g. fuzzy optimization [51, 52], fuzzy optimal control [53], etc.

In this section, we will briefly present the general principles of traditional FRB classifiers, predictors and controllers and will direct for more details in this area the interested readers to [5, 15].

3.1.3.2 FRB Predictors

A regression type FRB predictor is usually a first-order FRB system that employs the Takagi-Sugeno or *AnYa* type fuzzy rules in the following form [5, 6]:

Takagi-Sugeno type:

$$\begin{aligned} \boldsymbol{R}_i &: IF\left(x_{j,1}\ is\ LT_{i,1}\right) AND \left(x_{j,2}\ is\ LT_{i,2}\right) AND \ldots AND \left(x_{j,N}\ is\ LT_{i,N}\right) \\ & THEN\left(y_{j,i} = a_{i,0} + \sum_{l=1}^{N} a_{i,l} x_{j,l}\right) \end{aligned} \tag{3.5}$$

AnYa type:

$$\boldsymbol{R}_i :\ IF\left(\boldsymbol{x}_j \sim \boldsymbol{p}_i\right) \quad THEN\left(y_{j,i} = a_{i,0} + \sum_{l=1}^{N} a_{i,l} x_{j,l}\right) \tag{3.6}$$

where j denotes the current time instance, $j = 1, 2, \ldots, K$; $i = 1, 2, \ldots, M$; M denotes the number of fuzzy rules in the FRB.

For time series prediction, the model takes the following form:

$$y_j = f\left(\boldsymbol{x}_j\right) = x_{j+t,i} \tag{3.7}$$

where t is an integer indicating that the FRB predictor is for predicting the value of the ith input $x_{j,i}$ t time instances in advance; $f(\cdot)$ is the prediction model.

The overall output, y of the FRB predictors is usually determined by a weighted sum following the so-called "centre of gravity" principle [5, 21]:

$$y_j = \sum_{i=1}^{M} \bar{\lambda}_{j,i} y_{j,i}; \quad \bar{\lambda}_{j,i} = \frac{\lambda_{j,i}}{\sum_{l=1}^{M} \lambda_{j,l}} \tag{3.8}$$

or normalizing partial/local outputs (per rule) [16]:

$$y_j = \sum_{i=1}^{M} \lambda_{j,i} \bar{y}_{j,i}; \quad \bar{y}_{j,i} = \frac{y_{j,i}}{\sum_{l=1}^{M} y_{j,l}} \tag{3.9}$$

where $\lambda_{j,i}$ is the activation level/firing strength/score of confidence of the ith fuzzy rule.

3.1.3.3 FRB Classifiers

A. ***Zero-Order FRB Classifiers***

A zero-order FRB classifier consists of a set of Zadeh-Mamdani [27] type fuzzy rules in the following form [16]:

$$\boldsymbol{R}_i : \begin{array}{c} IF\left(x_{j,1}\ is\ LT_{i,1}\right)AND\left(x_{j,2}\ is\ LT_{i,2}\right)AND\ldots AND\left(x_{j,N}\ is\ LT_{i,N}\right) \\ THEN\left(y_{j,i}\ is\ LT_{i,out}\right) \end{array} \tag{3.10}$$

For a given input to the zero-order FRB classifier, $\boldsymbol{x}_j = \left[x_{j,1}, x_{j,2}, \ldots, x_{j,N}\right]^T$, which represents an instant vector of the feature values, its output represents the label, $LT_{i,out} = Label_i$. The label is quite often binary (for the so-called two-class classification problems), but in any case, integer.

It is well known that any multi-class classification problem (e.g. recognizing the ten digits or letters of the alphabet etc.) can be transformed into a number of two-class classification problems where this number is exactly the same as the number of classes in the original multi-class classification problem. The principle that is used is of "one against everything else". For example, an original problem of recognizing the ten digits (0, 1,. . ., 9) can easily be transformed into 10 two-class classification problems of the type "0 or else", "1 or else", etc.

The final label in a zero-order FRB classifier can be determined by the "winner takes all" (or more rarely, "few winners take all") principle using the normalized activation levels of each fuzzy rule, $\bar{\lambda}_{j,i}$ as follows:

$$Label(\boldsymbol{x}_j) = LT_{i^*,out}; \quad i^* = \underset{i=1,2,\ldots,M}{\arg\max}\left(\bar{\lambda}_{j,i}\right) \tag{3.11}$$

where $\bar{\lambda}_{j,i} = \frac{\lambda_{j,i}}{\sum_{l=1}^{M} \lambda_{j,l}}$; $\lambda_{j,i}$ is the activation level of the ith fuzzy rule.

The activation level itself is determined in Zadeh-Mamdani fuzzy rules as a product of the membership degrees to each fuzzy set (linguistic term) [54]:

$$\lambda_{j,i} = \prod_{l=1}^{N} \mu_{i,l}(x_{j,l}) \tag{3.12}$$

or by the minimum operator [54]:

$$\lambda_{j,i} = \min_{l=1,2,\ldots,N} (\mu_{i,l}(x_{j,l})) \tag{3.13}$$

where $\mu_{i,j}(\cdot)$ denotes the membership function of the jth fuzzy set of the ith fuzzy rule.

B. ***First-Order FRB Classifiers***

In 2007 it was proposed to use first-order FRB classifiers [16, 55, 56]. This allows a much higher level of flexibility and degrees of freedom which leads to much higher level of precision [16].

The overall output of the first-order FRB classifier can be determined by simple mechanism as follows (the example is for a two-class classifier, but can easily be generalized to multi-class classifiers). For example, as it was mentioned earlier, any multi-class classifier problem can be converted to a set of two-class classifier problems):

$$IF(y > 0.5) \quad THEN(Class\,0) \quad ELSE(Class\,1) \tag{3.14}$$

First order classifiers involve the use of a covariance matrix. A method for online selecting the most influential features was proposed in [57]. However, in comparison to the zero-order FRB classifiers, their training requires longer time and more data. In addition, while zero-order classifiers can be used in an unsupervised manner while the first-order classifiers at best can be semi-supervised [58], but not fully unsupervised.

3.1.3.4 FRB Controllers

The pioneering paper by Mamdani and Assilian [27] introduced the self-organizing fuzzy logic controllers and the Zadeh-Mamdani FRB systems as described earlier. In the following few decades a number of other types of fuzzy logic controllers were introduced, developed and applied to real engineering problems. The main issues are:

(1) the design, membership function, parameterization [30] and
(2) stability proof [59].

3.1.4 Evolving Fuzzy Systems

The concept of evolving fuzzy systems (EFS) was conceived around the turn of the 21st century [60, 61], and since then, it has been intensively researched and now matured [5]. EFSs are defined as self-organizing, self-developing FRB systems or neuro-fuzzy systems (NFS) that are capable of having both their structure and parameters self-updating online, and they are closely associated with streaming data processing [62]. In other words, EFSs are the dynamically evolving versions of the traditional fuzzy systems. EFSs aim at addressing the increasing demand of flexible, adaptive, robust, intelligent and interpretable systems for the advanced industry, autonomous systems, intelligent sensor networks, etc., and now have been widely applied to various real world applications [15, 63, 64].

A typical EFS is of "one pass" type. It is able to autonomously learn and extract knowledge from streaming data "on the fly" [56, 64–67], meanwhile, it keeps self-adapting parameters and self-evolving its structure. The basic idea behind it is to assume that the fuzzy system has an adjustable structure which allows the possible adaptation to follow the *shifts* and/or *drifts* of the ever-changing data pattern [68]. The system conducts both, structure evolution and parameter adaptation, based on the new observations from the data stream. However, usually, the structural changes appear much less frequently in comparison to the parameter adjustments, which take place at each processing round after each new data sample/observation is presented to the system.

Different EFSs use different online learning approaches to evolve their antecedent (IF) parts, but the majority of them use the recursive least squares (RLS) algorithm [69] or its extension, fuzzily weighted recursive least squares (FWRLS) introduced by the lead author of this book [6] to learn the consequent parameters (the THEN part).

Currently, the intensively-researched and widely-used EFSs include, but are not limited to, evolving Takagi-Sugeno (eTS) models [6, 65], dynamic evolving neural-fuzzy inference systems (DENFIS) [61], evolving fuzzy rule-based models [16, 66], sequential adaptive fuzzy inference system (SAFIS) [67, 70], flexible fuzzy inference systems (FLEXFIS) [71], parsimonious network based on fuzzy inference system (PANFIS) [72], generic evolving neuro-fuzzy inference system (GENFIS) [73], correntropy-based evolving fuzzy neural system (CENFS) [74], autonomous learning multi-model systems (*ALMMo*) [64], etc.

3.2 Artificial Neural Networks

Artificial neural networks (ANN) are computational structures formed of multiple interconnected computational neurons which are inspired by the biological neural network of the brain [75]. Computational ANNs are capable of approximating highly nonlinear and complex mathematical functions and relationships between inputs and outputs which is a consequence of the Kolmogorov theorem [76] and the pioneering publication by Hornik [77] formalized this significant theoretical result for any ANN with at least three layers (input, output and at least one hidden layer).

Indeed, the human brain (as well as the brain of mammals and other animals) is remarkable in its ability to process highly complex image, signal, textual and other information with extremely low power consumption (around 12 W) [78]. However, the biological brain has other aspects apart from the computational, such as biochemical, electromagnetic fields, genetic, etc. which are ignored in the ANNs. But, on the other hand, they are capable in some cases of solving problems that are hard for biological brain.

For example, recently, a series of GO matches between an algorithm performing so-called deep learning (DL) developed by the UK company specializing in games development named DeepMind and the World champion at that time (Lee Sedol) gained significant popularity and were also a part of the reported purchase of the company DeepMind by Google for £400 Million [79].

From computational point of view, ANNs are dynamic parallel information processing systems that perform embedded nonlinear transformations of the inputs. The main issue and shortcoming of the ANNs is their "black box" nature and lack of transparency as well as the long and computationally expensive training process [80].

The basic elements of an ANN are the computational neurons which mimic the biological neurons, see Fig. 3.2a. Computational neurons perform (usually non-linear) activation functions which map the sum of inputs into the output. They correspond to and mimic the somatic body which is also taking electric signals from the axons and map these to the synapses, see Fig. 3.2b.

Activation functions can be of different types. The most popular ones include:

(1) Binary function

$$f(x) = \begin{cases} 0 & x < 0 \\ 1 & x \geq 0 \end{cases} \tag{3.15}$$

(2) Linear function

$$f(x) = x \tag{3.16}$$

(3) Rectified linear function

$$f(x) = \begin{cases} 0 & x < 0 \\ x & x \geq 0 \end{cases} \tag{3.17}$$

(4) Sigmoid function

$$f(x) = \frac{1}{1 + e^{-x}} \tag{3.18}$$

Fig. 3.2 Comparison between computational and biological neurons

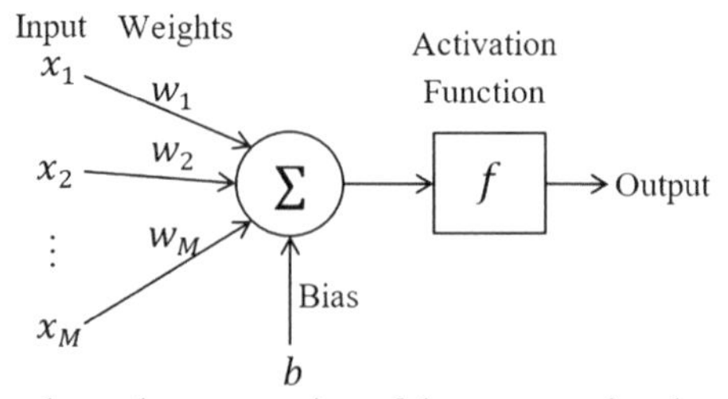

(a) A schematic presentation of the computational neuron

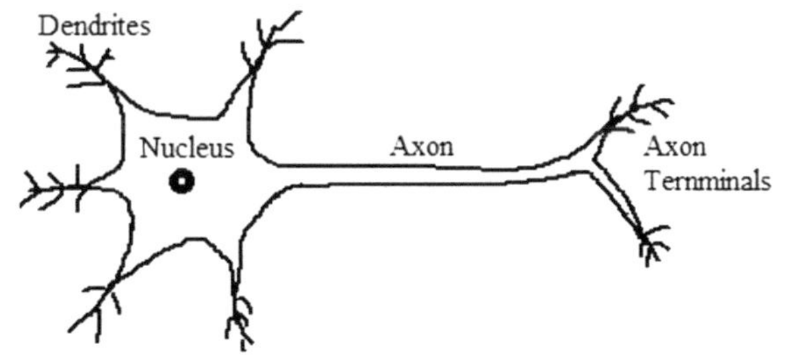

(b) A schematic presentation of the biological neuron

(5) Gaussian function

$$f(x) = e^{-x^2} \tag{3.19}$$

(6) Hyperbolic tangent function

$$f(x) = \tanh(x) = \frac{e^x - e^{-x}}{e^x + e^{-x}} \tag{3.20}$$

The curves of these functions are given in Fig. 3.3.

Historically, one of the earliest and best known computational models for neural networks based on mathematics and algorithms was introduced by Warren McCulloch and Walter Pitts in 1943 [75] called threshold logic. In 1970s and 1980s, with the development of the computational resources, a number of new, more complex ANNs started to emerge such as the multi-layer perceptron (MLP) [81], self-organizing map (SOM) [82], spiking neural network [83]. Around one decade ago, deep learning neural networks (DLNNs) started to gain a lot of popularity in both the academic circles and the general public [84, 85]. In 1974 Paul Werbos proposed (as part of one of his PhD thesis) the so-called *error back-propagation* (EBP) algorithm which is a layered form of the least mean square (LMS) gradient based optimization [86]. It became so popular that is still one of the benchmark methods to train ANN from data. P. Werbos later, in 1990 proposed the more advanced backpropagation through time (BPTT) learning method [87].

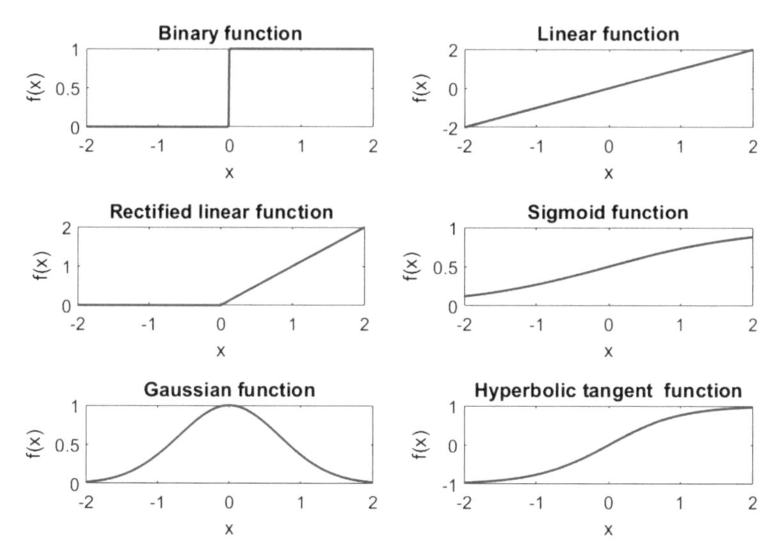

Fig. 3.3 Popular activation functions

The hype surrounding DL which goes beyond scientific research and spreads to publicity, investment and commerce [84, 85] helped resurrect the interest towards the ANN area as well as, more widely, to AI and machine learning. In what follows we will briefly present the basic principles of several specific types of ANN which will be used later in the book. For more details, the interested readers are referred, for example, to [84, 88, 89].

3.2.1 *Feedforward Neural Networks*

Feedforward neural networks are a class of ANNs whose neurons form an acyclic graph where information moves only in one direction from input to output. They are the simplest type of ANNs, and are extensively used in pattern recognition. The general architecture of a multilayer feedforward ANN also known as a multi-layer perceptron (MLP) is depicted in Fig. 3.4.

A typical MLP is composed of three types of layers:

(1) input layer,
(2) one or more hidden layers, and
(3) output layer.

Each layer is a group of neurons receiving connections from all neurons of the previous layer. Neurons from the same layer are not connected to each other (ANNs in which there is also a connection between the neurons of the same layer are called competitive ANNs, but the majority of ANNs are not competitive).

The input layer is the first layer of an MLP which is using input vector as its activation. Input layer receives no connections from other layers, but is fully connected to the first hidden layer. Each hidden layer is fully connected to the next hidden layer (if there is any), and the last hidden layer is fully connected to the output layer. The activations of the output units form the output of the overall MLP. The output is the result of the transformations of input data through neurons and layers in a form of distributed representation consisting of a huge number of weighted local representations across the entire networks, namely the weight matrices $\mathbf{W}^1$, $\mathbf{W}^2$ and $\mathbf{W}^3$ in Fig. 3.4.

The MLP is most often trained using the EBP algorithm which is the most widely used supervised learning algorithm for tuning the connection weights of feedforward neural networks [85]. EBP involves a two-phase cycle for weights optimization, namely:

(1) propagation, and
(2) weight update.

The input vector is firstly propagated forward layer by layer until it reaches the final layer, and the ANN generates the corresponding output. This results in an embedded function of Kolmogorov type [90] (here we use the MLP as depicted in Fig. 3.4, for example):

$$y_i = f\left(\sum_{j=1}^{Z_{H2}} w_{i,j}^3 f\left(\sum_{h=1}^{Z_{H1}} w_{j,h}^2 f\left(\sum_{l=1}^{Z_I} w_{h,l}^1 x_l\right)\right)\right) \tag{3.21}$$

where Z_I, Z_{H1}, Z_{H2} and Z_O are the sizes of the input, first hidden, second hidden and output layers, respectively; $\mathbf{W}^1$ is the $Z_{H1} \times Z_I$ dimensional matrix of weights of the

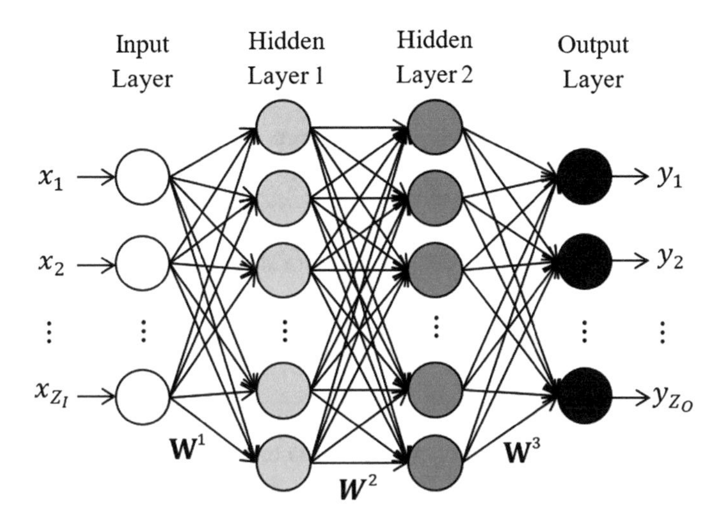

Fig. 3.4 General architecture of a multilayer feedforward NN (MLP) with two hidden layers

connections between the input and the first hidden layers; $\mathbf{W}^2$ is the $Z_{H2} \times Z_{H1}$ dimensional matrix of weights of the connections between the first and second hidden layers; $\mathbf{W}^3$ is the $Z_O \times Z_{H2}$ dimensional matrix of weights of the connections between the second hidden and the output layers; $w^1_{h,l} \in \mathbf{W}^1$ stands for the weight of the connection between the lth neuron of the input layer and the hth neuron of the first hidden layer; $w^2_{j,h} \in \mathbf{W}^2$ stands for the weight of the connection between the hth neuron of the first hidden layer and the jth neuron of the second hidden layer; $w^3_{i,j} \in \mathbf{W}^3$ stands for the weight of the connection between the jth neuron of the second hidden layer and the ith neuron of the output layer; $i = 1, 2, \ldots, Z_O$.

This is a superposition for which it was proven that it is able to approximate arbitrary continuous non-linear function normalized within the unit hypercube [90].

Weights of the network are, then, tuned by propagating the square error between the generated output and the target/reference value back through the network. The whole process repeats until the error (Eq. 3.22) reaches its minimum:

$$E_o = \sum_{k=1}^{K} (y_k - r_k)^2 \tag{3.22}$$

where E_o is the overall square error; y_k is the network output corresponding to the kth input and r_k is the respective reference value.

The general weight updating equations for EBP algorithm are given as follows [91, 92]:

$$\boldsymbol{\delta}^Q = \nabla_{\lambda} J \odot f'\left(\boldsymbol{b}^Q\right) \tag{3.23a}$$

$$\boldsymbol{\delta}^j = \left(\left(\mathbf{W}^{j+1}\right)^T \boldsymbol{\delta}^{j+1}\right) \odot f'\left(\boldsymbol{b}^j\right) \tag{3.23b}$$

$$\mathbf{W}^j \leftarrow \mathbf{W}^j - \beta \boldsymbol{\delta}^{j+1}\left(\boldsymbol{\lambda}^j\right)^T \tag{3.23c}$$

where $j = 1, 2, \ldots, Q$; Q is the number of layers; $\odot$ denotes the Hadamard product [93], which has the following form:

$$\begin{bmatrix} x_1 & x_2 \\ x_3 & x_4 \end{bmatrix} \odot \begin{bmatrix} y_1 & y_2 \\ y_3 & y_4 \end{bmatrix} = \begin{bmatrix} x_1 y_1 & x_2 y_2 \\ x_3 y_3 & x_4 y_4 \end{bmatrix} \tag{3.24}$$

$\boldsymbol{\lambda}^j$ and $\boldsymbol{b}^j$ are the vectors of the activation levels and inputs of the jth layer, respectively:

$$\boldsymbol{\lambda}^j = \left[\lambda^j_1, \lambda^j_2, \ldots, \lambda^j_{Z_j}\right]^T \tag{3.25a}$$

$$\boldsymbol{b}^j = \left[b^j_1, b^j_2, \ldots, b^j_{Z_j}\right]^T \tag{3.25b}$$

Z_j is the size of the jth layer; λ_i^j and b_i^j are the activation level and input of the ith neuron of the jth layer:

$$b_i^j = \sum_{l=1}^{Z_{j-1}} w_{i,l}^{j-1} \lambda_l^{j-1} \quad (3.26a)$$

$$\lambda_i^j = f\left(b_i^j\right) \quad (3.26b)$$

and there is $b_i^1 = x_i$ ($i = 1, 2, \ldots, Z_1$); J is the cost function, which usually takes the quadratic form given as:

$$J = \frac{1}{2} E_o = \frac{1}{2} \sum_{i=1}^{K} (y_i - r_i)^2 \quad (3.27)$$

$\nabla_\lambda J$ is the vector of partial derivatives of J expressed as:

$$\nabla_\lambda J = \left[\frac{\partial J}{\partial \boldsymbol{\lambda}_1^Q}, \frac{\partial J}{\partial \boldsymbol{\lambda}_2^Q}, \ldots, \frac{\partial J}{\partial \boldsymbol{\lambda}_{Z_Q}^Q} \right]^T \quad (3.28)$$

and $f'\left(\boldsymbol{b}^j\right)$ ($j = 1, 2, \ldots, Q$) is vector of partial derivatives of $f\left(\boldsymbol{b}^j\right)$ expressed as:

$$f'\left(\boldsymbol{b}^j\right) = \left[f'\left(b_1^j\right), f'\left(b_2^j\right), \ldots, f'\left(b_{Z_j}^j\right) \right]^T \quad (3.29)$$

$\mathbf{W}^j$ is the $Z_{j+1} \times Z_j$ weight matrix of the connections between the jth and $(j+1)$th layers, and the derivative of the cost function in regards to the weights is given as:

$$\frac{\partial J}{\partial w_{i,l}^j} = \lambda_i^j \delta_l^{j+1} \quad (3.30)$$

The EBP is, in fact, a practical application of the chain rule for derivatives and the least mean squares (LMS) algorithm [94], which itself is nothing else but a layered/chain form of steepest gradient algorithm to minimize the square error:

$$\Delta \boldsymbol{w}(j) = \beta \sum_{k=1}^{K} \left(r_k - \boldsymbol{x}_k^T \boldsymbol{w}(j-1) \right) \quad (3.31)$$

where $\Delta \boldsymbol{w}(j) = \boldsymbol{w}(j) - \boldsymbol{w}(j-1)$; $\boldsymbol{w}(j)$ is the $1 \times N$ dimensional coefficient vector calculated at the jth iteration round.

The key insight of this algorithm is that the derivative of the objective function in respect of the input of a layer can be computed by working backwards from the gradient in respect of the output of that layer. This is possible analytically because each activation function of the neurons is known by design and these are selected to be smooth and differentiable functions. The EBP equation can propagate gradients starting from the output layer all the way back to the input layer. Once these gradients have been computed, it is straightforward to compute the gradients with respect to the weights of each layer. The interested reader is directed for more details, for example, to [84, 88, 89, 92].

One can see the advantages and disadvantages of this approach: on one hand it is very slow and may fall into local minima; on the other hand, it guarantees smooth improvements and a clear theoretically proven rationale (unlike some *ad hoc* alternatives such as reinforcement learning, evolutionary algorithms, etc., which have no proven convergence).

One particular type of MLPs, namely the so-called radial basis function (RBF) type ANN, gained specific importance due to its links with the FRB systems and interpretability properties [95]. The general architecture of a RBF type MLP as depicted in Fig. 3.5 in which there is a single hidden layer and the activation functions of the neurons at the hidden layer are Gaussian with prototypes.

The radial basis function of the *i*th RBF neuron of the hidden layer in the ANN is formulated as follows:

$$f_i\left(\boldsymbol{x} = [x_1, x_2, \ldots, x_{Z_I}]^T\right) = e^{-(\boldsymbol{x}-\boldsymbol{c}_i)^T \boldsymbol{\Sigma}_i^{-1} (\boldsymbol{x}-\boldsymbol{c}_i)} \tag{3.32}$$

where $i = 1, 2, \ldots, Z_H$; Z_H is the size of the hidden layer; $\boldsymbol{c}_i$ and $\boldsymbol{\Sigma}_i$ are the respective mean vector and covariance matrix of the *i*th neuron, both of which can be obtained by supervised learning.

The output of the *i*th neuron in the final layer of the RBF NN is formulated as:

$$y_i = \sum_{j=1}^{Z_H} w_{i,j} f_j(\boldsymbol{x}) = \sum_{j=1}^{Z_H} w_{i,j} e^{-\left(\boldsymbol{x}-\boldsymbol{c}_j\right)^T \boldsymbol{\Sigma}_j^{-1} \left(\boldsymbol{x}-\boldsymbol{c}_j\right)} \tag{3.33}$$

where $i = 1, 2, \ldots, Z_O$.

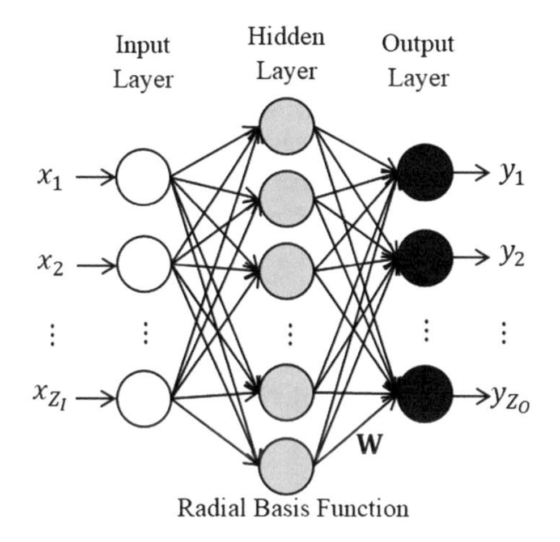

Fig. 3.5 General architecture of a RBF NN

3.2.2 *Deep Learning*

Deep learning (DL) architectures consist of multiple processing layers to learn high-level, abstract representations from raw data [85]. Representations at each layer are relatively simple and the power is in the deep (nested, hierarchical) architecture. A quintessential example of a DL model is the deep convolutional neural networks (DCNN).

In fact, the major reason for the recent success of DL models is the very large progress in hardware and software, which provides the essential computational resources for training such massive and complex computational models. The state-of-the-art performance achieved by the DL architectures sparked the interest of major technology companies such as Google, Facebook, Microsoft, Baidu, etc. to develop and offer commercial DL-based products and services [85].

3.2.2.1 Convolutional Neural Networks

DCNNs are the state-of-the-art approaches in the field of computer vision [85]. Recent DCNN architectures have 10–20 layers of rectified linear neurons, hundreds of millions of weights, and billions of connections between the units. A number of publications have demonstrated that DCNNs can produce highly accurate results in various image processing problems including, but not limited to, handwritten digits recognition [96–98], object recognition [99], human action recognition [100], remote sensing image classification [101], etc. Some publications suggest that the DCNNs can match the human performance on the handwritten digits recognition problems [96–98].

ANN in general, and MLP in particular, gained significant interest and were widely applied for prediction, classification and control problems, especially in 1990s. However, a particular interest to them surged again in this decade and especially, to particular type of multi-layered ANN that address image processing (as well as speech processing, games, human behavior, etc.) [84, 85].

If consider images, in particular, features of an image may change, be rotated or translated and traditional MLP cannot scale well for such particular problems, especially for large size images dense connections transforming input images through a series of hidden layers fully connected one-by-one do overfit and are practically impossible to train. The other important reason is that traditional MLP is translation-invariant, and thus, it is not able to learn the features that may change in the images.

At the same time, the human eye ignores some of the details and is focusing on the important, rotation- and translation- invariant features. Studying these phenomena, the Japanese researcher Kunihiko Fukushima introduced the so-called *Neucognitron* [102]. The *Neucognitron* was later generalized and significantly improved into the so-called convolutional neural network (CNN) [103]. CNNs are designed for solving image recognition problems. A typical structure is shown in

Fig. 3.6 [104]. A CNN receives *2D* inputs and extracts high-level features through multiple hidden layers which can be a combination of the following four types:

(1) Convolution layer;
(2) Pooling layer;
(3) Normalization layer;
(4) Fully connected layer.

Convolutional layer and pooling layers in CNNs are directly inspired by the classic notions of simple cells and complex cells in visual neuroscience [105].

Convolutional layers apply a convolution operation to the input, passing the result to the next layer. The role of the convolutional layer is to detect local conjunctions of features from the previous layer. It is at the core of a CNN and consists of a set of learnable parameters (filters). During the training process, the convolutional layer goes through all the training data samples and calculates the inner product between the inputs and the filter, which results in a feature map of the filter. The pooling layer operates on the feature maps and is for merging similar features into one. The main purposes of introducing the pooling layers into the CNN are:

(1) decreasing both the computational burden and the training time, and
(2) reducing the overfitting.

A typical pooling unit (max-pooling) calculates the maximum of a local patch of units in one feature map (or in a few feature maps). Neighboring pooling units take input from patches that are shifted by more than one row or column, results in the reduction of the dimension of the representation and the increase of the robustness to small shifts and distortions of the original image.

The normalization layer is useful when using neurons with unbounded activations (e.g. rectified linear neurons), because it permits the detection of high-frequency features with a big neuron response, while damping responses that are uniformly large in a local neighborhood. The introduction of this layer results in a faster training without significantly affecting the generalization ability of the network.

Finally, the fully connected layer connects every neuron in the previous layer to every neuron in it, which, in principle, is the same as the traditional feedforward

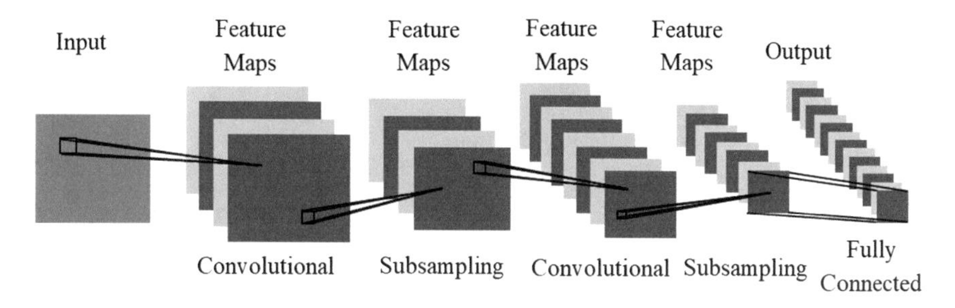

Fig. 3.6 A typical CNN

MLP as described in the previous subsection. A typical contemporary CNN such as AlexNet has hundreds of millions of trainable weights which requires significant computational resources and time (graphic processing units (GPU) and tens of hours of training) [85, 96, 97, 99].

3.2.2.2 Recurrent Neural Networks

Deep recurrent neural networks (DRNN) [106] and deep belief networks [107] can also be built by stacking multiple recurrent neural network (RNN) and restricted Boltzmann machine (RBM) hidden layers, respectively.

Feedforward MLPs assume independency between the input data samples. However, in many prediction tasks (e.g. time-series) the output is not only dependent on the current input, but also on several previous ones, and, thus, the sequences of inputs are important for a successful prediction. RNNs are designed to address this issue in sequential time series problems with various lengths.

The input to a RNN neuron is comprised by both, the current data sample and the previously observed one, which means that the output of the neuron at the current time instance is affected by the output of the previous one. To achieve this, each neuron is equipped with a feedback loop for returning the current output back for the next step. The typical structure of a RNN is depicted in Fig. 3.7.

To train the RNN, an extension of the EBP algorithm, named backpropagation through time (BPTT) [87] is used; this is because of the existence of cycles involving the RNN neurons. The original EBP algorithm works based on the error derivatives in respect of the weights in the upper layers, while RNNs do not have a stacked layer model. The core of the BPTT is a technique called *unrolling*, and this technique leads to a feedforward structure over time spans (see Fig. 3.7).

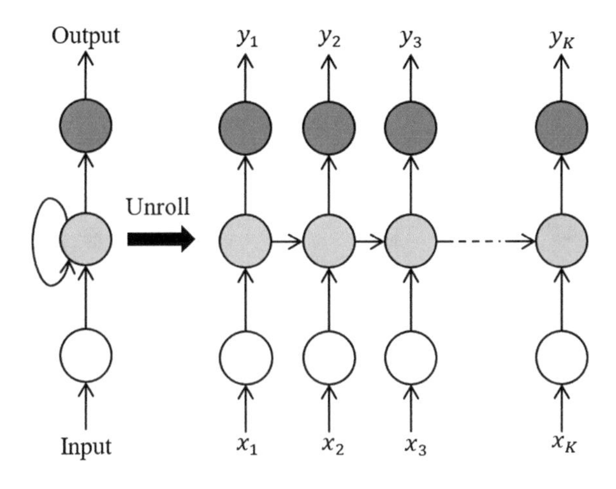

Fig. 3.7 The typical structure of a RNN

One of the most well-known extensions of RNN is the so-called *long short term memory* (LSTM) [108–110]. LSTM involves the concept of gates for its nodes. In addition to the original feedback loop in the RNN neuron, each LSTM neuron has multiplicative gates for the following functions:

Forget:

$$\boldsymbol{f}_k = f_1\left(\mathbf{W}_{f,x}\boldsymbol{x}_k + \mathbf{W}_{f,y}\boldsymbol{y}_{k-1} + \boldsymbol{b}_f\right) \tag{3.34}$$

Input:

$$\boldsymbol{i}_k = f_1\left(\mathbf{W}_{i,x}\boldsymbol{x}_k + \mathbf{W}_{i,y}\boldsymbol{y}_{k-1} + \boldsymbol{b}_i\right) \tag{3.35}$$

Output:

$$\boldsymbol{o}_k = f_1\left(\mathbf{W}_{o,x}\boldsymbol{x}_k + \mathbf{W}_{o,y}\boldsymbol{y}_{k-1} + \boldsymbol{b}_o\right) \tag{3.36}$$

where $\boldsymbol{f}_k$, $\boldsymbol{i}_k$ and $\boldsymbol{o}_k$ are the correspondingly activation vectors of the forget, input and output gates; $\mathbf{W}_{f,x}\mathbf{W}_{i,x}$ and $\mathbf{W}_{o,x}$, $\mathbf{W}_{f,y}\mathbf{W}_{i,y}$ and $\mathbf{W}_{o,y}$ are the weight matrices of the forget, input and output gates, respectively; $\boldsymbol{b}_f$, $\boldsymbol{b}_i$ and $\boldsymbol{b}_o$ are the biase vectors; $f_1(\cdot)$ is the sigmoid activation function as given by Eq. (3.18); $\boldsymbol{x}_k$ is the input vector to the LSTM neuron;$\boldsymbol{y}_{k-1}$ is the output of the neuron corresponding to $\boldsymbol{x}_{k-1}$, and the following equation takes place:

$$\boldsymbol{c}_k = \boldsymbol{f}_k \odot \boldsymbol{c}_{k-1} + \boldsymbol{i}_k \odot f_2\left(\mathbf{W}_{c,x}\boldsymbol{x}_k + \mathbf{W}_{c,y}\boldsymbol{y}_{k-1} + \boldsymbol{b}_c\right) \tag{3.37a}$$

$$\boldsymbol{y}_k = \boldsymbol{o}_k \odot f_3(\boldsymbol{c}_k) \tag{3.37b}$$

where $\boldsymbol{c}_k$ is the cell state vector; $\mathbf{W}_{c,x}$, $\mathbf{W}_{c,y}$ and $\boldsymbol{b}_c$ are the corresponding weight matrices and bias vector; $f_2(\cdot)$ is the hyperbolic tangent function (Eq. 3.20); $f_3(\cdot)$ is the linear function (Eq. 3.16).

The multiplicative forget, input and output gates are introduced for controlling the access to memory cells and prevent them from perturbation by irrelevant inputs. The typical structure of a so-called *peephole LSTM neuron* is given in Fig. 3.8 [109, 110].

Each LSTM gate will compute a value between 0 (turn off) and 1 (turn on) based on the input using an activation function (see Eqs. (3.15)–(3.20) and Fig. 3.3). The input gate is for controlling the input of the neuron, the output gate is for controlling the output, and the forget gate is for controlling the neuron to keep or forget its last content. An important difference between LSTMs and RNNs is that the LSTM neurons use forget gates to actively control the cell states and to prevent degradation. The differentiable function within the LSTM neurons is usually of sigmoid type. BPTT is a commonly used method in training LSTMs as well.

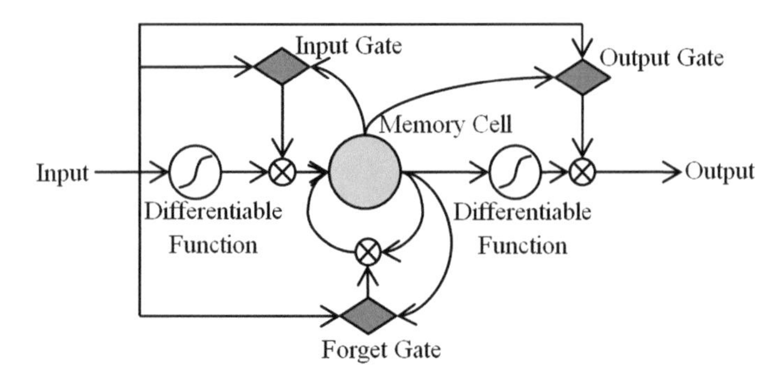

Fig. 3.8 The typical structure of a LSTM neuron

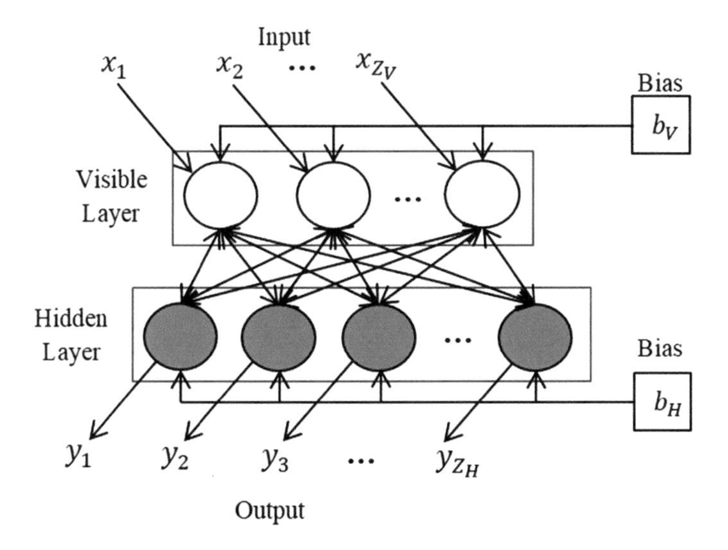

Fig. 3.9 The typical structure of a RBM

3.2.2.3 Restricted Boltzmann Machines

A restricted Boltzmann machine (RBM) is a stochastic ANN that can learn a probability function over the set of inputs. It is comprised of two layers:

(1) a visible layer that contains the input, and
(2) a hidden layer that contains the latent variables [111].

The typical structure of a RBM is given in Fig. 3.9.

RBMs are a variant of Boltzmann machines with the restriction applied to the connectivity of the neurons. The neurons of a RBM should form a bipartite graph: the neurons from the visible layer should be fully connected with the neurons from the hidden layer and vice versa, but there is no connection between the neurons of the same layer, while Boltzmann machines allow connections between the neurons in the hidden layer. This restriction allows for more efficient training algorithms to be developed. The RBM uses forward feeding to compute the latent variables and then, uses them to reconstruct the input with backward feeding. During the training stage, the training data is assigned to the visible neurons, and the weights optimization can be achieved using EBP and gradient decedent algorithms, and the goal is to maximize the product of all probabilities of the visible neurons.

3.3 Neuro-fuzzy Systems

ANN have proven computational advantages and same as the FRB systems have been proven universal approximator [25, 77]. FRB systems, however, have excellent interpretability, semantics and transparency properties. They are also able to incorporate human expertise, tolerate uncertainties. Therefore, the integration of both, ANN's layered architecture and the human interpretability properties of the FRB systems led in the early 1990s to the integration of the two approaches, which is usually called neuro-fuzzy systems (NFSs) [22, 112]. A typical NFS can be interpreted as a set of fuzzy rules [5, 28, 113], and it can also be represented as a special multi-layer feedforward neural network. NFSs are designed to realize the process of fuzzy reasoning. They identify fuzzy rules and learn membership functions by using learning algorithms that are typical for ANNs taking into account their layered structure.

Historically, the first NFSs, despite their names including terms such as "adaptive" and "evolving", e.g. adaptive-network-based fuzzy inference system (ANFIS) [22] and the evolving fuzzy neural networks (EFuNN) [114] and evolving FRB models [115, 116] were designed and trained offline and relied heavily on human expertise. ANFIS is the very first NFS architecture and is still widely applied in real applications [117, 118]. In this subsection, we use ANFIS [22] as an example. The general architecture of ANFIS is depicted in Fig. 3.10, where two fuzzy rules of Takagi-Sugeno type [28] are considered:

$$IF\left(x_1\ is\ LT_{1,1}\right)AND\left(x_2\ is\ LT_{1,2}\right)\quad THEN\left(y=a_{1,0}+a_{1,1}x_1+a_{1,2}x_2\right) \tag{3.38a}$$

$$IF\left(x_1\ is\ LT_{2,1}\right)AND\left(x_2\ is\ LT_{2,2}\right)\quad THEN\left(y=a_{2,0}+a_{2,1}x_1+a_{2,2}x_2\right) \tag{3.38b}$$

The first layer of ANFIS encodes the fuzzy sets ($x\ is\ LT$). The outputs of *Layer 1*, denoted by $\mu_{i,j}$ (in this case, $i,j=1,2$), are the fuzzy membership degrees (to what extent a particular value of the input variable (data) x_i is described by the linguistic term $LT_{1,1}$).

The second layer brings together the particular fuzzy sets of each fuzzy rules. It takes as inputs the fuzzy membership degrees, $\mu_{i,j}$ and generates as outputs the firing strengths, usually by product:

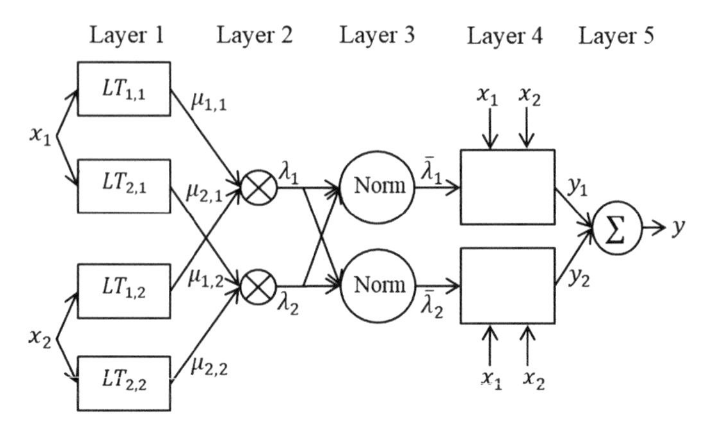

Fig. 3.10 General architecture of ANFIS

$$\lambda_i = \prod_j \mu_{i,j} \tag{3.39}$$

or "winner take all":

$$\lambda_i = \max_j \left(\mu_{i,j}\right) \tag{3.40}$$

The third layer normalizes the firing strengths from the previous layer, namely,

$$\bar{\lambda}_i = \frac{\lambda_i}{\sum_i \lambda_i} \tag{3.41}$$

The fourth layer produces the partial output per fuzzy rule as a first-order polynomial weighted by the normalized firing strength:

$$y_i = \bar{\lambda}_i \quad a_{i,0} + \sum_j a_{i,j}x_j \Bigg) \tag{3.42}$$

The final, fifth layer performs the summation of all incoming inputs and generates the overall output:

$$y = \sum_i y_i = \frac{\sum_i \lambda_i \left(a_{i,0} + \sum_j a_{i,j}x_j\right)}{\sum_i \lambda_i} \tag{3.43}$$

The training process of the ANFIS is a combination of gradient descent and least squares (LS) methods. In the forward pass, the outputs of the nodes within the network are passed forward up to *Layer 4* and the consequent parameters are determined by the LS method. In the backward pass, the error signals propagate backward and the premise parameters are updated using gradient descendent [14].

Towards the end of the last century as a result of an intensive research and development, much more sophisticated NFSs were proposed which offer capability of self-developing, self-learning both their parameters and structure online [6, 61, 67, 72, 119]. Currently, the most popular NFSs include, but are not limited to, the evolving Takagi-Sugeno systems (eTS) [6, 60], DENFIS [61], SAFIS [67, 70], PANFIS [72], GENFIS [73] as well as the recently introduced CENFS [74]. There are also different versions of eTS, and these algorithms include, e.g. simpl_eTS [65], eTS+ [57].

3.4 Conclusions

In this chapter, we focus on the two main forms of *computational intelligence*, namely FRB systems and ANN. Both of them have been proven universal approximators avoiding the need to have first principles or a statistical model. They can be designed based on data. FRB systems can also express its internal model in a form understandable by humans and can use human domain expert knowledge (if such knowledge exists). Furthermore, it has been proven [120] that:

(1) any FRB system may be approximated by a feedforward ANN, and
(2) any feedforward ANN may be approximated by a FRB system.

On the other hand, there are major differences between the two concepts as well. Fuzzy systems are excellent in handling uncertainties, and they also have transparent, human-interpretable internal representation. They are capable of self-organizing, self-updating both their structures and parameters in an online, dynamic environment. Neural networks, in contrast, are excellent in providing high precision in most cases but are fragile when facing new data patterns. They lack transparency ("black box" type) and the training process is usually limited to offline, and requires huge amount of computation resources and data.

DL type ANNs are the latest and most successful example. They demonstrated excellent level of performance in terms of precision (often comparable with that of a human, especially in image processing). Nonetheless, the DL-based approaches still have a number of major drawbacks:

(1) The training process is computationally expensive and requires a huge amount of data;
(2) The training process is opaque and lack of human interpretability (black box type);
(3) The training process is limited to offline and requires re-training for samples with feature properties different than the observed samples.

3.5 Questions

(1) Which are the main components of the computational intelligence?
(2) What is the inspiration for and what the main types of computational intelligence do emulate?
(3) What do FRB systems offer that makes them attractive?
(4) What are the main advantages of the mainstream deep learning?
(5) What are the main shortcomings of the mainstream deep learning?

3.6 Take Away Items

- Computational Intelligence was coined as a term two decades ago as an umbrella term to cover several research areas that were already developed and had in common the nature-inspired computational methods that offer alternative to traditional deterministic and stochastic approaches and mimic/emulate the brain, human reasoning, natural evolution; these include FRB Systems (reasoning), ANNs (brain) and evolutionary computation whereby the first two are used in the book.
- Both, FRB systems and ANNs have been proven universal approximators which is very important property.
- FRB systems, in addition, are also transparent and semantically interpretable; they are a powerful tool to deal with uncertainties.
- The mainstream Deep Learning offers high precision solutions, however, they have many drawbacks, e.g.,

  (1) The training process is computational expensive and requires a huge amount of data;
  (2) The training process is opaque, and model architecture lacks the human interpretability (black box type);
  (3) The training process is limited to offline and requires re-training for samples with feature properties different than the observed samples.

References

1. J.C. Bezdek, What is computational intelligence?, *Computational Intelligence Imitating Life* (IEEE Press, New York, 1994), pp. 1–12
2. W. Duch, What is computational intelligence and what could it become?, *Computational Intelligence, Methods and Applications Lecture Notes* (NAnYang Technological University, Singapour, 2003)
3. L.A. Zadeh, Fuzzy sets. Inf. Control **8**(3), 338–353 (1965)
4. L.A. Zadeh, Outline of a new approach to the analysis of complex systems and decision processes. IEEE Trans. Syst. Man Cybern. **1**, 28–44 (1973)

5. P. Angelov, *Autonomous Learning Systems: From Data Streams to Knowledge in Real Time* (Wiley, New York, 2012)
6. P.P. Angelov, D.P. Filev, An approach to online identification of Takagi-Sugeno fuzzy models. IEEE Trans. Syst. Man, Cybern. Part B Cybern. **34**(1), 484–498 (2004)
7. P. Angelov, R. Ramezani, X. Zhou, Autonomous novelty detection and object tracking in video streams using evolving clustering and Takagi-Sugeno type neuro-fuzzy system, in *IEEE International Joint Conference on Neural Networks*, 2008, pp. 1456–1463
8. D. Chakraborty, N.R. Pal, Integrated feature analysis and fuzzy rule-based system identification in a neuro-fuzzy paradigm. IEEE Trans. Syst. Man Cybern. Part B Cybern. **31**(3), 391–400 (2001)
9. A. Lemos, W. Caminhas, F. Gomide, Adaptive fault detection and diagnosis using an evolving fuzzy classifier. Inf. Sci. (Ny) **220**, 64–85 (2013)
10. C.F. Juang, C.T. Lin, An on-line self-constructing neural fuzzy inference network and its applications. IEEE Trans. Fuzzy Syst. **6**(1), 12–32 (1998)
11. P.P. Angelov, *Evolving Rule-Based Models: A Tool for Design of Flexible Adaptive Systems* (Springer, Berlin Heidelberg, 2002)
12. P. Angelov, A fuzzy controller with evolving structure. Inf. Sci. (Ny) **161**(1–2), 21–35 (2004)
13. R.E. Precup, H.I. Filip, M.B. Rədac, E.M. Petriu, S. Preitl, C.A. Dragoş, Online identification of evolving Takagi-Sugeno-Kang fuzzy models for crane systems. Appl. Soft Comput. J. **24**, 1155–1163 (2014)
14. A. Al-Hmouz, J. Shen, R. Al-Hmouz, J. Yan, Modeling and simulation of an adaptive neuro-fuzzy inference system (ANFIS) for mobile learning. IEEE Trans. Learn. Technol. **5** (3), 226–237 (2012)
15. E. Lughofer, *Evolving Fuzzy Systems-Methodologies, Advanced Concepts and Applications* (Springer, Berlin, 2011)
16. P. Angelov, X. Zhou, Evolving fuzzy-rule based classifiers from data streams. IEEE Trans. Fuzzy Syst. **16**(6), 1462–1474 (2008)
17. D. Leite, P. Costa, F. Gomide, Interval approach for evolving granular system modeling, *Learning in Non-stationary Environments* (Springer, New York, 2012), pp. 271–300
18. O. Cordón, F. Herrera, P. Villar, Generating the knowledge base of a fuzzy rule-based system by the genetic learning of the data base. IEEE Trans. Fuzzy Syst. **9**(4), 667–674 (2001)
19. W. Pedrycz, F. Gomide, *Fuzzy Systems Engineering: Toward Human-Centric Computing* (Wiley, New York, 2007)
20. L. Liu, M. Tamer Özsu, *Encyclopedia of Database Systems* (Springer, Berlin, 2009)
21. P. Angelov, R. Yager, A new type of simplified fuzzy rule-based system. Int. J. Gen Syst. **41** (2), 163–185 (2011)
22. J.S.R. Jang, ANFIS: adaptive-network-based fuzzy inference system. IEEE Trans. Syst. Man Cybern. **23**(3), 665–685 (1993)
23. http://www.worldweatheronline.com
24. G.W.F. Hegel, *Science of Logic* (Humanities Press, New York, 1969)
25. L.X. Wang, J.M. Mendel, Fuzzy basis functions, universal approximation, and orthogonal least-squares learning. IEEE Trans. Neural Netw. **3**(5), 807–814 (1992)
26. E.H. Mamdani, Application of fuzzy algorithms for control of simple dynamic plant. Proc. Inst. Electr. Eng. **121**(12), 1585 (1974)
27. E.H. Mamdani, S. Assilian, An experiment in linguistic synthesis with a fuzzy logic controller. Int. J. Man Mach. Stud. **7**(1), 1–13 (1975)
28. T. Takagi, M. Sugeno, Fuzzy identification of systems and its applications to modeling and control. IEEE Trans. Syst. Man. Cybern. **15**(1), 116–132 (1985)
29. P. Angelov, An approach for fuzzy rule-base adaptation using on-line clustering. Int. J. Approx. Reason. **35**(3), 275–289 (2004)
30. P.P. Angelov, X. Gu, Empirical fuzzy sets. Int. J. Intell. Syst. **33**(2), 362–395 (2017)

31. P. Angelov, X. Gu, D. Kangin, Empirical data analytics. Int. J. Intell. Syst. **32**(12), 1261–1284 (2017)
32. C.C. Lee, Fuzzy logic in control systems: fuzzy logic controller—Part 1. IEEE Trans. Syst. Man Cybern. **20**(2), 404–418 (1990)
33. A. Okabe, B. Boots, K. Sugihara, S.N. Chiu, *Spatial tessellations: concepts and applications of Voronoi diagrams*, 2nd edn. (Wiley, Chichester, 1999)
34. R.R. Yager, D.P. Filev, Approximate clustering via the mountain method. IEEE Trans. Syst. Man. Cybern. **24**(8), 1279–1284 (1994)
35. R.R. Yager, D.P. Filev, *Essentials of Fuzzy Modeling and Control*, vol. 388 (Wiley, New York, 1994)
36. C. Von Altrock, B. Krause, H. Zimmermann, Advanced fuzzy logic control technologies in automotive applications, in *IEEE International Conference on Fuzzy Systems*, 1992, pp. 835–842
37. L.I. Larkin, A fuzzy logic controller for aircraft flight control, in *IEEE Conference on Decision and Control*, 1984, pp. 894–897
38. R. Hoyer, U. Jumar, Fuzzy control of traffic lights, in *IEEE Conference on Fuzzy Systems*, 1994, pp. 1526–1531
39. X. Zhou, P. Angelov, Real-time joint landmark recognition and classifier generation by an evolving fuzzy system. IEEE Int. Conf. Fuzzy Syst. **44**(1524), 1205–1212 (2006)
40. P. Angelov, P. Sadeghi-Tehran, R. Ramezani, An approach to automatic real-time novelty detection, object identification, and tracking in video streams based on recursive density estimation and evolving Takagi-Sugeno fuzzy systems. Int. J. Intell. Syst. **29**(2), 1–23 (2014)
41. M. Sugeno, *Industrial Applications of Fuzzy Control* (Elsevier Science Inc., 1985)
42. J.J. Macias-Hernandez, P. Angelov, X. Zhou, Soft sensor for predicting crude oil distillation side streams using Takagi Sugeno evolving fuzzy models, vol. 44, no. 1524, pp. 3305–3310, 2007
43. X. Gu, P.P. Angelov, A.M. Ali, W.A. Gruver, G. Gaydadjiev, Online evolving fuzzy rule-based prediction model for high frequency trading financial data stream, in *IEEE Conference on Evolving and Adaptive Intelligent Systems (EAIS)*, 2016, pp. 169–175
44. J. Trevisan, P.P. Angelov, A.D. Scott, P.L. Carmichael, F.L. Martin, IRootLab: a free and open-source MATLAB toolbox for vibrational biospectroscopy data analysis. Bioinformatics **29**(8), 1095–1097 (2013)
45. J.A. Iglesias, P. Angelov, A. Ledezma, A. Sanchis, Human activity recognition based on evolving fuzzy systems. Int. J. Neural Syst. **20**(5), 355–364 (2010)
46. J. Andreu, P. Angelov, Real-time human activity recognition from wireless sensors using evolving fuzzy systems, in *IEEE International Conference on Fuzzy Systems*, 2010, pp. 1–8
47. J. Casillas, O. Cordón, M.J. Del Jesus, F. Herrera, Genetic feature selection in a fuzzy rule-based classification system learning process for high-dimensional problems. Inf. Sci. (Ny) **136**, 135–157 (2001)
48. H. Ishibuchi, T. Yamamoto, Rule weight specification in fuzzy rule-based classification systems. IEEE Trans. Fuzzy Syst. **13**(4), 428–435 (2005)
49. X. Zhou, P. Angelov, Autonomous visual self-localization in completely unknown environment using evolving fuzzy rule-based classifier, in *IEEE Symposium on Computational Intelligence in Security and Defense Applications*, 2007, pp. 131–138
50. J.C. Bezdek, R. Ehrlich, W. Full, FCM: the fuzzy c-means clustering algorithm. Comput. Geosci. **10**(2–3), 191–203 (1984)
51. H.J. Zimmermann, Description and optimization of fuzzy systems. Int. J. Gen. Syst. **2**(1), 209–215 (1975)
52. P. Angelov, A generalized approach to fuzzy optimization. Int. J. Intell. Syst. **9**(4), 261–268 (1994)
53. D. Filev, P. Angelov, Fuzzy optimal control. Fuzzy Sets Syst. **47**(2), 151–156 (1992)
54. D.J. Dubois, *Fuzzy Sets and Systems: Theory and Applications* (Academic Press, 1980)
55. P. Angelov, X. Zhou, F. Klawonn, Evolving fuzzy rule-based classifiers, in *IEEE Symposium on Computational Intelligence in Image and Signal Processing*, 2007, pp. 220–225

56. P. Angelov, E. Lughofer, X. Zhou, Evolving fuzzy classifiers using different model architectures. Fuzzy Sets Syst. **159**(23), 3160–3182 (2008)
57. P. Angelov, Evolving Takagi-Sugeno fuzzy systems from streaming data (eTS+), in *Evolving Intelligent Systems: Methodology and Applications* (Wiley, New York, 2010)
58. X. Gu, P.P. Angelov, Semi-supervised deep rule-based approach for image classification. Appl. Soft Comput. **68**, 53–68 (2018)
59. H.-J. Rong, P. Angelov, X. Gu, J.-M. Bai, Stability of evolving fuzzy systems based on data clouds. IEEE Trans. Fuzzy Syst. (2018). https://doi.org/10.1109/TFUZZ.2018.2793258
60. P. Angelov, R. Buswell, Evolving rule-based models: a tool for intelligent adaption, in *IFSA World Congress and 20th NAFIPS International Conference*, 2001, pp. 1062–1067
61. N.K. Kasabov, Q. Song, DENFIS: dynamic evolving neural-fuzzy inference system and its application for time-series prediction. IEEE Trans. Fuzzy Syst. **10**(2), 144–154 (2002)
62. P. Angelov, Evolving fuzzy systems. Scholarpedia **3**(2), 6274 (2008)
63. L. Maciel, R. Ballini, F. Gomide, Evolving possibilistic fuzzy modeling for realized volatility forecasting with jumps. IEEE Trans. Fuzzy Syst. **25**(2), 302–314 (2017)
64. P.P. Angelov, X. Gu, J.C. Principe, Autonomous learning multi-model systems from data streams. IEEE Trans. Fuzzy Syst. **26**(4), 2213–2224 (2016)
65. P. Angelov, D. Filev, Simpl_eTS: a simplified method for learning evolving Takagi-Sugeno fuzzy models, in *IEEE International Conference on Fuzzy Systems*, 2005, pp. 1068–1073
66. R.D. Baruah, P.P. Angelov, J. Andreu, Simpl_eClass: simplified potential-free evolving fuzzy rule-based classifiers, in *IEEE International Conference on Systems, Man, and Cybernetics (SMC)*, 2011, pp. 2249–2254
67. H.J. Rong, N. Sundararajan, G. Bin Huang, P. Saratchandran, Sequential adaptive fuzzy inference system (SAFIS) for nonlinear system identification and prediction. Fuzzy Sets Syst. **157**(9), 1260–1275 (2006)
68. E. Lughofer, P. Angelov, Handling drifts and shifts in on-line data streams with evolving fuzzy systems. Appl. Soft Comput. **11**(2), 2057–2068 (2011)
69. R.M. Johnstone, C. Richard Johnson, R.R. Bitmead, B.D.O. Anderson, Exponential convergence of recursive least squares with exponential forgetting factor. Syst. Control Lett. **2**(2), 77–82 (1982)
70. H.J. Rong, N. Sundararajan, G. Bin Huang, G.S. Zhao, Extended sequential adaptive fuzzy inference system for classification problems. Evol. Syst. **2**(2), 71–82 (2011)
71. E.D. Lughofer, FLEXFIS: a robust incremental learning approach for evolving Takagi-Sugeno fuzzy models. IEEE Trans. Fuzzy Syst. **16**(6), 1393–1410 (2008)
72. M. Pratama, S.G. Anavatti, P.P. Angelov, E. Lughofer, PANFIS: a novel incremental learning machine. IEEE Trans. Neural Netw. Learn. Syst. **25**(1), 55–68 (2014)
73. M. Pratama, S.G. Anavatti, E. Lughofer, Genefis: toward an effective localist network. IEEE Trans. Fuzzy Syst. **22**(3), 547–562 (2014)
74. R. Bao, H. Rong, P.P. Angelov, B. Chen, P.K. Wong, Correntropy-based evolving fuzzy neural system. IEEE Trans. Fuzzy Syst. (2017). https://doi.org/10.1109/tfuzz.2017.2719619
75. W.S. McCulloch, W. Pitts, A logical calculus of the ideas immanent in nervous activity. Bull. Math. Biophys. **5**(4), 115–133 (1943)
76. A.N. Kolmogorov, Grundbegriffe der wahrscheinlichkeitsrechnung. Ergebnisse der Math. **3** (1933)
77. K. Hornik, M. Stinchcombe, H. White, Multilayer feedforward networks are universal approximators. Neural Netw. **2**(5), 359–366 (1989)
78. L.C. Aiello, P. Wheeler, The expensive-tissue hypothesis: the brain and the digestive system in human and primate evolution. Curr. Anthropol. **36**(2), 199–221 (1995)
79. http://uk.businessinsider.com/googles-400-million-acquisition-of-deepmind-is-looking-good-2016-7
80. S. Mitra, Y. Hayashi, Neuro-fuzzy rule generation: survey in soft computing framework. IEEE Trans. Neural Netw. **11**(3), 748–768 (2000)

81. D.E. Rumelhart, J.L. McClell, *Parallel Distributed Processing: Explorations in the Microstructure of Cognition, vol. 1: Foundations* (MIT Press, Cambridge, 1986)
82. T. Kohonen, The self-organizing map. Neurocomputing **21**(1–3), 1–6 (1998)
83. W. Maass, Networks of spiking neurons: the third generation of neural network models. Neural Netw. **10**(9), 1659–1671 (1997)
84. I. Goodfellow, Y. Bengio, A. Courville, *Deep Learning* (MIT Press, Crambridge, 2016)
85. Y. LeCun, Y. Bengio, G. Hinton, Deep learning. Nat. Methods **13**(1), 35 (2015)
86. P. Werbos, *Beyond Regression: New Fools for Prediction and Analysis in the Behavioral Sciences* (Harvard University, 1974)
87. P.J. Werbos, Backpropagation through time: what it does and how to do it. Proc. IEEE **78** (10), 1550–1560 (1990)
88. C.M. Bishop, *Neural Networks for Pattern Recognition* (Oxford University Press, 1995)
89. C.M. Bishop, *Pattern Recognition and Machine Learning* (Springer, New York, 2006)
90. A.N. Kolmogorov, On the representation of continuous functions of many variables by superposition of continuous functions of one variable and addition, in *Dokl. Akad. Nauk. SSSR.* vol. 114 (1957)
91. D.E. Rumelhart, G.E. Hinton, R.J. Williams, Learning representations by back-propagating errors. Nature **323**(6088), 533–536 (1986)
92. M. Nielsen, *Neural Networks and Deep Learning* (Determination Press, 2015)
93. R.A. Horn, C.R. Johnson, *Matrix Analysis* (Cambridge University Press, 1990)
94. B. Widrow, S.D. Stearns, *Adaptive Signal Processing* (Prentice-Hall, Englewood Cliffs, 1985)
95. D. Lowe, D. Broomhead, Multivariable functional interpolation and adaptive networks. Complex Syst. **2**(3), 321–355 (1988)
96. D.C. Cireşan, U. Meier, L.M. Gambardella, J. Schmidhuber, Convolutional neural network committees for handwritten character classification, in *International Conference on Document Analysis and Recognition*, vol. 10, 2011, pp. 1135–1139
97. D. Ciresan, U. Meier, J. Schmidhuber, Multi-column deep neural networks for image classification, in *Conference on Computer Vision and Pattern Recognition*, 2012, pp. 3642–3649
98. P.Y. Simard, D. Steinkraus, J.C. Platt, Best practices for convolutional neural networks applied to visual document analysis, *Seventh International Conference on Document Analysis* and. *Recognition* (Proceedings, 2003), pp. 958–963
99. A. Krizhevsky, I. Sutskever, G.E. Hinton, ImageNet classification with deep convolutional neural networks, in *Advances in Neural Information Processing Systems*, 2012, pp. 1097–1105
100. K. Charalampous, A. Gasteratos, On-line deep learning method for action recognition. Pattern Anal. Appl. **19**(2), 337–354 (2016)
101. L. Zhang, L. Zhang, V. Kumar, Deep learning for remote sensing data. IEEE Geosci. Remote Sens. Mag. **4**(2), 22–40 (2016)
102. K. Fukushima, S. Miyake, Neocognitron: a self-organizing neural network model for a mechanism of visual pattern recognition, *Competition and Cooperation in Neural Nets* (Springer, Berlin, 1982), pp. 267–285
103. Y. LeCun, B. Boser, J.S. Denker, D. Henderson, R.E. Howard, W. Hubbard, L.D. Jackel, Backpropagation applied to handwritten zip code recognition. Neural Comput. **1**(4), 541–551 (1989)
104. S. Lawrence, C.L. Giles, A.C. Tsoi, A.D. Back, Face recognition: a convolutional neural-network approach. IEEE Trans. Neural Netw. **8**(1), 98–113 (1997)
105. D.H. Hubel, T.N. Wiesel, Receptive fields, binocular interaction and functional architecture in the cat's visual cortex. J. Physiol. **160**(1), 106–154 (1962)
106. S.C. Prasad, P. Prasad, *Deep Recurrent Neural Networks for Time Series Prediction*, vol. 95070, pp. 1–54, 2014
107. A. Mohamed, G.E. Dahl, G. Hinton, Acoustic modeling using deep belief networks. IEEE Trans. Audio. Speech. Lang. Processing **20**(1), 14–22 (2012)
108. S. Hochreiter, J. Urgen Schmidhuber, Long short-term memory. Neural Comput. **9**(8), 1735–1780 (1997)

109. F. Gers, Long short-term memory in recurrent neural networks (2001)
110. F.A. Gers, N.N. Schraudolph, J. Schmidhuber, Learning precise timing with LSTM recurrent networks. J. Mach. Learn. Res. **3**(1), 115–143 (2002)
111. G.E. Hinton, R.R. Salakhutdinov, Reducing the dimensionality of data with neural networks. Science (80.) **313**(5786), 504–507 (2006)
112. C.T. Lin, C.S.G. Lee, Neural-network-based fuzzy logic control and decision system. IEEE Trans. Comput. **40**(12), 1320–1336 (1991)
113. K.S.S. Narendra, J. Balakrishnan, M.K.K. Ciliz, Adaptation and learning using multiple models, switching, and tuning. IEEE Control Syst. Mag. **15**(3), 37–51 (1995)
114. N. Kasabov, Evolving fuzzy neural networks for supervised/unsupervised online knowledge-based learning. IEEE Trans. Syst. Man Cybern. Part B **31**(6), 902–918 (2001)
115. P.P. Angelov, Evolving fuzzy rule-based models, in *International Fuzzy Systems Association World Congress*, 1999, pp. 19–23
116. P.P. Angelov, Evolving fuzzy rule-based models. J. Chinese Inst. Ind. Eng. **17**(5), 459–468 (2000)
117. U. Çaydaş, A. Hasçalik, S. Ekici, An adaptive neuro-fuzzy inference system (ANFIS) model for wire-EDM. Expert Syst. Appl. **36**(3 PART 2), 6135–6139 (2009)
118. I. Yilmaz, O. Kaynar, Multiple regression, ANN (RBF, MLP) and ANFIS models for prediction of swell potential of clayey soils. Expert Syst. Appl. **38**(5), 5958–5966 (2011)
119. P.P. Angelov, D.P. Filev, N.K. Kasabov, *Evolving Intelligent Systems: Methodology and Applications* (2010)
120. Y. Hayashi, J.J. Buckley, Approximations between fuzzy expert systems and neural networks. Int. J. Approx. Reason. **10**(1), 63–73 (1994)

Part II
Theoretical Fundamentals of the Proposed Approach

Chapter 4
Empirical Approach—Introduction

In this chapter, we will describe the fundamentals of the proposed new "*empirical*" approach as a systematic methodology with its non-parametric quantities derived entirely from the actual data [1–3] with no subjective and/or problem-specific assumptions made. It has a potential to be a powerful extension of (and/or alternative to) the traditional probability theory, statistical learning and computational intelligence methods.

4.1 Principles and Concept

The theoretical foundations of traditional data analysis approaches is based on the probability theory and statistical learning as an essential and widely used tool for quantitative analysis of stochastic type of uncertainties. However, as discussed in Chap. 3, probability theory and statistical learning rely heavily on a number of restrictive assumptions that usually do not hold in reality, i.e. pre-defined smooth and "convenient to use" probability density function (PDF), infinite amount of observations, independent and identically distributed (*iid*) data, etc. PDF is the derivative of the cumulative distribution function (CDF), and the differentiation operation often creates numerical problems. It is also a challenge for functions that are not analytically defined or are complex.

The ***general problem of probability theory*** was defined by Kolmogorov as follows:

"***Given*** *a CDF, F(x), describe outcomes of* ***random experiments*** *for a* ***given theoretical model***" [4].

Vapnik and Izmailov define the ***general problem of statistics*** as follows: "***Given iid*** *observations of outcomes of the* ***same random experiments****, estimate the statistical model that defines these observations*" [5].

P. P. Angelov and X. Gu, *Empirical Approach to Machine Learning*, Studies in Computational Intelligence 800,
https://doi.org/10.1007/978-3-030-02384-3_4

Both, traditional probability theory and statistics rely on strong (usually impractical) requirements and assumptions. They also assume a random nature of the variables, which is, in fact, only correct for few problems such as entirely independent experts or the so-called "fair games", i.e. gambling. However, real processes of interest (such as climate, economic, social, mechanical, electronic, biological, etc.) are complex and not always with a clear nature (deterministic or stochastic).

The appeal of the traditional statistical approach comes from its solid mathematical foundation and the ability to provide guarantees of performance on condition that there is plenty of data. Another assumption is that the data is generated from the same distribution as being assumed in the probability theory. These strong *prior* assumptions imposed on the data, make traditional statistical approaches fragile in real-world applications. A more detailed analysis was made in Sect. 2.1.6.

In this book, we propose a novel computational methodology based entirely on the *empirical* observations of discrete data samples and the relative proximity of these points in the data space. More importantly, unlike traditional approaches, the proposed one does not impose any *prior* assumptions on the data generation model and does not require any *prior* knowledge about the problem. It is free from any user- or problem-specific parameters and free from the restrictions about the amount or the random nature of data.

The theoretical core of the book, the *empirical approach to machine learning*, is formulated as follows [2]:

> "***Given observations of outcomes*** *of* **real** *processes/experiments* **alone**, *estimate the ensemble properties of the data and, furthermore, estimate these for any feasible outcome*".

It is using the following non-parametric ensemble functions [2, 3]:

(1) Cumulative proximity which is an objective measure of farness and inverse of centrality;
(2) Eccentricity which is a measure of anomaly and tail of the distribution;
(3) *Data density* which is the central paradigm in pattern recognition and is derived here from the actual data;
(4) *Typicality* which is a fundamentally new measure in pattern recognition that resembles PDF, but is designed differently and from the data alone.

We further consider the discrete and continuous, the unimodal and multimodal versions of these fundamental measures that quantify the data pattern as well as their recursive calculation forms, which serve an instrumental role for streaming data processing.

4.2 Cumulative Proximity—A Measure of the Data Mutual Position

Cumulative proximity, q,was firstly introduced in [6–8] as one of the key concepts of the newly proposed *empirical* approach which is closely related to the similar quantities of farness and centrality introduced much earlier in the graph theory and networks analysis indicating the most important vertices within a graph [9, 10].

Cumulative proximity is the first association measure that can quite easily be derived from the actual data and, thus, plays an important role in the proposed *empirical* (centered at data, not assumptions) approach. Its importance comes from the fact that it provides *centrality* information for each particular data item without making any *prior* assumptions about the data generation model. Data samples (items) are represented by a N dimensional vector (a point in the N dimensional data/feature space, $\mathbf{R}^N$). The cumulative proximity at $\boldsymbol{x}_i \in \mathbf{R}^N$, denoted by $q_K(\boldsymbol{x}_i)$, is expressed as [3]:

$$q_K(\boldsymbol{x}_i) = \sum_{j=1}^{K} d^2(\boldsymbol{x}_i, \boldsymbol{x}_j); \quad i = 1, 2, \ldots, K \tag{4.1}$$

where $\boldsymbol{x}_i, \boldsymbol{x}_j \in \{\boldsymbol{x}\}_K; d(\boldsymbol{x}_i, \boldsymbol{x}_j)$ denotes the distance/dissimilarity between $\boldsymbol{x}_i$ and $\boldsymbol{x}_j$, which can be of any type, i.e. Euclidean, Mahalanobis, or cosine dissimilarity.

It is also worth to notice that the average square distance/dissimilarity between any two data samples within $\{\boldsymbol{x}\}_K$ can be expressed as [11]:

$$\bar{d}_K^2 = \frac{1}{K^2} \sum_{i=1}^{K} q_K(\boldsymbol{x}_i) \tag{4.2}$$

An illustrative example regarding the calculation of cumulative proximity at $\boldsymbol{x}_i$ and $\boldsymbol{x}_j$ using the Euclidean distance is given in Fig. 4.1. The values of $q_K(\boldsymbol{x}_i)$ and $q_K(\boldsymbol{x}_i)$ equal to the sum of Euclidean distances from $\boldsymbol{x}_i$ and $\boldsymbol{x}_j$ to all the other data samples in the data space, which are the lines in green for $\boldsymbol{x}_i$ and the dash lines in grey for $\boldsymbol{x}_j$.

The cumulative proximity converts the N dimensional data vector into a scalar and, importantly, can be calculated recursively as follows [12]:

$$q_K(\boldsymbol{x}_i) = q_{K-1}(\boldsymbol{x}_i) + d^2(\boldsymbol{x}_i, \boldsymbol{x}_K); \quad i = 1, 2, \ldots, K-1 \tag{4.3}$$

A visual example of the cumulative proximity values (without normalization or standardization) calculated from the real climate data (temperature and wind speed) measured in Manchester, UK for the period 2010–2015 [13] is given in Fig. 4.2. In this example, for visual clarity we use Euclidean type of distance metric.

From this figure one can see that the closer to the global mean a data sample is, the lower its value of cumulative proximity will be, which is quite logical.

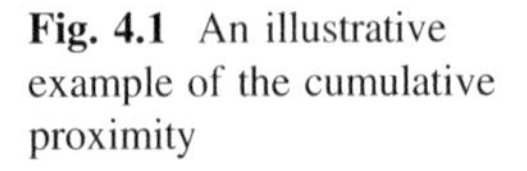

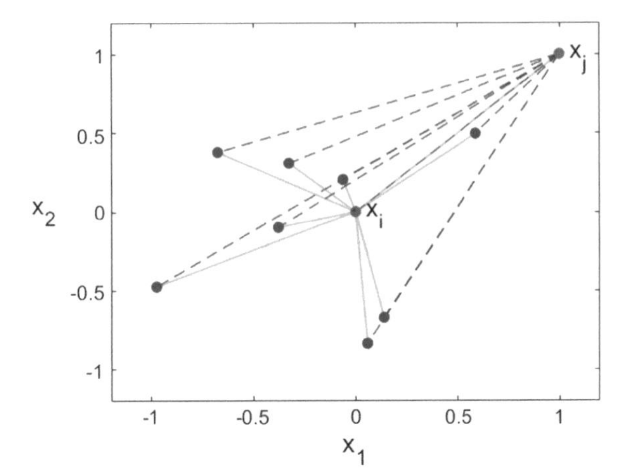

Fig. 4.1 An illustrative example of the cumulative proximity

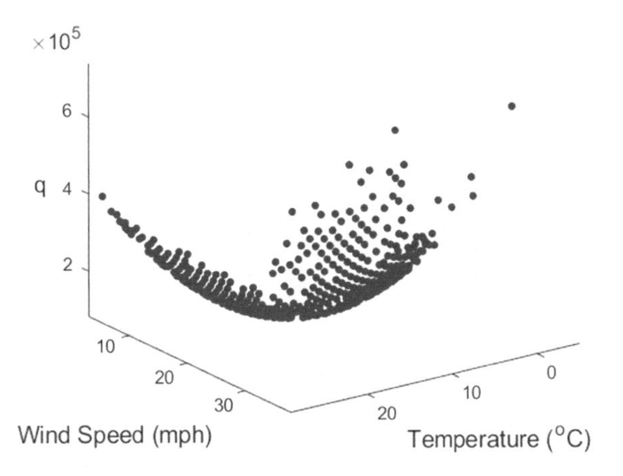

Fig. 4.2 Cumulative proximity of the real climate data

The sum of cumulative proximities of all the data samples can also be updated recursively as follows [6–8]:

$$\sum_{i=1}^{K} q_K(\boldsymbol{x}_i) = \sum_{i=1}^{K-1} q_{K-1}(\boldsymbol{x}_i) + 2q_K(\boldsymbol{x}_K) \tag{4.4}$$

For many types of distance measures, the cumulative proximity, $q_K(\boldsymbol{x}_i)$ and the sum of cumulative proximities, $\sum_{i=1}^{K} q_K(\boldsymbol{x}_i)$ can be recursively calculated in a more elegant form. In this book, we give the recursive expressions for the most commonly used types of distance metrics, such as Euclidean and Mahalanobis as well as for the cosine dissimilarity. But the cumulative proximity as defined in Eqs. (4.1), (4.3) and (4.4) are valid for any types of distance/dissimilarity.

The detailed mathematical derivations are presented in the Appendix A.

A. ***Euclidean distance***

Let us recall that the Euclidean type of distance metric is defined as follows:

$$d(\boldsymbol{x}_i, \boldsymbol{x}_j) = \sqrt{(\boldsymbol{x}_i - \boldsymbol{x}_j)^T(\boldsymbol{x}_i - \boldsymbol{x}_j)} = \|\boldsymbol{x}_i - \boldsymbol{x}_j\| \tag{4.5}$$

The recursive expressions of $q_K(\boldsymbol{x}_i)$ and $\sum_{i=1}^{K} q_K(\boldsymbol{x}_i)$ are then given as follows (the detailed mathematical derivations are given by equations (A.1) and (A.2) in Appendix A.1, respectively):

$$q_K(\boldsymbol{x}_i) = K\left(\|\boldsymbol{x}_i - \boldsymbol{\mu}_K\|^2 + X_K - \|\boldsymbol{\mu}_K\|^2\right) \tag{4.6}$$

$$\sum_{i=1}^{K} q_K(\boldsymbol{x}_i) = 2K^2\left(X_K - \|\boldsymbol{\mu}_K\|^2\right) \tag{4.7}$$

where $\boldsymbol{\mu}_K$ and X_K are the means of $\{\boldsymbol{x}\}_K$ and $\{\boldsymbol{x}^T\boldsymbol{x}\}_K$, respectively, and both of them can be updated recursively as follows [12]:

$$\boldsymbol{\mu}_K = \frac{K-1}{K}\boldsymbol{\mu}_{K-1} + \frac{1}{K}\boldsymbol{x}_K \tag{4.8}$$

$$X_K = \frac{K-1}{K}X_{K-1} + \frac{1}{K}\|\boldsymbol{x}_K\|^2 \tag{4.9}$$

B. ***Mahalanobis distance***

Similarly, it is well known that the Mahalanobis distance is defined as follows [14]:

$$d(\boldsymbol{x}_i, \boldsymbol{x}_j) = \sqrt{(\boldsymbol{x}_i - \boldsymbol{x}_j)^T \Sigma_K^{-1}(\boldsymbol{x}_i - \boldsymbol{x}_j)} \tag{4.10}$$

In this case, the recursive expressions for calculating $q_K(\boldsymbol{x}_i)$ and $\sum\limits_{i=1}^{K} q_K(\boldsymbol{x}_i)$ are formulated as follows (the detailed mathematical derivations are presented in equations (A.3) and (A.4a) in Appendix A.2, respectively) [14]:

$$q_K(\boldsymbol{x}_i) = K\left((\boldsymbol{x}_i - \boldsymbol{\mu}_K)^T \Sigma_K^{-1}(\boldsymbol{x}_i - \boldsymbol{\mu}_K) + X_K - \boldsymbol{\mu}_K^T \Sigma_K^{-1}\boldsymbol{\mu}_K\right) \tag{4.11}$$

$$\sum_{i=1}^{K} q_K(\boldsymbol{x}_i) = 2K^2\left(X_K - \boldsymbol{\mu}_K^T \Sigma_K^{-1}\boldsymbol{\mu}_K\right) \tag{4.12}$$

where $\boldsymbol{\mu}_K$ is the mean of $\{\boldsymbol{x}\}_K$; $\boldsymbol{\Sigma}_K$ is the covariance matrix, $\Sigma_K = \frac{1}{K}\sum\limits_{j=1}^{K}(\boldsymbol{x}_j - \boldsymbol{\mu}_K)(\boldsymbol{x}_j - \boldsymbol{\mu}_K)^T$; $X_K = \frac{1}{K}\sum\limits_{j=1}^{K}\boldsymbol{x}_j^T\Sigma_K^{-1}\boldsymbol{x}_j$.

Alternatively, $\sum_{i=1}^{K} q_K(\boldsymbol{x}_i)$ can be further simplified by using the covariance matrix symmetricity property leading to [15]:

$$\sum_{i=1}^{K} q_K(\boldsymbol{x}_i) = 2K^2 N \tag{4.13}$$

The detailed mathematical derivation of this expression can be found in equation (A.4b) of the Appendix A.2.

One can conclude from Eqs. (4.12) and (4.13) that: $X_K - \boldsymbol{\mu}_K^T \Sigma_K^{-1} \boldsymbol{\mu}_K = N$.

The covariance matrix $\boldsymbol{\Sigma}_K$ can also be updated recursively in an elegant form as it was demonstrated in [12]:

$$\mathbf{X}_K = \frac{K-1}{K}\mathbf{X}_{K-1} + \frac{1}{K}\boldsymbol{x}_K \boldsymbol{x}_K^T \tag{4.14}$$

$$\Sigma_K = \mathbf{X}_K - \boldsymbol{\mu}_K \boldsymbol{\mu}_K^T \tag{4.15}$$

C. ***Cosine dissimilarity***

As it was discussed in Sect. 2.2.1.E the cosine dissimilarity is not a full metric and is defined as a complement to the cosine. For example, we use the following form of a cosine dissimilarity [16, 17]:

$$d(\boldsymbol{x}_i, \boldsymbol{x}_j) = \sqrt{2\left(1 - \cos\left(\theta_{\boldsymbol{x}}^{i,j}\right)\right)} = \left\| \frac{\boldsymbol{x}_i}{\|\boldsymbol{x}_i\|} - \frac{\boldsymbol{x}_j}{\|\boldsymbol{x}_j\|} \right\| \tag{4.16}$$

It has been demonstrated in [16, 17] that $q_K(\boldsymbol{x}_i)$ and $\sum_{i=1}^{K} q_K(\boldsymbol{x}_i)$ can be calculated recursively [similarly to Eqs. (4.6) and (4.7)]:

$$q_K(\boldsymbol{x}_i) = K\left(\left\| \frac{\boldsymbol{x}_i}{\|\boldsymbol{x}_i\|} - \bar{\boldsymbol{\mu}}_K \right\|^2 + \bar{X}_K - \|\bar{\boldsymbol{\mu}}_K\|^2 \right) \tag{4.17}$$

$$\sum_{i=1}^{K} q_K(\boldsymbol{x}_i) = 2K^2\left(\bar{X}_K - \|\bar{\boldsymbol{\mu}}_K\|^2\right) \tag{4.18}$$

where $\bar{\boldsymbol{\mu}}_K$ and $\bar{X}_K$ are the respective means of $\left\{\frac{\boldsymbol{x}}{\|\boldsymbol{x}\|}\right\}_K$ and $\left\{\left\|\frac{\boldsymbol{x}}{\|\boldsymbol{x}\|}\right\|^2\right\}_K$, and both of them can be updated recursively in a similar way as $\boldsymbol{\mu}_K$ (Eq. 4.8) and X_K (Eq. 4.9) [17]:

$$\bar{\boldsymbol{\mu}}_K = \frac{K-1}{K}\bar{\boldsymbol{\mu}}_{K-1} + \frac{1}{K}\frac{\boldsymbol{x}_K}{\|\boldsymbol{x}_K\|} \tag{4.19}$$

$$\bar{X}_K = \frac{K-1}{K}\bar{X}_K + \frac{1}{K}\left\|\frac{\boldsymbol{x}_K}{\|\boldsymbol{x}_K\|}\right\|^2 \equiv 1 \tag{4.20}$$

4.3 Eccentricity—A Measure of Anomaly

Eccentricity, ξ is defined as the normalized cumulative proximity [6, 7] and is an important measure of the ensemble properties of the data samples that are away from the mode/peak. It is very useful to disclose the tails of the data distribution and to detect anomalies/outliers.

The eccentricity at $\boldsymbol{x}_i$, denoted by $\xi_K(\boldsymbol{x}_i)$, is expressed as follows [6, 7]:

$$\xi_K(\boldsymbol{x}_i) = \frac{2q_K(\boldsymbol{x}_i)}{\sum_{j=1}^{K} q_K(\boldsymbol{x}_j)}; \quad i = 1, 2, 3, \ldots, K \tag{4.21}$$

where the coefficient 2 is used to balance the numerator and denominator because the distance between any two data samples is counted twice.

It is easy to demonstrate that [6, 7]:

$$\sum_{i=1}^{K} \xi_K(\boldsymbol{x}_i) = 2; \quad 0 \le \xi_K(\boldsymbol{x}_i) \le 1 \tag{4.22}$$

An illustrative example of the eccentricity calculated for the same data used in Fig. 4.2 is depicted in Fig. 4.3. It can be seen from this figure that the further away a data sample is from the global mean, the higher its eccentricity is.

When $K \to \infty$, the value of eccentricity is very small because of the external constraint (Eq. 4.22). To avoid this problem, the standardized eccentricity, ε was also introduced as follows [6, 7]:

$$\varepsilon_K(\boldsymbol{x}_i) = K\xi_K(\boldsymbol{x}_i) = \frac{2q_K(\boldsymbol{x}_i)}{\frac{1}{K}\sum_{j=1}^{K} q_K(\boldsymbol{x}_j)}; \quad i = 1, 2, \ldots, K \tag{4.23}$$

Compared with $\xi_K(\boldsymbol{x}_i)$, the value of ε does not decrease with the increase of the number of observations, and similarly to Eq. (4.22), the following property holds for the sum of all standardized eccentricities:

$$\sum_{i=1}^{K} \varepsilon_K(\boldsymbol{x}_i) = 2K \tag{4.24}$$

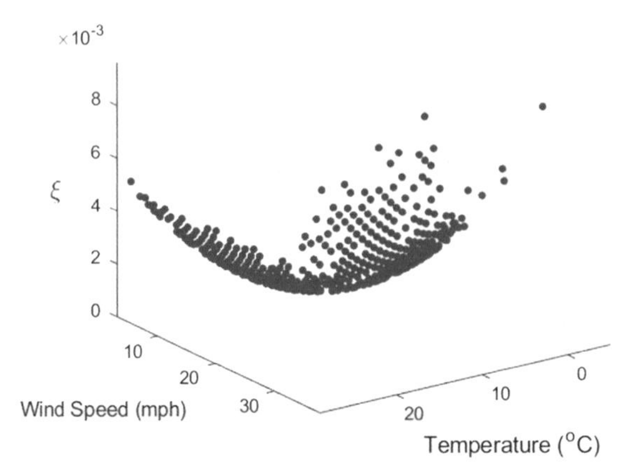

Fig. 4.3 Eccentricity of the real climate data

Fig. 4.4 Standardized eccentricity of the real climate data

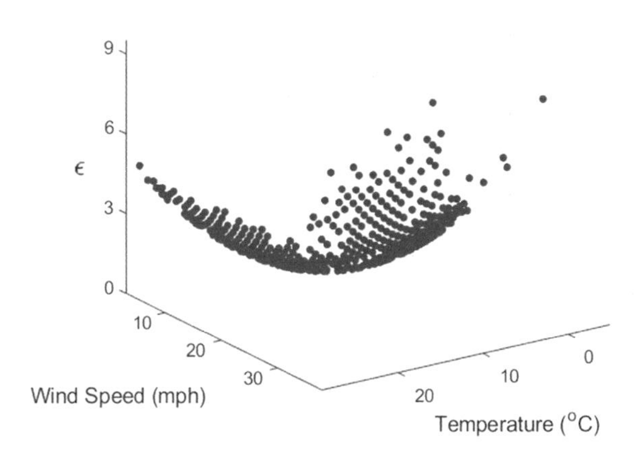

The standardized eccentricity of the same data used in Figs. 4.2 and 4.3 is depicted in Fig. 4.4, where the Euclidean distance is used.

The expression given by Eq. (4.23) is generic and applicable for any type of distance metrics. However, specifically, for the case when the distance matrix is Euclidean, the standardized eccentricity, ε can be simplified using Eqs. (4.6) and (4.7) resulting in:

$$\varepsilon_K(\boldsymbol{x}_i) = K \cdot \frac{2K\left(\|\boldsymbol{x}_i - \boldsymbol{\mu}_K\|^2 + X_K - \|\boldsymbol{\mu}_K\|^2\right)}{2K^2\left(X_K - \|\boldsymbol{\mu}_K\|^2\right)} \tag{4.25}$$

or further:

$$\varepsilon_K(\boldsymbol{x}_i) = 1 + \frac{\|\boldsymbol{x}_i - \boldsymbol{\mu}_K\|^2}{\sigma_K^2} \tag{4.26}$$

where σ_K denote the standard deviation, $\sigma_K^2 = X_K - \|\boldsymbol{\mu}_K\|^2$.

This form is particularly convenient because it lends itself to a recursive form of update which is computationally very efficient and allows real-time applications with very light requirements to the hardware.

Standardized eccentricity, ε is a very useful non-parametric measure for detecting anomalies, which simplifies the Chebyshev inequality, significantly.

Let us recall that the Chebyshev inequality for the case of Euclidean distance can be expressed in the following form [18]:

$$\begin{array}{l} P\left(\|\boldsymbol{\mu}_K - \boldsymbol{x}_i\|^2 \leq n^2\sigma_K^2\right) \geq 1 - \frac{1}{n^2} \\ P\left(\|\boldsymbol{\mu}_K - \boldsymbol{x}_i\|^2 > n^2\sigma_K^2\right) < \frac{1}{n^2} \end{array}; \quad i = 1, 2, \ldots, K \tag{4.27}$$

where n denotes the number multiplied by the standard deviation σ_K; $\sigma_K^2 = X_K - \|\boldsymbol{\mu}_K\|^2$. In practice, $n = 3$ is the most often used, and sometimes, $n = 6$, also known as "3σ" (or, respectively, "6σ") principle.

Using the standardized eccentricity, the Chebyshev inequality can be reformulated in a significantly more elegant form as follows [2]:

$$\begin{aligned} P\left(\varepsilon_K(\boldsymbol{x}_i) \leq n^2 + 1\right) &\geq 1 - \frac{1}{n^2} \\ P\left(\varepsilon_K(\boldsymbol{x}_i) > n^2 + 1\right) &< \frac{1}{n^2} \end{aligned} \tag{4.28}$$

From the above equation one can see that for a particular data sample $\boldsymbol{x}_i$, if $\varepsilon_K(\boldsymbol{x}_i) \geq 1 + n^2$, it is $n\sigma_K$ away from the global mean.

For example, for $n = 3$, it reduces to simply checking if $\varepsilon_K(\boldsymbol{x}_i) \geq 10$, which can be done visually from Fig. 4.4. It means that there is a guarantee that no more than $\frac{1}{9}$ of the data samples/points are anomalous for any (arbitrary) data distribution (if the data distribution is Gaussian/normal, then, no more than 0.3% of the data samples can be outliers). In the example depicted on Fig. 4.4, the standardized eccentricity at all the samples is below this threshold.

The attractiveness of the reformulated Chebyshev inequality compared with its original form is that no *prior* assumptions are made on the nature of the data (random or deterministic), the generation model, the amount of data and their independence and the standard deviation, sigma is not necessary to calculate.

It is also quite interesting that this also applies to other types of distance/dissimilarity, i.e. Mahalanobis distance, cosine dissimilarity, etc. Moreover, the eccentricity and the standardized eccentricity can be both updated recursively for different types of distance/dissimilarity.

4.4 *Data Density*—A Central Paradigm in Pattern Recognition

Data density is a term that is widely used in machine learning. It is a central paradigm in pattern recognition. Usually, it is used in the context of PDF and/or clustering [19, 20]. More generally, density is a term that is defined as a number of items (in this case, number of data points) per unit of volume/area.

In the *empirical* approach to machine learning proposed in this book we define the *data density* through the mutual proximity of the individual data points/samples which represent observations or measurements [6]. It is convenient to consider the density, D calculated at $\boldsymbol{x}_i$ as the inverse of standardized eccentricity [1, 2]:

$$D_K(\boldsymbol{x}_i) = \frac{1}{\varepsilon_K(\boldsymbol{x}_i)}; \quad i = 1, 2, \ldots, K \tag{4.29}$$

Obviously, $0 < D_K(\boldsymbol{x}_i) \leq 1$. *Data density* can only be defined for two or more data points/samples. The *data density* of a single data point is postulated as $D_1(\boldsymbol{x}_1) \equiv 1$.

We use the same illustrative example for the *data density* of the real climate data as used in Fig. 4.2, see Fig. 4.5.

One can see from this figure that the data samples that are closer to global mean have higher *data density* values.

The *data density* value evaluated at a particular data sample is inversely proportional to the sum of distances between this data samples and all the other data samples.

For a static dataset $\{\boldsymbol{x}\}_K$, the numerator of $D_K(\boldsymbol{x}_i), i = 1, 2, \ldots, K$ is $\sum_{j=1}^{K} q_K(\boldsymbol{x}_j)$, which is the same for all the data samples and therefore, can be viewed as a normalizing constant. The denominator of $D_K(\boldsymbol{x}_i)$, namely, $q_K(\boldsymbol{x}_i)$ is

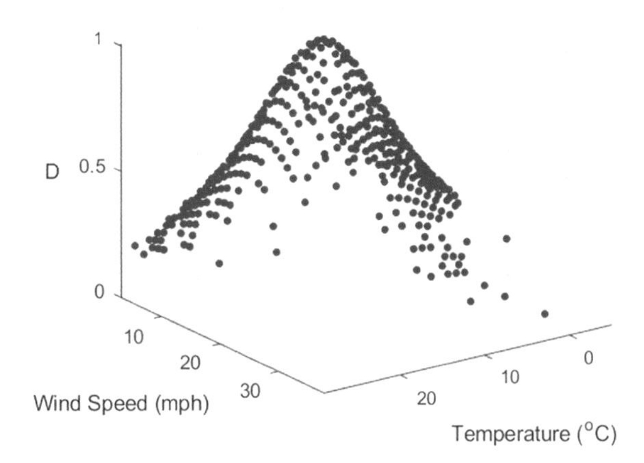

Fig. 4.5 *Data density* of the climate data

exactly the sum of distances between $\boldsymbol{x}_i$ to all the data samples within the data space. We illustrate the concept of the *data density*, D in Fig. 4.6. The dots are the data samples and the circles around them indicate the *data density* calculated at these data samples. The higher the *data density* is, the larger the circle will be.

From Fig. 4.6 one can see that the data point that is closest to the global mean, $\boldsymbol{\mu}$ of them has the largest *data density*.

Similar to the standardized eccentricity, the *data density* can be used to reformulate the Chebyshev inequality in an elegant form as follows:

$$
\begin{aligned}
&P\left(D_K(\boldsymbol{x}_i) = \varepsilon_K^{-1}(\boldsymbol{x}_i) \geq \frac{1}{n^2+1}\right) \geq 1 - \frac{1}{n^2} \\
&P\left(D_K(\boldsymbol{x}_i) = \varepsilon_K^{-1}(\boldsymbol{x}_i) < \frac{1}{n^2+1}\right) < \frac{1}{n^2}
\end{aligned}
\tag{4.30}
$$

For example, for the most popular values of value of $n = 3$, we get:

$$
\begin{aligned}
&P(D_K(\boldsymbol{x}_i) \geq 0.1) \geq \frac{8}{9} \\
&P(D_K(\boldsymbol{x}_i) < 0.1) < \frac{1}{9}
\end{aligned}
\tag{4.31}
$$

which is valid for any type of data generation model.

Data density can also be updated on a sample-by-sample basis in an online environment using Eqs. (4.3)–(4.20). It is quite interesting to notice that when Euclidean distance is used, the *data density* reduces to a Cauchy type function [2, 3]:

$$
D_K(\boldsymbol{x}_i) = \frac{1}{1 + \frac{\|\boldsymbol{x}_i - \boldsymbol{\mu}_K\|^2}{\sigma_K^2}}; \quad i = 1, 2, \ldots, K \tag{4.32}
$$

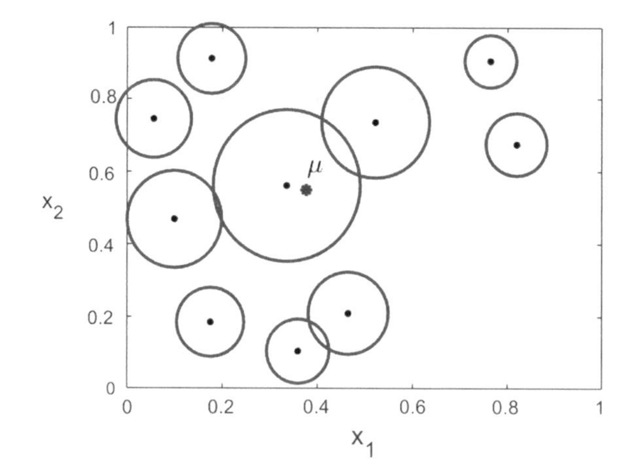

Fig. 4.6 A simple illustrative example of the *data density*, D (the size of the circles is proportional to D)

where $\sigma_K^2 = X_K - \|\boldsymbol{\mu}_K\|^2$.

It is also interesting to notice that there are no parameters involved and there was no *prior* assumption of the type of the distribution or even randomness or impudence of the data.

The *data density* formulated in this way has a very interesting resemblance with the Newton's law of universal gravitation which is formulated as follows:

$$f_{AB} = \frac{g_o m_A m_B}{r_{AB}^2} \quad (4.33)$$

where f_{AB} is the attracting force between two objects, denoted by A and B; g_o is the gravitational constant ($6.674 \times 10^{-11}\ \mathrm{N} \cdot \mathrm{m}^2 \cdot \mathrm{kg}^{-2}$); m_A and m_B are the respective masses of the two objects; r_{AB} is the distance between the centres of the mass of them. That is, if we ignore the constant and the masses and use the notation for distance we use in this book ($d(A, B)$ instead of r_{AB}), we can state:

$$f_{AB} \sim \frac{1}{d^2(A, B)} \quad (4.34)$$

Indeed, the value of the *data density* evaluated at a particular data sample indicates how strongly this particular data sample is influenced by the other data samples in the data space due to their mutual proximity and attraction. It is also inversely proportional to the square distance between these two data samples.

In fact, in physics there is a more wider law called *inverse square law* which applies to any physical quantity whose intensity is inversely proportional to the square of the distance to the source/emitter. The *data density*, *D*, formulated in this book has similar properties but applied to the data points in the data space.

4.5 *Typicality*—A Fundamentally New Measure in Data Analysis and Machine Learning

Typicality is a fundamentally new measure in pattern recognition resembling the traditional unimodal probability mass function (PMF). *Typicality*, τ, calculated at $\boldsymbol{x}_i$, denoted by $\tau_K(\boldsymbol{x}_i)$ was defined as the normalized *data density* [1–3, 6]:

$$\tau_K(\boldsymbol{x}_i) = \frac{D_K(\boldsymbol{x}_i)}{\sum_{j=1}^{K} D_K(\boldsymbol{x}_j)}; \quad j = 1, 2, \ldots, K \quad (4.35)$$

It has the following properties shared with the PMF:

$$0 < \tau_K(\boldsymbol{x}_i) \leq 1 \quad (4.36)$$

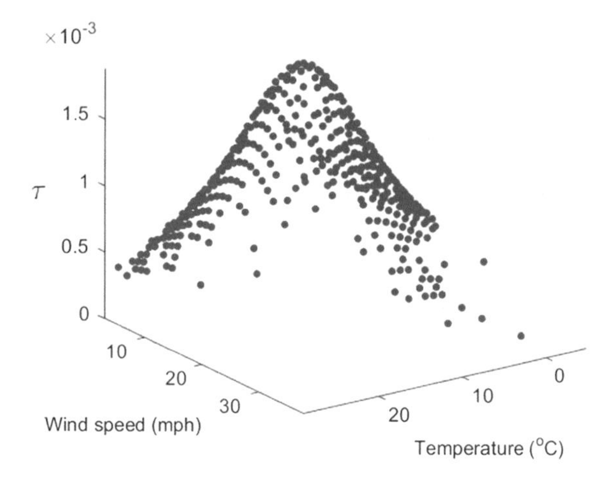

Fig. 4.7 *Typicality* of the real climate data

$$\sum_{i=1}^{K} \tau_K(\boldsymbol{x}_i) = 1 \tag{4.37}$$

However, the main difference between them is that *typicality* is automatically defined by the data unlike the PMF that may have non-zero values for infeasible values of the variable. In addition, for the definition of the *typicality*, τ, the randomness of which is not necessarily assumed [3].

The *typicality* of the same dataset plotted in Fig. 4.2 is depicted in Fig. 4.7, from which one can see that, the data samples that are closer to the global mean have higher *typicality* values.

4.6 Discrete Multimodal *Data Density* and *Typicality*

In the previous sections of this chapter, the *data density* and *typicality* were defined *empirically* (from the actual/real observable data). The power of this new, *empirical* approach is that no restrictive *prior* assumptions were made, no user- or problem-specific thresholds or other parameters were used. In many cases, however, a particular data point may have been observed multiple times. A simple example is a climate/weather observation when we may have not just a single day but multiple days when the temperature is, for example, 12 °C.

Indeed, while the *typicality* (and *data density*) resembles and corresponds to PDF in the traditional probability theory, there is another (historically older and easier to explain, thus, used in schools and at the beginning of each text book) form of probability- the frequentistics, based on counting the frequencies of occurrence of an event or a particular value of the random variable (see Chap. 3). All considerations so far in this chapter resulted in unimodal discrete distributions centered at the global mean, $\boldsymbol{\mu}_K$.

In practice, however, unimodal representations are, usually, insufficient to describe the actual experimental/observable data [21–24]. To address this problem, the traditional probability theory often involves mixture of unimodal distributions, which requires identifying the local modes first. This is always a challenging task solved as a separate problem (usually, clustering, optimization or using domain expert knowledge) [24].

In the *empirical* approach presented in this book, we provide the following non-parametric discrete quantities that are derived directly from the data, which provide multimodal distributions automatically without the need of user decisions [3]:

(1) *Multimodal data density*, D^M;
(2) *Multimodal typicality*, τ^M.

For the data samples which repeat more than once we can state that:

$$\exists \boldsymbol{x}_i = \boldsymbol{x}_j, \quad i \neq j \tag{4.38}$$

Let us denote the unique data samples (each distinct value taken only once) by $\{\boldsymbol{u}\}_L = \{\boldsymbol{u}_1, \boldsymbol{u}_2, \ldots, \boldsymbol{u}_L\}$, where L is the number of unique data samples. It is obvious that $\{\boldsymbol{u}\}_L \subseteq \{\boldsymbol{x}\}_K$. Let us also denote the corresponding numbers of occurrences by $\{F\}_L = \{F_1, F_2, \ldots, F_L\}; \sum_{i=1}^{L} F_i = K$. These can be obtained automatically from $\{\boldsymbol{x}\}_K$ in online mode.

For example, for $\{\boldsymbol{x}\}_{10} = \{1, 2, 2, 4, 5, 7, 7, 7, 8, 8\}$, one can easily identify the unique samples from $\{\boldsymbol{x}\}_{10}$, which are $\{\boldsymbol{u}\}_6 = \{1, 2, 4, 5, 7, 8\}$ and the corresponding numbers of occurrences, $\{F\}_6 = \{1, 2, 1, 1, 3, 2\}$.

The multimodal *data density*, D^M, estimated at $\boldsymbol{u}_i$ ($i = 1, 2, \ldots, L$) is formulated as the weighted (by the number of occurrences of each particular data sample) *data density* [1]:

$$D_K^M(\boldsymbol{u}_i) = F_i D_K(\boldsymbol{u}_i) \tag{4.39}$$

The multimodal *data density* of the real climate data used in the previous subsection for illustration is depicted in Fig. 4.8. Again, Euclidean type of distance metric is used.

The multimodal *typicality* estimated at the unique points, $\boldsymbol{u}_i$ ($i = 1, 2, \ldots, L$) is expressed as follows [1]:

$$\tau_K^M(\boldsymbol{u}_i) = \frac{F_i D_K(\boldsymbol{u}_i)}{\sum_{j=1}^{L} F_j D_K(\boldsymbol{u}_j)}; \quad i = 1, 2, \ldots, L \tag{4.40}$$

Despite its simplicity, this expression is a very powerful generalization of the two basic types of probability forms:

(1) **the frequency-of-occurrence-based** (the one used in "fair games" examples, like throwing dices, tossing a coin etc.), and

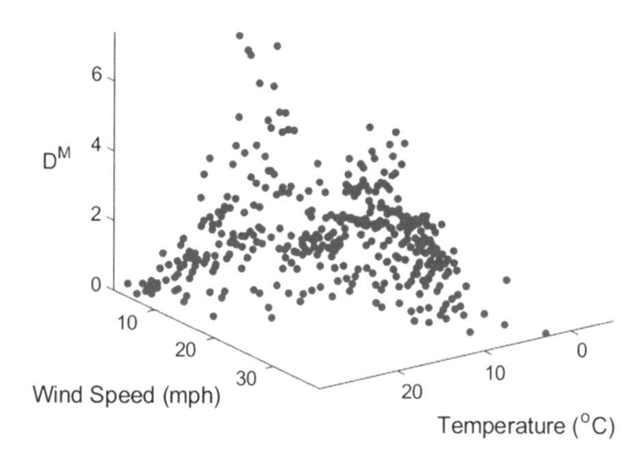

Fig. 4.8 Multimodal data density of the real climate data

(2) **the distribution-based**.

These two types of defining the probabilities are well known from each textbook and course: the former is usually used in schools and in the first tens of pages of each book on probabilities and often exemplified with "fair games", such as tossing coins, dices, colored balls in urns, etc. Indeed, if we ignore the *data density*, D (consider it equal to 1, which is true for "fair games" as further discussed at the end of this section), we will get the same result as Eq. (2.1) staring from Eq. (4.40):

$$\tau_K^M(\boldsymbol{u}_i) = \frac{F_i D_K(\boldsymbol{u}_i)}{\sum_{j=1}^{L} F_j D_K(\boldsymbol{u}_j)} = \frac{F_i}{\sum_{j=1}^{L} F_j}; \quad i = 1, 2, \ldots, L \tag{4.41}$$

The latter (distribution-based) definition of the probability is the main one used in practice and it occupies most of the university-level books and courses. It is covered by the normalized *data density*, see Eq. (4.35) and Fig. 4.7.

As a result of the fusion in the form of a weighted sum, which we introduced in Eq. (4.40), we get a multimodal distribution of the *typicality*, τ^M automatically from the observed experimental data without imposing *prior* assumptions about:

(1) the number of modes/peaks;
(2) the nature (random as in "fair games" or deterministic) of the data;
(3) the quantity (infinite or not) of the available data (in fact, this expression works for as little as a couple of data points);
(4) the mutual dependence or independence of the data;
(5) the data generation model.

The above properties combined with the computational simplicity for the cases where recursive calculations are possible makes it powerful. It lies at the core of the proposed new *empirical* approach to machine learning presented in this book.

An illustrative example of it is depicted in Fig. 4.9a using the same data that was used in Fig. 4.2. This was compared with the histogram generated from the same data presented in Fig. 4.9b.

If we analyze these results we can see that the multimodal *typicality*, τ^M provides quite similar information to the histogram, but directly from the data without any pre-processing techniques such as clustering or the need of decisions by the user regarding the size of the histogram cells. It has to be stressed that histograms can create problems in cases when the dimensionality of the data, N is high (in this example only $N = 2$ dimensions are used for visualization purposes, but in real problems this number may be high and the number of data points, K required to get a meaningful histogram is in the order of $K = \lceil 2^{H-1} \rceil$ based on the Sturges' rule [25], where H is the number of bins. This is because the histogram on the vertical axis (in this case—τ^M; in other cases—probability) can only take values from a

Fig. 4.9 Multimodal *typicality* and histogram based on the real climate data

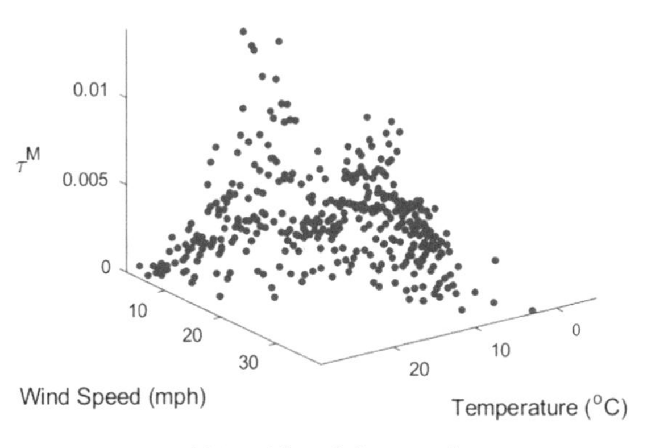

(a) Multimodal *typicality*

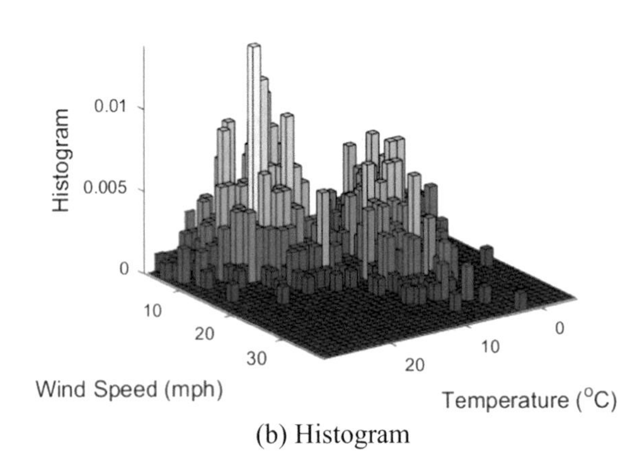

(b) Histogram

finite set $\left\{0, \frac{1}{K}, \frac{2}{K}, \ldots, 1\right\}$. The choice of the size of the cells is subjective. In practical problems, the minimum number of data samples required for a meaningful histogram may run into many thousands or more for the reasons mentioned above.

In contrast, τ^M does not require any user- or problem-specific parameters and it can take any real values. The proposed approach can work with as little as couple of data samples and does not require any input from the user. One can see that the multimodal *typicality* takes into full account not only the frequencies of the repeated data samples, but also their mutual positions.

When all the data samples in the dataset have different values, namely, $F_i = 1, \forall i$, and the quantization step of the histogram is not properly set, the histogram is unable to show any useful information, see Fig. 4.10b. However, the multimodal *typicality* is still able to show the mutual distribution information of the data and is automatically being reduced/transformed into a unimodal distribution, see Fig. 4.10a. In this figure, we use the real climate data that was used in the previous illustrative examples in this chapter to demonstrate the idea.

We further consider a synthetic dataset with only the temperature measured in one place for 60 days, which has only 3 unique data samples $\{u\}_3 = \{8, 14, 16\}$ °C with the frequencies of occurrence $\{F\}_3 = \{30, 20, 10\}$. With this synthetic example, we aim to compare:

(1) the discrete multimodal *typicality*, τ^M;
(2) the probability mass function (PMF), and
(3) the probability density function (PDF).

The comparison is depicted in Fig. 4.11.

From the above example one can clearly conclude that when compared with the PMF, τ^M contains the extra spatial/similarity information concerning the mutual distribution of the data (the data sample in the middle has higher value than the PMF).

For the equally distant data samples, the multimodal *typicality* is exactly the same as the PMF:

$$\tau_K^M(\boldsymbol{u}_i) = \frac{F_i}{\sum_{j=1}^{L} F_j}; \quad i = 1, 2, \ldots, L \tag{4.42}$$

Such equally distant data samples are, for example, results of "fair games", e.g. throwing dices, etc. For example, let us consider tossing a coin for 200 times. Let 105 of the outcomes are "head" (the rest will, obviously, be "tail"). Then we can state that $\{\boldsymbol{u}\}_2 = \{head, tail\}$ and $\{F\}_2 = \{105, 95\}$. Then one can obtain the multimodal *typicality* of the 200 observations as follows:

$$\tau_K^M(head) = \frac{105}{200} = 52.5\% \text{ and } \quad \tau_K^M(tail) = \frac{95}{200} = 47.5\%.$$

Fig. 4.10 Multimodal *typicality* and histogram of the unique data samples

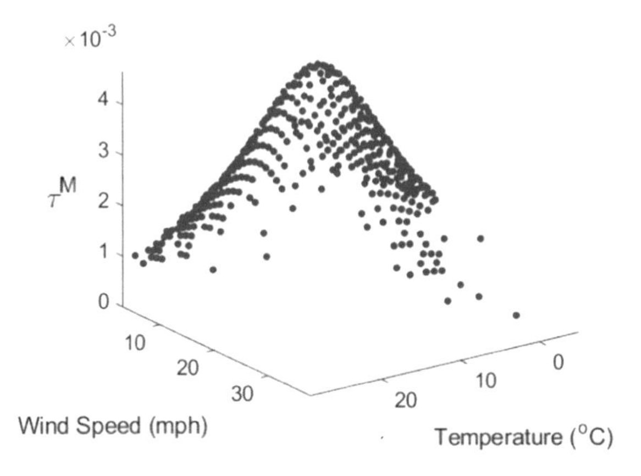

(a) Multimodal *typicality*

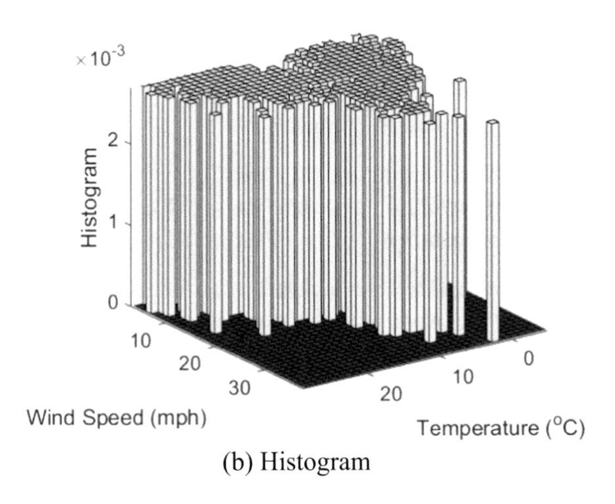

(b) Histogram

Fig. 4.11 Multimodal *typicality* versus PMF versus PDF

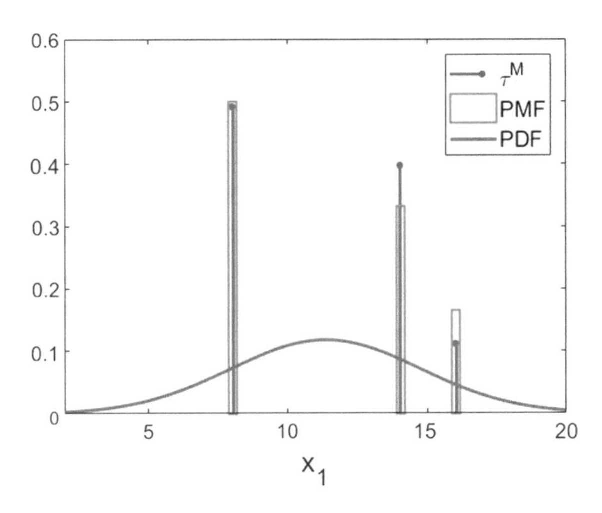

The discrete multimodal *data density* and *typicality* considered in the previous section are valid at the unique data samples, $\{\boldsymbol{u}\}_L$ only.

4.7 Continuous Form of *Data Density* and *Typicality*

All the non-parametric empirically derived quantities described in Sects. 4.2–4.6 are discrete and calculated at the actual data points in the data space and are valid at the unique data samples, $\{\boldsymbol{u}\}_L$ only.

These discrete quantities are entirely based on the ensemble properties and mutual position of the empirical observations of actual data points. The individual data samples/observations are not assumed to be independent or identically distributed (*iid*); instead, their mutual dependence is taken into consideration directly through the mutual distance between them. They also do not require infinite number of observations and can work with minimum two data samples. They are free from the well-known paradoxes of the traditional probability theory and statistics [2, 3]. Last, but not least, they can be calculated recursively for various types of distance/similarity measure.

4.7.1 *Continuous Unimodal* Data Density *and* Typicality

Discrete quantities are useful to describe data sets or streams, however, for making inference, the continuous forms of *data density* and *typicality* are needed. This is the case because the discrete quantities are valid only for the existing data points.

To address this let us consider the continuous form of the *data density* (in its *unimodal* form) [3]:

$$D_K^C(\boldsymbol{x}) = \frac{1}{\varepsilon_K(\boldsymbol{x})}; \quad \boldsymbol{x} \in \mathbf{R}^N \tag{4.43}$$

and it can expressed in terms of the cumulative proximity as follows:

$$D_K^C(\boldsymbol{x}) = \frac{\sum_{j=1}^{K} q_K(\boldsymbol{x}_j)}{2Kq_K(\boldsymbol{x})} \tag{4.44}$$

where $\boldsymbol{x}_j \in \{\boldsymbol{x}\}_K$.

The continuous forms of *data density* and *typicality* are introduced as envelops of the discrete forms.

In the case when Euclidean distance is used, the continuous *data density* is simplified to a continuous Cauchy type function covering the whole data space, ($\boldsymbol{x} \in \mathbf{R}^N$) [3]:

$$D_K^C(\boldsymbol{x}) = \frac{1}{1 + \frac{\|\boldsymbol{x} - \boldsymbol{\mu}_K\|^2}{\sigma_K^2}} \tag{4.45}$$

Compared with Eq. (4.28), the main difference in Eq. (4.45) is that $\boldsymbol{x}$ does not have to be a physically existing data point in the data space. The illustrative example of the continuous *data density* of the real climate data as used in the previous example is given in Fig. 4.12.

As described in Sect. 4.5, the *typicality*, in its discrete form, is the normalized *data density*. Following the same concept, the continuous *typicality* is defined as the normalized continuous *data density* using integral instead of the sum [3]:

$$\tau_K^C(\boldsymbol{x}) = \frac{D_K^C(\boldsymbol{x})}{\int_{\boldsymbol{x}} D_K^C(\boldsymbol{x}) d\boldsymbol{x}} \tag{4.46}$$

As one can see from Eq. (4.46), by solving the integral of the continuous *data density* $D_K^C(\boldsymbol{x})$ within the data space and dividing $D_K^C(\boldsymbol{x})$ by it, the unit integral of $\tau_K^C(\boldsymbol{x})$ can always be guaranteed for all types of distances [3]:

$$\int_{x} \tau_K^C(\boldsymbol{x}) d\boldsymbol{x} = 1 \tag{4.47}$$

This is a very important result, because it makes the continuous *typicality*, τ^C comparable with the traditional PDF in terms of this condition. However, as stressed already, the *typicality* has a number of advantages (less restrictions) due to the way it was defined and generated in a constructivist way from data directly.

For example, if use Euclidean distance without loss of generality, we can transform $D_K^C(\boldsymbol{x})$[21–23] by considering the well-known expression of the multivariate Cauchy distribution as follows:

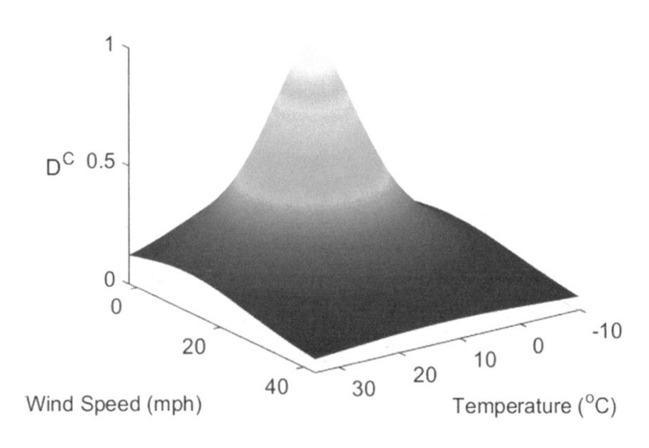

Fig. 4.12 Continuous *data density* of the real climate data

$$f(\boldsymbol{x}) = \frac{\Gamma\left(\frac{N+1}{2}\right)}{\pi^{\frac{N+1}{2}} \sigma_o^N \left(1 + \frac{(\boldsymbol{x}-\boldsymbol{\mu})^T(\boldsymbol{x}-\boldsymbol{\mu})}{\sigma_o^2}\right)^{\frac{N+1}{2}}} \tag{4.48}$$

where $\boldsymbol{x} = [x_1, x_2, \ldots, x_N]^T$; $\boldsymbol{\mu} = E[\boldsymbol{x}]$; σ_o is the scalar parameter; $\Gamma(\cdot)$ is the gamma function and π is the well-known constant.

It is very important that Eq. (4.48) also guarantees that:

$$\int_{x_1}\int_{x_2}\cdots\int_{x_N} f(x_1, x_2, \ldots, x_N)dx_1\,dx_2\ldots dx_N = 1 \tag{4.49}$$

Based on Eq. (4.48), we introduce the normalized continuous *data density* for the Euclidean distance as follows [3]:

$$\bar{D}_K^C(\boldsymbol{x}) = \frac{\Gamma\left(\frac{N+1}{2}\right)}{\pi^{\frac{N+1}{2}}\sigma_K^N}\left(D_K^C(\boldsymbol{x})\right)^{\frac{N+1}{2}} = \frac{\Gamma\left(\frac{N+1}{2}\right)}{\pi^{\frac{N+1}{2}} \sigma_K^N \left(1 + \frac{\|\boldsymbol{x}-\boldsymbol{\mu}_K\|^2}{\sigma_K^2}\right)^{\frac{N+1}{2}}} \tag{4.50}$$

where $\sigma_K^2 = X_K - \|\boldsymbol{\mu}_K\|^2$.

Due to the fact that $\int_{\boldsymbol{x}} \bar{D}_K^C(\boldsymbol{x})d\boldsymbol{x} = 1$, for the Euclidean distance, it follows that [3]:

$$\tau_K^C(\boldsymbol{x}) = \bar{D}_K^C(\boldsymbol{x}) = \frac{\Gamma\left(\frac{N+1}{2}\right)}{\pi^{\frac{N+1}{2}} \sigma_K^N \left(1 + \frac{\|\boldsymbol{x}-\boldsymbol{\mu}_K\|^2}{\sigma_K^2}\right)^{\frac{N+1}{2}}} \tag{4.51}$$

The continuous *typicality*, τ^C of the same real climate dataset as used in the previous example is depicted in Fig. 4.13. We also compare the continuous *typicality* with the unimodal PDF per attribute, and depict the comparison in Fig. 4.14.

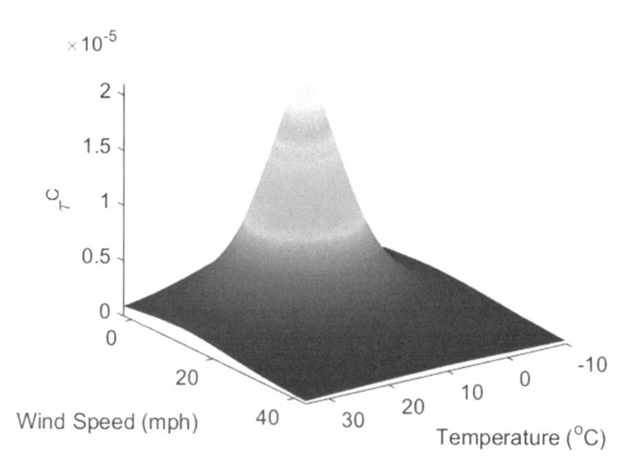

Fig. 4.13 Continuous *typicality* of the real climate data

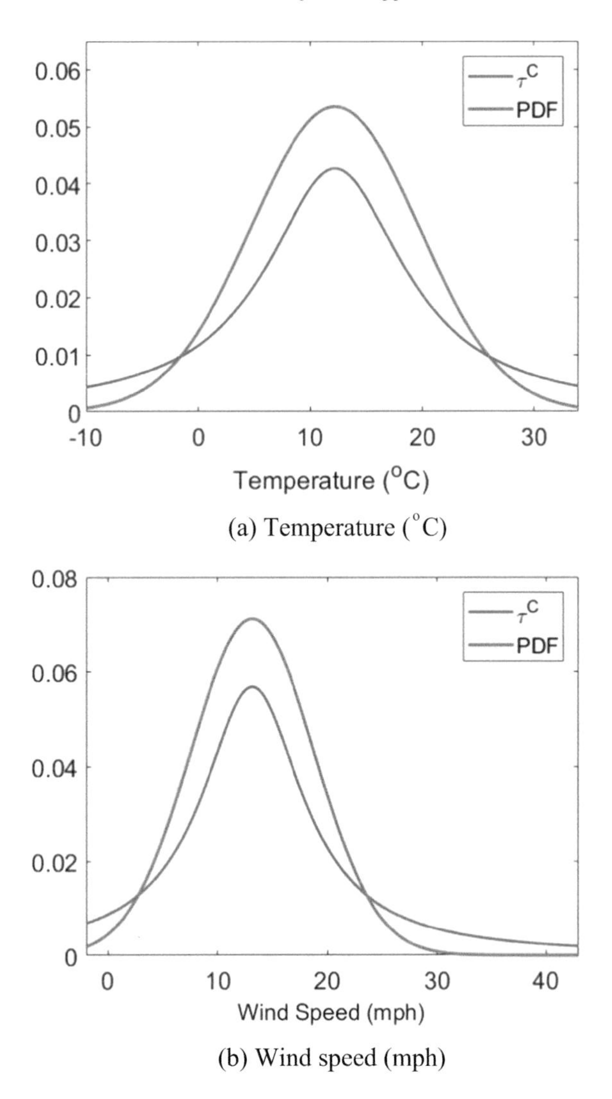

Fig. 4.14 Continuous *typicality* versus unimodal PDF per attribute

From the above comparison one can see that, the continuous *typicality* shares the most important properties of the continuous unimodal PDF, namely

(1) integrates to 1;
(2) provides a closed analytical form, and
(3) represents the probability of a particular value of the data, $\boldsymbol{x}_i$.

However, it has the advantage that there is no requirement for *prior* assumptions as well as user- or problem-specific parameters, and it is derived entirely from the *empirical* observations. This makes the τ^C an effective and efficient alternative to the traditional (frequentistic, distribution-based) as well as other alterative (i.e. belief-based [26]) forms of probability.

Compared with the discrete *typicality*, τ, which is only valid at the observed values, the inference for arbitrary values can be easily made using the continuous *typicality*, τ^C. To illustrate this, let us use the first attribute of the real climate data (temperature), denoted by x_1. If the Euclidean distance is used, the value of τ^C at $x_1 = 2.75$ and $x_1 = 14.25$ can be directly obtained using Eq. (4.51) as $\tau_K^C(x_1 = 2.75) = 0.0164$ and $\tau_K^C(x_1 = 14.25) = 0.0396$, which means that $x_1 = 14.25$ is (twice) more likely to happen than $x_1 = 2.75$ (see also Fig. 4.15).

Similar to the unimodal PDF, if we want to obtain the continuous *typicality* of all the data samples larger than $\boldsymbol{y}$, we can integrate τ^C as follows [3]:

$$T(\boldsymbol{x} > \boldsymbol{x}_o) = 1 - \int_{\boldsymbol{x}=-\infty}^{\boldsymbol{x}_o} \tau_K^C(\boldsymbol{x})d\boldsymbol{x} \tag{4.52}$$

Following the previous example, if we want to obtain the value of the continuous *typicality*, τ^C for all $x_1 > x_o$, we have:

$$\begin{aligned} T(x_1 > x_o) &= 1 - \int_{x_1=-\infty}^{x_o} \tau_K^C(x_1)dx_1 \\ &= 1 - \left(\frac{1}{\pi}\arctan\left(\frac{x_o - \mu_{K,1}}{\sigma_{K,1}}\right) + \frac{1}{2}\right) \end{aligned} \tag{4.53}$$

where $\mu_{K,1}$ is the mean of the first attribute of the data; $\sigma_{K,1}$ is the corresponding standard deviation. If $x_o = 20$ (see the shadowy segment in Fig. 4.15), then $T(x_1 > 20) = 0.2429$.

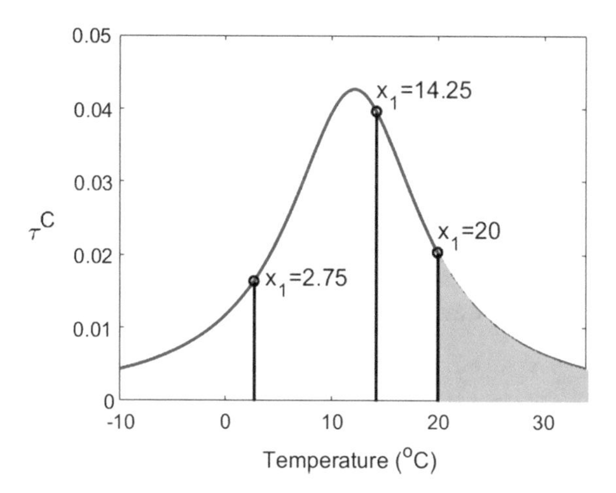

Fig. 4.15 Example of making inference using the continuous *typicality*, τ^C

4.7.2 *Continuous Multimodal* Data Density *and* Typicality

Traditional approaches make assumptions and approximate the actual data with smooth distribution functions which are usually unimodal and globally valid or multimodal, but with pre-defined number of modes/peaks determined in some cases by procedures like clustering, expectation maximization (EM) or using subjective knowledge provided by domain experts.

Instead, in the proposed *empirical* approach we start from the data, use no *prior* assumptions and user- or problem-specific thresholds, parameters, etc. to determine the continuous *data density*, D^C and *typicality*, τ^C, respectively.

4.7.2.1 Continuous Multimodal *Data Density*

The continuous version of the multimodal *data density* is a mixture that arises from the distance/dissimilarity metric used to measure the mutual distances and the density of the data samples in the data space. For any $\boldsymbol{x}$, the continuous multimodal *data density*, D^{CM}, is defined in a general form very similar to the mixture distributions [3]:

$$D_K^{CM}(\boldsymbol{x}) = \frac{1}{K}\sum_{i=1}^{N} S_{K,i} D_{K,i}^{C}(\boldsymbol{x}) \tag{4.54}$$

where $D_{K,i}^{C}(\boldsymbol{x})$ is the continuous *data density* of $\boldsymbol{x}$ calculated locally in the ith cluster/*data cloud* denoted by $\mathbf{C}_{K,i}, D_{K,i}^{C}(\boldsymbol{x}) = \frac{1}{\varepsilon_{K,i}(\boldsymbol{x})}; S_{K,i}$ is the corresponding support (number of members) of $\mathbf{C}_{K,i}$, and $\sum_{i=1}^{N} S_{K,i} = K; N$ is the total number of the clusters/*data clouds* existing in the data space.

Equation (4.54) is a weighted combination of continuous data densities, D^{CM} calculated locally from each *data cloud.* Equation (4.54) is valid for all types of distance/dissimilarity.

The continuous multimodal *data density*, D^{CM}, is defined non-parametrically from the local modes/peaks of the data. It is a very good representation of the multimodal nature of the data distribution and behaves differently in the trough regions in contrast with the unimodal expressions (Eq. 4.32). When using Euclidean distance, the continuous multimodal *data density*, D^{CM}, can be reformulated as a mixture of Cauchy distributions as follows [3]:

$$D_K^{CM}(\boldsymbol{x}) = \frac{1}{K}\sum_{i=1}^{N} \frac{S_{K,i}}{1 + \frac{\left\|\boldsymbol{x}-\boldsymbol{\mu}_{K,i}\right\|^2}{\sigma_{K,i}^2}} \tag{4.55}$$

where $\boldsymbol{\mu}_{K,i}$ and $\sigma_{K,i}$ are the mean and standard deviation of the data samples within the ith *data cloud*.

The continuous multimodal *data density* of the same climate dataset as used in the previous examples considering Euclidean distance is depicted in Fig. 4.16. In Fig. 4.16a, we separate the data based on the ground truth (the true labels of the observations) into two clusters. In Fig. 4.16b, we use the autonomous data partitioning (ADP) algorithm which will be detailed in Chap. 7, Sect. 7.4 to find the local modes in a fully data-driven, non-parametric way.

As we can see from Fig. 4.16, the continuous multimodal *data density*, D^{CM} is able to detect the natural multimodal structure of the data distribution.

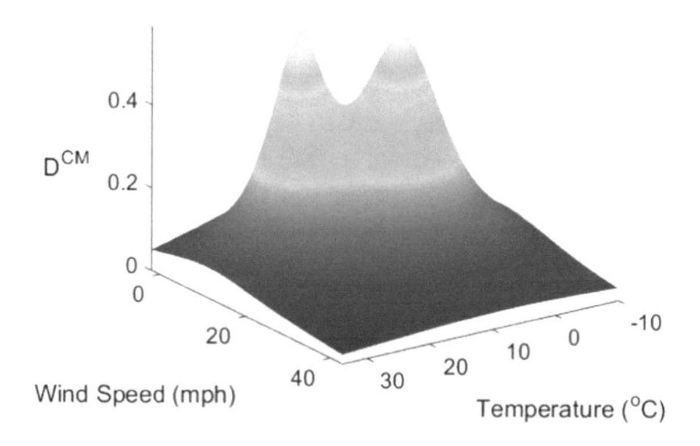

(a) Ground truth based D^{CM}

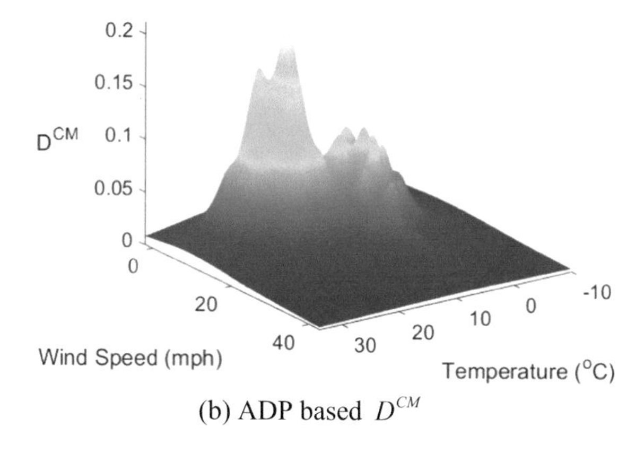

(b) ADP based D^{CM}

Fig. 4.16 Continuous multimodal *data density*, D^{CM} of the climate data

4.7.2.2 Continuous Multimodal *Typicality*

In a similar manner, we also introduce the continuous form of the multimodal *typicality*, τ^{CM}, which is formulated as a normalized continuous multimodal *data density* [3]:

$$\tau_K^{CM}(\boldsymbol{x}) = \frac{D_K^{CM}(\boldsymbol{x})}{\int_{\boldsymbol{x}} D_K^{CM}(\boldsymbol{x}) d\boldsymbol{x}} \tag{4.56}$$

In more detail, combining Eqs. (4.54) and (4.56), we get:

$$\tau_K^{CM}(\boldsymbol{x}) = \frac{\sum_{i=1}^{N} S_{K,i} D_{K,i}^{C}(\boldsymbol{x})}{\sum_{i=1}^{N} S_{K,i} \int_{\boldsymbol{x}} D_{K,i}^{C}(\boldsymbol{x}) d\boldsymbol{x}} \tag{4.57}$$

It can also be proven that the continuous multimodal *typicality,* τ^{CM} is integrated to 1.

The multimodal *typicality* is very close to the multimodal PDF by its nature, but it is derived entirely from the data and is free from *prior* assumptions as well as from user- or problem-specific thresholds and parameters. The continuous multimodal *typicality* τ^{CM} that is derived from D^{CM} of the real climate data given in Fig. 4.16 is depicted in Fig. 4.17.

We further compare the continuous multimodal *typicality* with multimodal PDF, unimodal PDF and PMF plotted in Fig. 4.1, and show the comparison in Fig. 4.18, where τ^{CM} is generated from the data using the proposed *empirical* approach is used.

As one can see from this figure that, the continuous multimodal *typicality* τ^{G} is able to objectively identify the main modes/peaks of the data pattern and compares favorably in regards to the unimodal and multimodal PDFs.

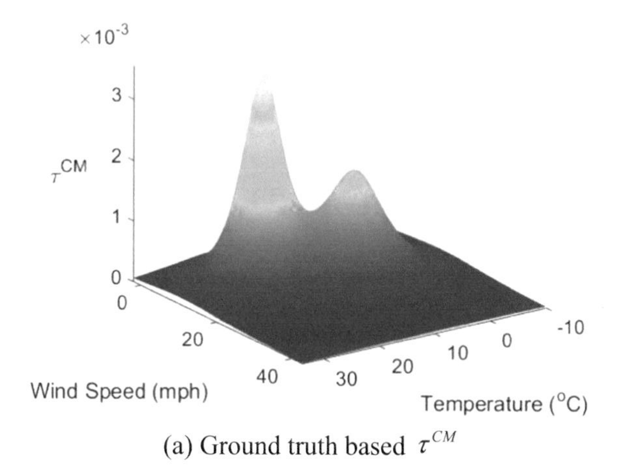

(a) Ground truth based τ^{CM}

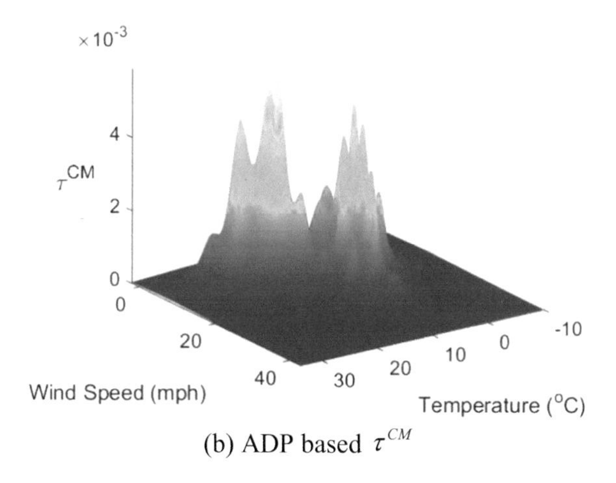

(b) ADP based τ^{CM}

Fig. 4.17 Continuous multimodal *typicality* of the real climate data

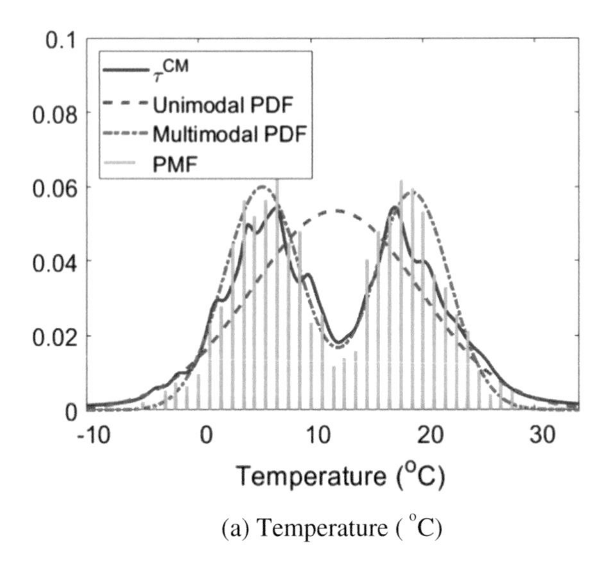

(a) Temperature (°C)

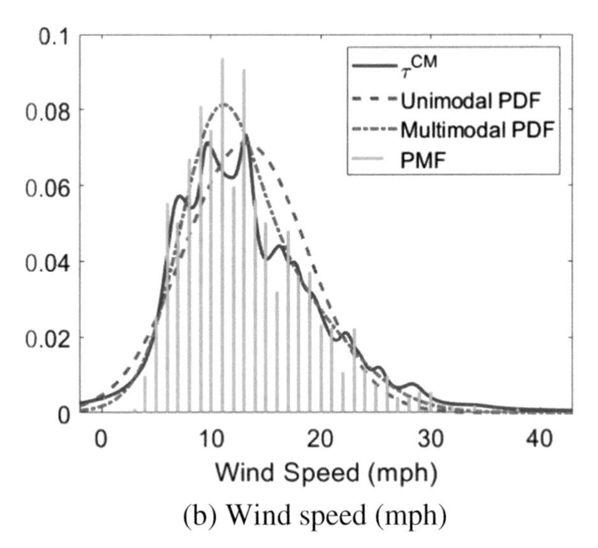

(b) Wind speed (mph)

Fig. 4.18 Continuous multimodal *typicality*, τ^{CM} versus unimodal and multimodal PDFs versus PMF

4.8 Conclusions

In this chapter, we systematically presented the non-parametric quantities of the proposed new *empirical* approach [1–3], including:

(1) the *cumulative proximity*, q;
(2) the *eccentricity*,ξ and *the standardized eccentricity*, ε;
(3) the *data density*, D, and
(4) the *typicality*, τ.

We also discussed their properties and presented their recursive calculation expressions. The discrete and continuous as well as unimodal and multimodal versions of *data density* (D^C, D^M, D^{CM}) and *typicality* ($\tau^C, \tau^M, \tau^{CM}$) were described and analyzed. The non-parametric quantities introduced in this chapter are of great importance and serve as the theoretical foundation of the new *empirical* approach proposed in this book.

As a fundamentally new measure in pattern recognition, the discrete version of the *typicality* τ resembles the unimodal PDF, but is in a discrete form. The discrete multimodal *typicality* τ^M resembles the PMF, but is derived from data directly and is free from the paradoxes and problems that the traditional approaches are suffering from [2, 3]. The continuous *typicality* τ^C and multimodal *typicality* τ^{CM} share many properties with the unimodal and multimodal PDFs, but is free from *prior* assumptions and user- and problem-specific parameters.

The most distinctive feature of the proposed new *empirical* approach is that it is not limited by restrictive impractical *prior* assumptions about the data generation model as the traditional probability theory and statistical learning approaches are. Instead, it is fully based on the ensemble properties and mutual distribution of the *empirical* discrete data. Moreover, it does not require an explicit assumption of randomness or determinism of the data, their independence, or even their amount. This new *empirical* approach touches the very foundation of data analysis, and thus, there are a wide range of applications including, but not limited to, anomaly detection, clustering/data partitioning, classification, prediction, fuzzy rule-based (FRB) system, deep rule-based (DRB) system, etc. We will describe the main ones in the remainder of this book.

4.9 Questions

(1) Which are the non-parametric quantities of the proposed new *empirical* approach?
(2) What are the similarities and differences between the *typicality* and the PDF and PMF?

(3) What are the differences between the *data density* as defined in this book and the density in terms of "density based clustering approaches" used in the literature, e.g. DBSCAN (density-based spatial clustering of applications with noise) or between the kernel density estimation (KDE) vs recursive density estimation (RDE)?

4.10 Take Away Items

- The non-parametric quantities of the proposed new *empirical* approach include:
  (1) the *cumulative proximity*,
  (2) the *eccentricity*, and the *standardized eccentricity*;
  (3) the *data density*, and
  (4) the *typicality*.
- They have recursive versions, unimodal and multimodal, discrete and continuous forms/versions.
- They are based on ensemble properties of the data and not by *prior* restrictive assumptions.
- The discrete version of the *typicality* resembles the unimodal PDF, but is in a discrete form. The discrete multimodal *typicality* resembles the PMF.

References

1. P.P. Angelov, X. Gu, J. Principe, D. Kangin, Empirical data analysis—a new tool for data analytics, in *IEEE International Conference on Systems, Man, and Cybernetics*, 2016, pp. 53–59
2. P. Angelov, X. Gu, D. Kangin, Empirical data analytics. Int. J. Intell. Syst. **32**(12), 1261–1284 (2017)
3. P.P. Angelov, X. Gu, J. Principe, A generalized methodology for data analysis. IEEE Trans. Cybern. **48**(10), 2981–2993 (2018).
4. A.N. Kolmogorov, *Foundations of the Theory of Probability* (Chelsea, Oxford, England, 1950)
5. V. Vapnik, R. Izmailov, Statistical inference problems and their rigorous solutions. Stat. Learn. Data Sci. **9047**, 33–71 (2015)
6. P. Angelov, Outside the box: an alternative data analytics framework. J. Autom. Mob. Robot. Intell. Syst. **8**(2), 53–59 (2014)
7. P.P. Angelov, Anomaly detection based on eccentricity analysis, in *2014 IEEE Symposium Series in Computational Intelligence, IEEE Symposium on Evolving and Autonomous Learning Systems, EALS, SSCI 2014*, 2014, pp. 1–8
8. P. Angelov, *Typicality* distribution function—a new density-based data analytics tool," in *IEEE International Joint Conference on Neural Networks (IJCNN)*, 2015, pp. 1–8
9. G. Sabidussi, The centrality index of a graph. Psychometrika **31**(4), 581–603 (1966)

10. L.C. Freeman, Centrality in social networks conceptual clarification. Soc. Netw. **1**(3), 215–239 (1979)
11. X. Gu, P.P. Angelov, J.C. Principe, A method for autonomous data partitioning. Inf. Sci. (Ny) **460–461**, 65–82 (2018)
12. P. Angelov, *Autonomous Learning Systems: From Data Streams to Knowledge in Real Time* (Wiley, 2012)
13. http://www.worldweatheronline.com
14. R. De Maesschalck, D. Jouan-Rimbaud, D.L.L. Massart, The mahalanobis distance. Chemometr. Intell. Lab. Syst. **50**(1), 1–18 (2000)
15. D. Kangin, P. Angelov, J.A. Iglesias, Autonomously evolving classifier TEDAClass. Inf. Sci. (Ny) **366**, 1–11 (2016)
16. X. Gu, P.P. Angelov, D. Kangin, J.C. Principe, A new type of distance metric and its use for clustering. Evol. Syst. **8**(3), 167–178 (2017)
17. X. Gu, P. Angelov, D. Kangin, J. Principe, Self-organised direction aware data partitioning algorithm. Inf. Sci. (Ny) **423**, 80–95 (2018)
18. J.G. Saw, M.C.K. Yang, T.S.E.C. Mo, Chebyshev inequality with estimated mean and variance. Am. Stat. **38**(2), 130–132 (1984)
19. M. Ester, H.P. Kriegel, J. Sander, X. Xu, A density-based algorithm for discovering clusters in large spatial databases with noise, in *International Conference on Knowledge Discovery and Data Mining*, 1996, vol. 96, pp. 226–231
20. P.P. Angelov, D.P. Filev, An approach to online identification of Takagi-Sugeno fuzzy models, IEEE Trans. Syst. Man, Cybern. Part B Cybern. **34**(1), 484–498 (2004)
21. S.Y. Shatskikha, Multivariate Cauchy distributions as locally gaussian distributions. J. Math. Sci. **78**(1), 102–108 (1996)
22. C. Lee, Fast simulated annealing with a multivariate Cauchy distribution and the configuration's initial temperature. J. Korean Phys. Soc. **66**(10), 1457–1466 (2015)
23. S. Nadarajah, S. Kotz, Probability integrals of the multivariate t distribution. Can. Appl. Math. Q. **13**(1), 53–84 (2005)
24. A. Corduneanu, C.M. Bishop, Variational Bayesian model selection for mixture distributions, in *Proceedings of the Eighth International Conference on Artificial Intelligence and Statistics*, pp. 27–34, 2001
25. H.A. Sturges, The choice of a class interval. J. Am. Stat. Assoc. **21**(153), 65–66 (1926)
26. T. Bayes, An essay towards solving a problem in the doctrine of chances. Philos. Trans. R. Soc. **53**, 370 (1763)

Chapter 5
Empirical Fuzzy Sets and Systems

Fuzzy sets are described mathematically by defining their membership functions. Historically, these were first designed subjectively [1, 2]. In 1990s, the so-called *data-driven* approach offered to use clustering to objectively design fuzzy sets including their membership functions [3]. This was first done offline and then online methods were also offered which allowed to process data streams [4, 5].

The flowcharts of the existing approaches (both, subjective and objective) for designing a fuzzy rule-based (FRB) system are depicted in Fig. 5.1.

We proposed recently so-called *empirical* fuzzy sets [6], which are derived from the *empirical* observations. The important difference from the *data-driven* approach (which was also using the actual data as a source to generate traditional membership functions) is the fact that *empirical* fuzzy sets are prototype-based, non-parametric and multimodal instead of unimodal functions of pre-defined, pre-selected type (most often triangular, Gaussian, trapezoidal). This will be further detailed in this chapter, but from practical, pragmatic point of view, the design of *empirical* fuzzy sets is much easier since it only requires selecting few prototypes and does not require to pre-select the type of membership functions and their parameters. This is a huge difference, because typically the number of fuzzy sets (respectively, linguistic terms) per variable used in fuzzy systems is more than 3 (often 5 or 7) and the number of variables in some problems may be tens or even hundreds and more (the total number of membership functions, respectively, fuzzy sets, is exponential: T^N, where T is the number of linguistic terms/fuzzy sets; N is the number of variables). Therefore, instead of the need to define and parameterize possibly many thousands or millions subjectively pre-determined functions what can be done is to simply identify a small number of actual data that constitute prototypes. This can be done either:

(1) subjectively, by asking a human domain expert, or
(2) objectively (refer to the autonomous data partitioning (ADP) algorithm presented in Chap. 7) based on the *typicality* and *data density*.

P. P. Angelov and X. Gu, *Empirical Approach to Machine Learning*, Studies in Computational Intelligence 800,
https://doi.org/10.1007/978-3-030-02384-3_5

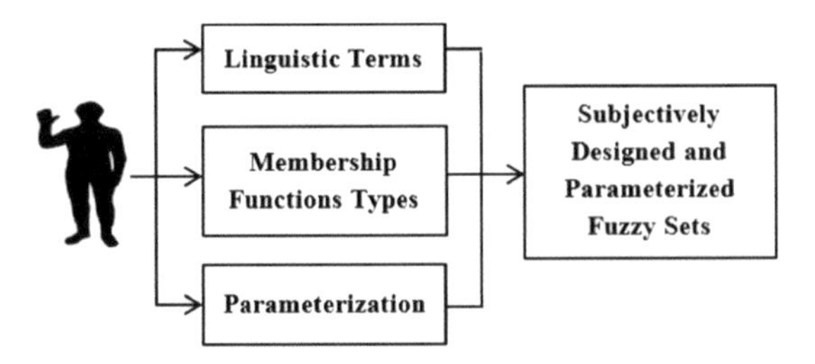

(a) Subjective approach to designing fuzzy sets

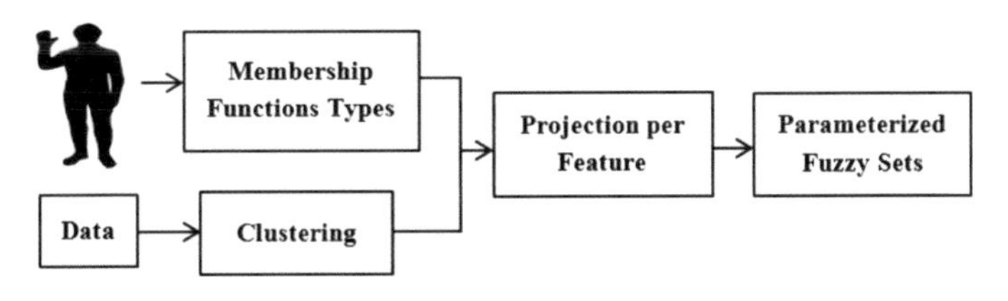

(b) Objective approach of designing fuzzy sets

Fig. 5.1 The flowcharts of the existing approaches for FRB system identification

Empirical fuzzy sets offer a very interesting link between the objective data and the subjective linguistic terms. Furthermore, *empirical* fuzzy sets can deal with heterogeneous data, e.g. data that have text, image/video as well as signals together. In addition, *empirical* fuzzy rule-based (FRB) systems can be formed in the same way as with the traditional fuzzy sets. These can also be used as a tool to represent expert knowledge.

In this chapter, the concepts and principles of *empirical* fuzzy sets and rule-based systems, including both, the objective and subjective approaches for their design as well as examples of real-world applications are presented.

5.1 The Concept

As it was mentioned above, the main issue in the design of fuzzy sets is to define the membership functions. The traditional and historically first approach of designing fuzzy sets, involved the human domain expert defining the type of the membership function, e.g. triangular, trapezoidal, Gaussian, etc. as well as parameterizing it based on subjective judgement. This approach has its own rationale in the two-way process of:

(1) formalizing expert knowledge and representing it in a mathematical form through membership functions, and
(2) extracting human-intelligible and understandable, transparent linguistic information from data and representing it in the form of IF…THEN rules.

However, to handcraft a FRB system for a specific problem in the subjective way, human experts are expected to incorporate their expertise about the problem and make many ad hoc decisions. This is usually very cumbersome in practice.

Let us consider a typical problem of medical decision support system based on a traditional FRB system.

Firstly, the domain expert (medical doctor) will have to define the variables/attributes/features of interest, e.g.

- systolic blood pressure,
- body temperature,
- oxygen concentration in the blood, and
- pulse, etc.

(S)he will have to after that choose a type of membership functions, e.g.

- bell shaped,
- Gaussian,
- triangular, and
- trapezoidal, etc.

After that, the expert must select the number of linguistic terms (respectively, fuzzy sets) to be used per variable. These can be the same number per variable or not; usually, people use an odd number, e.g. 3, 5 or 7. It has been proven that humans cannot distinguish more than nine linguistic labels, e.g. *Low*, *Medium* and *High* are the most used ones and *Very*, *Extremely* as well as *Rather* can be added to *Low* and *High* as so called linguistic hedges. As a result, for problems with tens or hundreds of variables the total number of membership functions may spiral out of control—a problem known as "curse of dimensionality" [7].

With the proposed *empirical* fuzzy sets, the expert (in this case, medical doctor) is only required to provide few examples of ill and healthy patients (prototypes of the two classes, in this case).

A very similar problem would have a sports coach/trainer when trying to determine good and bad players. The traditional approach of using fuzzy sets and systems would require, first, the coach to identify the various attributes/features such as

- number of passes made,
- distance run in a match or in training,
- time spent or efforts made during the training, etc.

It will require afterwards a, possibly, huge number of membership functions to be defined, their type (e.g. bell shaped, Gaussian, triangular, trapezoidal, or other convenient to use functions) to be determined. Instead, it is so much easier to ask the coach to simply, say "players X, Y, Z are *good*" and "players V, W are *bad*" without even explaining why.

5.2 Objective Method for *Empirical* Membership Functions Design

In 1990s, the so-called data-driven approach to designing fuzzy sets and systems was developed [3] and started to gain popularity. It is based on clustering the data and is the first attempt to move towards a more objective and *empirical* approach. Nonetheless, there are some problems left with this approach, i.e.

(1) ad hoc decisions are still needed for defining the membership functions;
(2) significant differences between the real data distribution and the membership functions still exists.

The proposed objective approach for *empirical* fuzzy sets starts with the actual data and, unlike the previous data-driven approach, does not involve or require any user- or problem-specific assumptions, parameters, thresholds or choices to be made. Its flowchart is depicted in Fig. 5.2.

The objective approach involves the ADP algorithm described in more detail in Chap. 7. This approach is based on the multimodal *data density* and *typicality* described in the previous chapter. The local peaks of the *typicality* (respectively, the *data density*) are used as prototypes because they are the most representative data points/vectors.

The ADP method [8] described in Chap. 7 has both, an offline and an online versions. In this section we use the offline version of the ADP method for illustration. We consider the real climate data (temperature and wind speed) measured in Manchester, UK for the period 2010–2015 (downloadable from [9]) as used in the numerical examples in the previous chapters for illustration. Within this dataset, roughly half (480) of the data samples were recorded during the winter and the remaining (459) data samples were recorded during the summer. The "winter" and "summer" are the two classes.

The ADP method, which is described in Chap. 7 and the respective algorithm, which is described in Chap. 11 does identify automatically (without any information required from the user) 15 prototypes which define 29 *empirical* fuzzy membership functions and sets, respectively which are depicted in Fig. 5.3 where the black asterisks "*" denote the identified prototypes and the dots "·" in different colors denote the data samples associated with them based on the "nearest neighbor" principle forming Voronoi tessellations [10]:

$$\mathbf{C}_{n*} \leftarrow \mathbf{C}_{n*} + \boldsymbol{x}; \quad n* = \underset{i=1,2,\ldots,M}{\arg\min}\,(d(\boldsymbol{x},\boldsymbol{p}_i)) \tag{5.1}$$

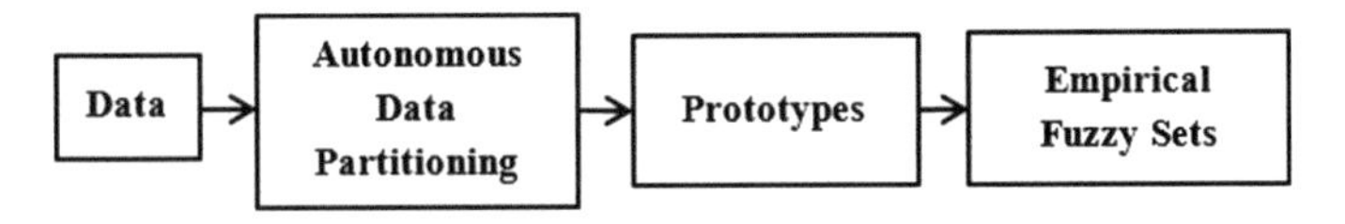

Fig. 5.2 The flowchart of the objective approach for *empirical* fuzzy sets design

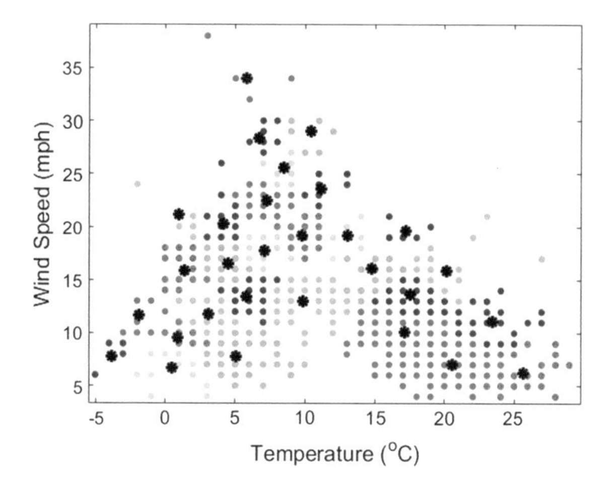

Fig. 5.3 The identified prototypes by the ADP algorithm with the climate data

where M denotes the number of *data clouds*; $\boldsymbol{p}_i$ denotes the prototype of the ith *data cloud*, $\mathbf{C}_i$; $\mathbf{C}_{n*}$ denotes the *data cloud* with the prototype closest to $\boldsymbol{x}$.

The *empirical* fuzzy membership functions can be defined through the *data density* estimated at the prototype. Remember that as defined in the previous chapter,

$$\mu_i^{\varepsilon}(\boldsymbol{x}) = D_i(\boldsymbol{x}) \tag{5.2}$$

where $\mu_i^{\varepsilon}(\boldsymbol{x})$ is the *empirical* membership function of the ith data cloud, $\mathbf{C}_i$; $D_i(\boldsymbol{x})$ is the local *data density* calculated within $\mathbf{C}_i$.

If Euclidean type of distance is used, then the expression reduces to a Cauchy type function defined around the prototype, $\boldsymbol{p}_i$:

$$\mu_i^{\varepsilon}(\boldsymbol{x}) = \frac{1}{1 + \frac{\|\boldsymbol{x}-\boldsymbol{p}_i\|^2}{X_i - \|\boldsymbol{\mu}_i\|}} \tag{5.3}$$

where $\boldsymbol{p}_i$, $\boldsymbol{\mu}_i$ and X_i are the prototype, mean and average scalar product of the data samples of $\mathbf{C}_i$.

The corresponding *3D empirical* fuzzy memberships in both, discrete and continuous form are given in Fig. 5.4.

5.3 Subjective Method of Expressing Expert Opinions

Empirical fuzzy sets allow the subjective specifics that the fuzzy logic theory is strong with to be easier incorporated and formalized into a mathematical form. However, instead of spending significant efforts on handcrafting a traditional (Zadeh-Mamdani [1] type or Takagi-Sugeno [2, 11, 12] type fuzzy sets and FRB

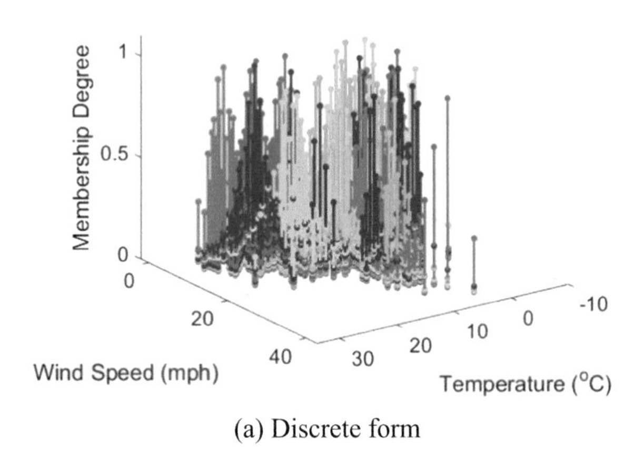

(a) Discrete form

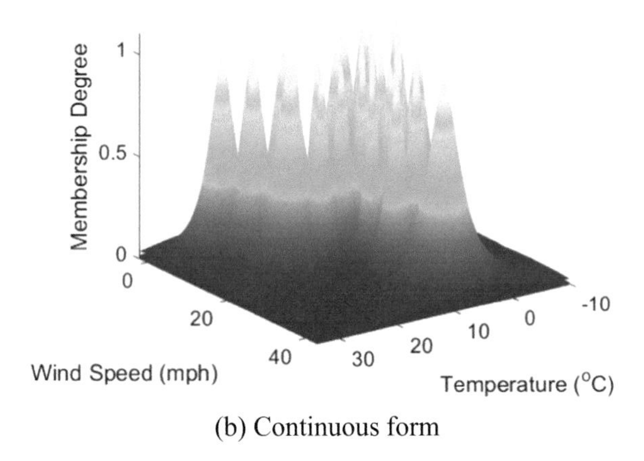

(b) Continuous form

Fig. 5.4 Visualization of the *3D empirical* fuzzy sets of the climate data

systems, respectively) experts only need to select a number of most typical samples as prototypes, and the *empirical* fuzzy sets will be built automatically based on them. Optionally, human experts can also help to define the labels/names of the linguistic terms, classes (if any).

The flowchart of the subjective approach for *empirical* fuzzy sets design is given in Fig. 5.5.

Let us consider a simple illustrative example using the same climate data (with two attributes: temperature and wind speed). The design of an *empirical* FRB system requires simply selecting at least one prototype per class (in this case, the number of classes is 2). For example, if the user selects two prototypes to be the data samples $\boldsymbol{p}_1 = [1\,^\circ\mathrm{C}, 7\,\mathrm{mph}]^T = \boldsymbol{x}_{14}$ and $\boldsymbol{p}_2 = [11\,^\circ\mathrm{C}, 20\,\mathrm{mph}]^T = \boldsymbol{x}_{357}$ recorded in the winter. The meaning is that on one of the two most *typical* winter days, temperature in Manchester is 1 °C and the wind speed is 7 mph, and on the other

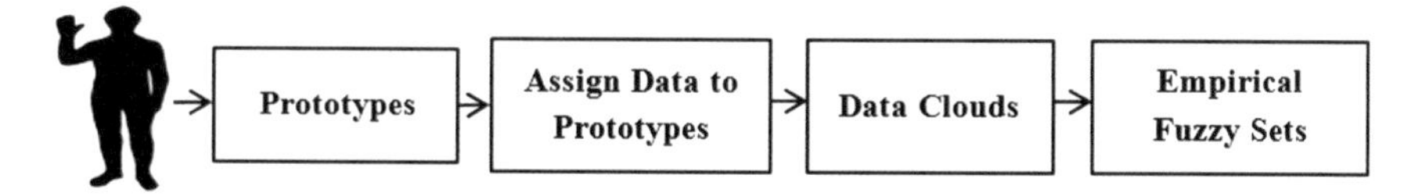

Fig. 5.5 The subjective approach for *empirical* fuzzy sets design

most *typical* winter day, the temperature in Manchester is 11 °C and the wind speed is 20 mph. "The most *typical*" means "with the highest value of the *typicality* (and *data density*). Similarly, let us assume that, the summer prototype selected by the user is $\boldsymbol{p}_3 = [20\,^{\circ}\mathrm{C}, 12\,\mathrm{mph}]^T = \boldsymbol{x}_{88}$. In general, the user may select as many prototypes per class as (s)he wants, but this will not change the principle and the further considerations.

All remaining data points/vectors are associated with the nearest prototype forming in this way a so-called Voronoi tessellation [10]. In this way, the *empirical* fuzzy membership functions are being based on the per class *data density*. Naturally, the prototypes will have the highest value of membership (equal to 1 because the distance from each one of them to itself is zero). The membership of any other data point to the respective class (e.g. "winter" or "summer") will be less than 1 and inversely proportional to the square distance to the prototype (remember, the analogy with the planets in the outer space and the law of gravity).

The *empirical* fuzzy sets built around the three prototypes we considered as a simple example is visualized in a *3D* form in Fig. 5.6.

Compared with the efforts spent on handcrafting traditional fuzzy sets, the *empirical* approach only requires human experts to define couple of prototypes, which is much simpler, easier, and highly transparent. The prototypes have a clear meaning unlike the parameters of traditional membership functions which may or may not have a clear meaning. For the simple illustrative example considered earlier, prototype $\boldsymbol{p}_1$ represents a cold, but calm day; prototype $\boldsymbol{p}_2$ represents a moderate, but windy day and prototype $\boldsymbol{p}_3$ represents a warm day with a moderate wind. The simplification in terms of the human involvement with the proposed approach can play a very important role in the collaboration between computer scientists and experts from different areas. Indeed, domain experts often find it difficult to understand mathematical models, computing terminology, etc. Instead of asking the domain experts (e.g. biologists, market or climate experts, etc.) to explicitly define the variables of interest (e.g. temperature, wind speed for this simple example or many other variables in other possible problems), to then define how many linguistic terms and how to label and parametrize them (e.g. *Low*, *Medium*, *High*, etc.) with the proposed *empirical* approach everything the domain expert has to do is to say "this is a *typical* summer/winter day" at least for one day for each class. Moreover, the domain expert can be (optionally) involved to further improve/refine the membership functions interactively and has a full interpretability and understanding of the internal structure and functioning of the model.

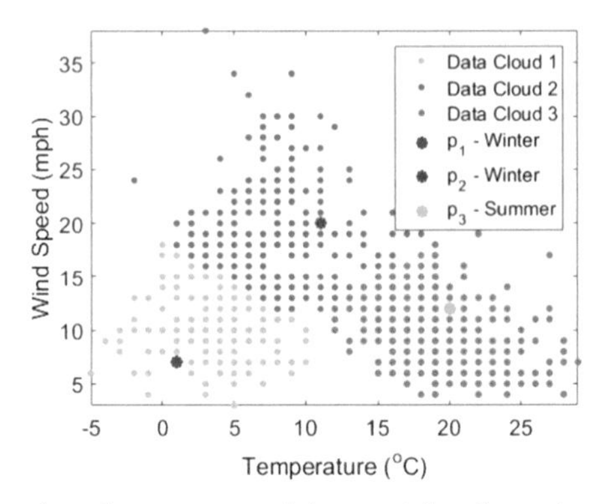

(a) The three selected prototypes and the remaining data points associated with them

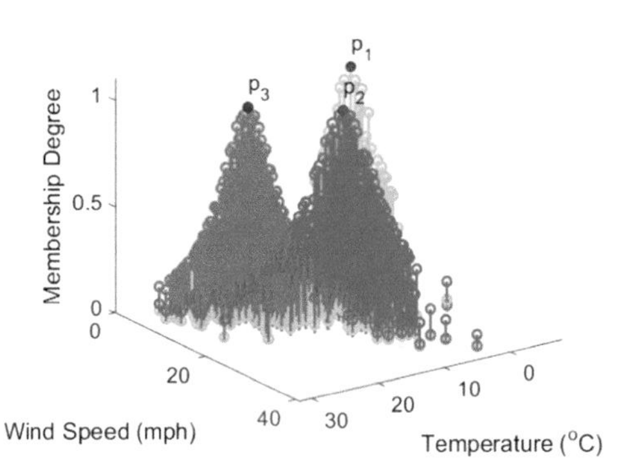

(b) *3D* discrete *empirical* membership functions

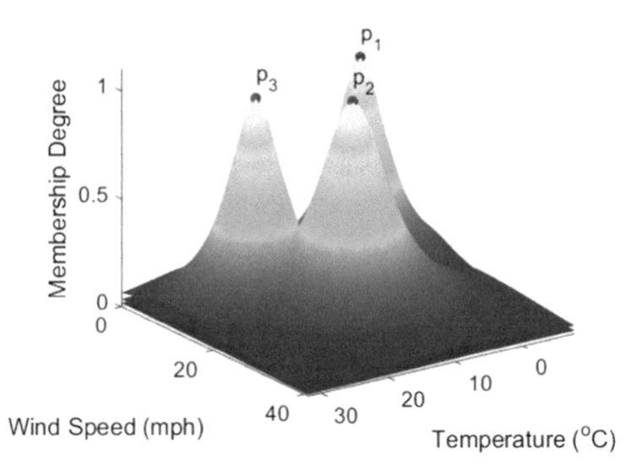

(c) *3D* continuous *empirical* membership functions

Fig. 5.6 The *empirical* fuzzy sets designed via the subjective approach

5.4 *Empirical* Fuzzy Rule-Based Systems

Based on the *empirical* fuzzy sets we can now define the *empirical* FRB systems of zero- and first-orders as follows:

(1) Zero-order (integer/singleton consequents, for example, indicating a class label):

$$\boldsymbol{R}_i : IF(\boldsymbol{x} \sim \boldsymbol{p}_i) \quad THEN\left(y \sim LT_{out,i}\right) \tag{5.4}$$

where $LT_{i,out}$ is linguistic term of the output of the ith fuzzy rule.

(2) First-order with functional (usually, linear) consequents:

$$\boldsymbol{R}_i : IF(\boldsymbol{x} \sim \boldsymbol{p}_i) \quad THEN(y = f(\boldsymbol{x}, \boldsymbol{a}_i)) \tag{5.5a}$$

$$\boldsymbol{R}_i : IF(\boldsymbol{x} \sim \boldsymbol{p}_i) \quad THEN\left(y = \bar{\boldsymbol{x}}^T \boldsymbol{a}_i\right) \tag{5.5b}$$

where $\bar{\boldsymbol{x}} = \left[1, \boldsymbol{x}^T\right]^T$.

Let us consider the same climate dataset as used in the previous illustrative examples in order to visualize and compare with the traditional approach. The following rules were generated based on the example presented in Fig. 5.6, where the climate data is re-grouped based on the selected earlier prototypes before into three classes "*Cold, but Calm*", "*Moderate, but Windy*" and "*Warm with Moderate Wind*".

$$\boldsymbol{R}_1: \begin{array}{c} IF(x_T \text{ is } Low) AND(x_W \text{ is } Low) \\ THEN(y_D \text{ is } "Cold,\ but\ Calm") \end{array} \tag{5.6a}$$

$$\boldsymbol{R}_2: \begin{array}{c} IF\ (x_T \text{ is } Medium) AND(x_W \text{ is } High) \\ THEN(y_D \text{ is } "Moderate,\ but\ Windy") \end{array} \tag{5.6b}$$

$$\boldsymbol{R}_3: \begin{array}{c} IF(x_T \text{ is } High) AND(x_W \text{ is } Moderate) \\ THEN(y_D \text{ is } "Warm\ with\ Moderate\ Wind") \end{array} \tag{5.6c}$$

where x_T and x_W represent the temperature and wind speed of a particular day, respectively; y_D is the weather of that day.

As a next step, based on the prototypes of the data samples of the three classes, for the variable x_T, namely, temperature, we can interpret the linguistic term "*Low*" as "*around* 1 °C", "*Medium*" as "*around* 11 °C" and "*High*" as "*around* 20 °C"; for the variable x_W, namely, wind speed, the linguistic term "*Low*" as "*around* 7 mph", "*Moderate*" as "*around* 12 mph" and "*High*" as "*around* 20 mph".

The next step of the traditional method is to select the type of the membership function (e.g. Gaussian, triangular, etc.) and to decide its parameters.

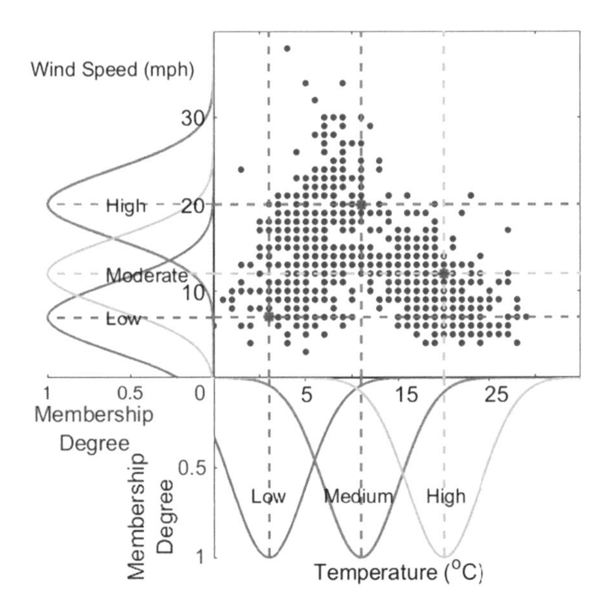

Fig. 5.7 Example of a traditional FRB system (the red asterisks "*" denote the means and the blue dots "." denote data samples)

After this, one can finally define the required fuzzy sets. In Fig. 5.7, we depict the FRB system with Gaussian type membership functions. FRB systems with other types of membership functions can also be built in a similar way.

Alternatively, the *empirical* FRB system has the following form:

$$\boldsymbol{R}_1\text{: } IF\left(\boldsymbol{x} \sim [1\,^{\circ}\mathrm{C}, 7\,\mathrm{mph}]^T\right) \quad THEN(y \,\mathrm{is}\, ''Cold,\ but\ Calm'') \tag{5.7a}$$

$$\boldsymbol{R}_2\text{: } IF\left(\boldsymbol{x} \sim [11\,^{\circ}\mathrm{C},\ 20\,\mathrm{mph}]^T\right) \quad THEN(y \,\mathrm{is}\, ''Moderate,\ but\ Windy'') \tag{5.7b}$$

$$\boldsymbol{R}_3\text{: } IF\left(\boldsymbol{x} \sim [20\,^{\circ}\mathrm{C},\ 12\,\mathrm{mph}]^T\right) \quad THEN(y \,\mathrm{is}\, ''Warm\ with\ Moderate\ Wind'') \tag{5.7c}$$

where $\boldsymbol{x} = [x_T, x_W]^T$.

The visualization of the *empirical* FRB systems (using discrete or continuous *empirical* membership functions) based on the same dataset as Fig. 5.7 is given in Fig. 5.8, where the red asterisks "*" denote the prototypes; the dots "." in different colors denote data samples belonging to different *data clouds* [4].

5.5 Dealing with Categorical Variables

One common practice for the traditional machine learning approaches to process categorical variables is to map them to different integer numbers. For example, one may use

- digit "*1*" to represent occupation category "driver";
- digit "2" to represent occupation category "industry worker";
- digit "*3*" to represent "cook", and so on.

Alternatively, one can use the *1-of-C* encoding method [13] to map the categorical variables into a series of orthogonal binary vectors like using:

- "*001*" to represent job category "driver";
- "*010*" to represent "worker";
- "*100*" to represent "cook", and so on.

However, no matter what kind of mapping is used, the encoding process always minimizes the true differences between data samples from different categories. This minimization is more obvious in high-dimensional problems. In many cases, different categories are inconsistent and, in fact, incomparable. The best way for handling different categories is to process them separately and, thus, avoid the manual interferences between each other during the processing.

For example, using the encoding method described above implies that, the "driver" is closer to the "worker" than to the "cook" which may not be the case in reality.

One unique characteristic of the *empirical* fuzzy sets is the ability to deal with problems containing categorical variables. Let us consider a *Z*-dimensional vector of categorical variables, $v_l = \left[v_{l,1}, v_{l,2}, \ldots, v_{l,Z}\right]^T$ $(l = 1, 2, \ldots, K)$; $v_{l,j}$ is the *j*th categorical variable of $\boldsymbol{v}_l$; $\{\boldsymbol{\chi}\}_K = \{\boldsymbol{\chi}_1, \boldsymbol{\chi}_2, \ldots, \boldsymbol{\chi}_K\}$ is the data set/stream that consists of both the categorical and numerical variables in each data sample, $\boldsymbol{\chi}_l = \left[\boldsymbol{v}_l^T, \boldsymbol{x}_l^T\right]^T$; $\upsilon_i = \left[\upsilon_{i,1}, \upsilon_{i,2}, \ldots, \upsilon_{i,Z}\right]^T$ is the corresponding vector of categorical variables of the

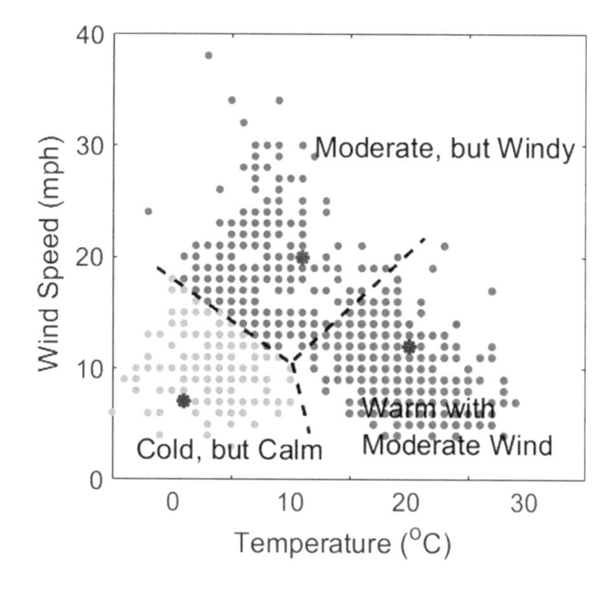

Fig. 5.8 Example of an *empirical* FRB system for the same data

ith numerical prototype, $\boldsymbol{p}_i$; the set of possible values of the jth categorical variable is denoted by $\mathbb{C}_j$, and $v_{l,j}$ and $\upsilon_{i,j}$ can only take one value from $\mathbb{C}_j$. Examples of categorical variables may include:

- gender;
- occupation;
- brand, and so on.

By considering both, the numerical and categorical variables, we introduce the *empirical* fuzzy rules in a more general form as follows [6]:

$$IF(\boldsymbol{v}_l \sim \boldsymbol{\upsilon}_i) \quad THEN\, IF\, (\boldsymbol{x}_l \sim \boldsymbol{p}_i) \quad THEN\left(y_l \sim LT_{out,i}\right) \tag{5.8}$$

Similarly, the first-order *empirical* fuzzy rules can be formulated in the following more general form [6]:

$$IF(\boldsymbol{v}_l \sim \boldsymbol{\upsilon}_i) \quad THEN\, IF(\boldsymbol{x}_l \sim \boldsymbol{p}_i) \quad THEN\left(y_l = \bar{\boldsymbol{x}}_l^T \boldsymbol{a}_i\right) \tag{5.9}$$

where $\bar{\boldsymbol{x}}_l = \left[1, \boldsymbol{x}_l^T\right]^T$.

The output of the categorical (*THEN IF*) part in the general *empirical* fuzzy rule is defined by a Boolean function ("true" or "false" only) expressed as:

$$B_i(\boldsymbol{v}_l) = \begin{cases} 1 & \boldsymbol{v}_l = \boldsymbol{\upsilon}_i \\ 0 & \boldsymbol{v}_l \neq \boldsymbol{\upsilon}_i \end{cases} \tag{5.10}$$

At least one prototype is required for each category in order to build the general *empirical* fuzzy rule. For data that contain multiple categorical variables, that is $v_{l,1}, v_{l,2}, \ldots, v_{l,Z}$, at least $\prod_{j=1}^{Z} b_j$ prototypes are needed, where b_j is the cardinality of $\mathbb{C}_j$.

As the *empirical* FRB system requires at least one prototype for each category. Therefore, the data will be split per category and processed separately. This is very different from the traditional approaches, which ignore the real differences between categorical variables. This is, however, very convenient for parallelization.

In order to illustrate this, we use the same climate dataset used in the previous examples for visualization. As the climate dataset consists of data samples from two categories (let us call these "*0*" and "*1*" rather than "*Winter*" and "*Summer*"). We, firstly, separate the two categories and, then, randomly select three data samples from each category as the prototypes to build the *empirical* fuzzy sets. The randomly selected prototypes are visualized in Fig. 5.9.

Using the identified prototypes, six *empirical* fuzzy sets in total with the respective *empirical* membership functions are identified. The structure of the *empirical* FRB system is then developed based on these *empirical* fuzzy rules. The *3D* visualization of the *empirical* fuzzy sets derived from data is depicted in Fig. 5.10, and the corresponding *empirical* fuzzy rules are given by Eqs. (5.11a) and (5.11b).

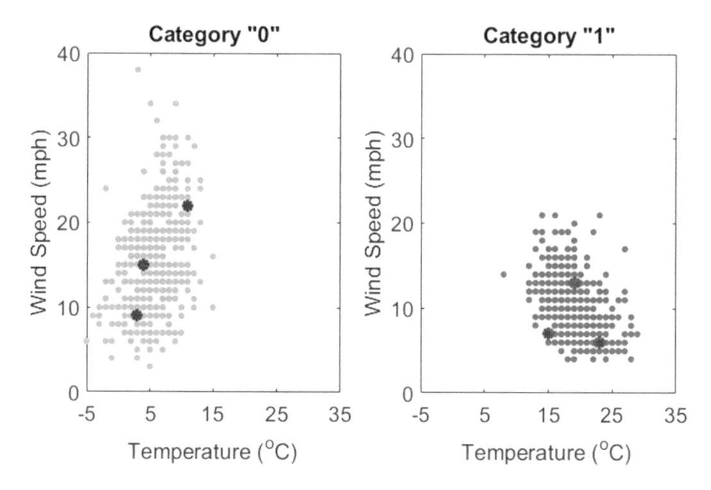

Fig. 5.9 Prototypes identified from the data samples from the two categories (the red asterisks "*" are the prototypes)

$$IF\,(v \sim "0")\quad THEN\; IF\,\left(\boldsymbol{x} \sim [3\,^{\circ}\mathrm{C}, 9\,\mathrm{mph}]^T\right)\; OR\; \left(\boldsymbol{x} \sim [4\,^{\circ}\mathrm{C}, 15\,\mathrm{mph}]^T\right)\; OR\; \left(\boldsymbol{x} \sim [11\,^{\circ}\mathrm{C}, 22\,\mathrm{mph}]^T\right)\quad THEN\,(y\,\mathrm{is}\; "Winter") \tag{5.11a}$$

$$IF\,(v \sim "1")\quad THEN\; IF\,\left(\boldsymbol{x} \sim [15\,^{\circ}\mathrm{C}, 7\,\mathrm{mph}]^T\right)\; OR\; \left(\boldsymbol{x} \sim [19\,^{\circ}\mathrm{C}, 13\,\mathrm{mph}]^T\right)\; OR\; \left(\boldsymbol{x} \sim [23\,^{\circ}\mathrm{C}, 6\,\mathrm{mph}]^T\right)\quad THEN\,(y\,\mathrm{is}\; "Summer") \tag{5.11b}$$

5.6 Comparative Analysis

In comparison with the traditional FRB systems, the *empirical* membership function in the case of using Euclidean type of distance has the form of a Cauchy function which is quite similar to the usually used Gaussian function [12]. However, it has no parameters that should be pre-defined by the user and this type of a function is not a result of a subjective choice (which in the case of traditional membership functions could have also been a triangular, trapezoidal, etc.) but an objective result which only depends on the type of distance used and the actual data.

Moreover, the way to define the value of the membership function (Eqs. 5.2 and 5.3) for the *empirical* FRB systems is quite different from the traditional FRB systems. To build a traditional (Zadeh-Mamdani or Takagi-Sugeno type) FRB systems, human experts need to define multiple parameters as described in Chap. 3. In comparison to this, the *empirical* fuzzy sets simplify the process of designing fuzzy sets. They are defined by multivariate data points/vectors representing the focal points of the non-parametric, shape-free *data clouds* consisting of data

Fig. 5.10 *3D* visualization of the *empirical* fuzzy functions per numerical variable

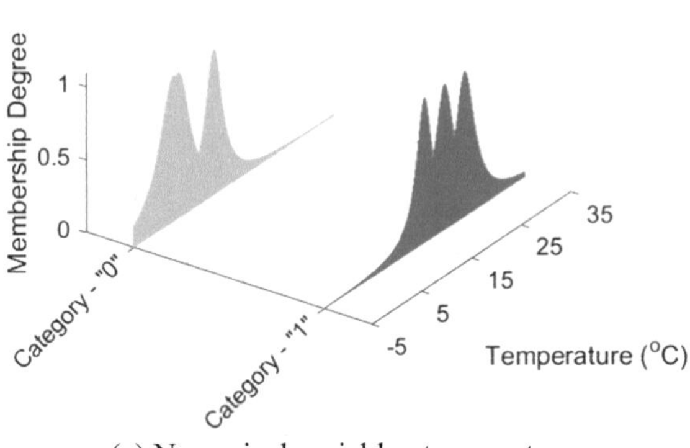

(a) Numerical variable : temperature

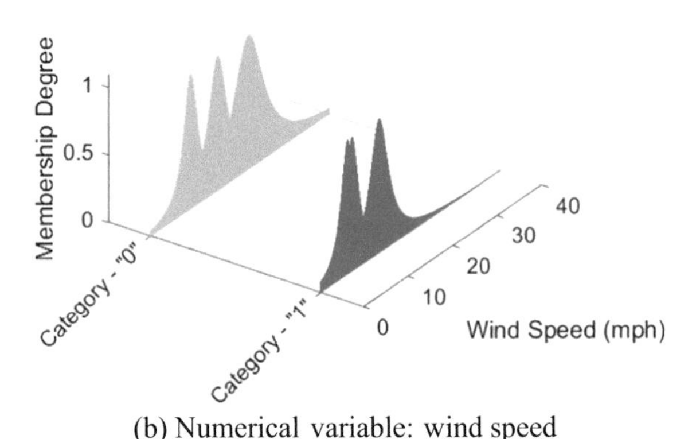

(b) Numerical variable: wind speed

samples associated with the nearest focal point forming Voronoi tessellation [10]. These focal points (prototypes) are used as the antecedent (IF) part of the *empirical* fuzzy sets, which significantly reduces the required efforts of human experts and, at the same time, largely enhances the objectiveness of FRB systems.

The *empirical* FRB systems do not need to assume that the *empirical* membership functions are in the form of continuous functions like the two traditional types of FRB systems in Fig. 5.7. The *empirical* membership functions derived from data are in discrete form [14, 15], see Fig. 5.11a.

However, considering that the two attributes, namely temperature and wind speed, are from a continuous domain, continuous membership functions of Cauchy form can also be defined around the prototype as depicted in Fig. 5.11b forming envelops around the actual discrete data points. It has to be noticed that, in Fig. 5.11, the *empirical* membership functions are visualized per feature.

Fig. 5.11 *Empirical* FRB systems

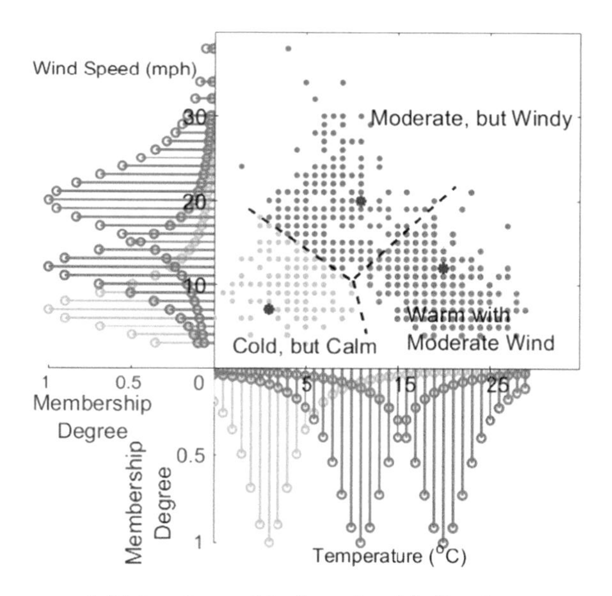

(a) Discrete *empirical* membership functions

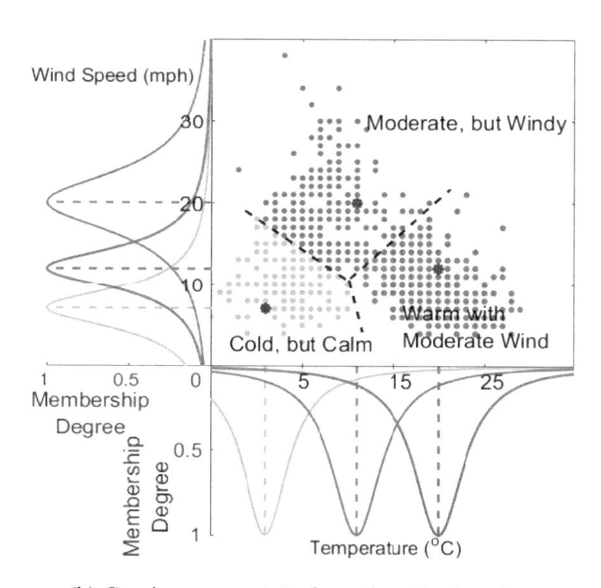

(b) Continuous *empirical* membership functions

Unlike the traditional membership functions used in the Zadeh-Mamdani and Takagi-Sugeno types of FRB systems which are defined per attribute, the *empirical* fuzzy sets are extracted from the data directly in a vector form [6]. Nonetheless, one can still draw $(n+1)$ dimensional *empirical* membership functions based on the particular n $(1 \leq n \leq N)$ features/attributes of the data resembling the $(n+1)$

dimensional probability distribution function (PDF) [14, 15]. Additionally, the $(n+1)$ dimensional *empirical* fuzzy sets can be either discrete or continuous or a combination of discrete and continuous. The $n + 1$ $(n = 2)$ dimensional discrete and continuous *empirical* fuzzy sets are shown in Fig. 5.4.

The difference between the *empirical* membership functions and the PDFs is that the peaks (maxima) of the former are *1*: $\mu_i^{\varepsilon}(\boldsymbol{x}) = 1$ and they can be linguistically interpreted, for example as: "*Low*", "*Medium*", "*High*" and so on, per feature/attribute based on projections in the same way as the traditional fuzzy sets or as "close to prototype $\boldsymbol{p}_i$" in the *empirical* fuzzy sets.

The *empirical* FRB systems are much more convenient and computationally simpler in high-dimensional problems in comparison to the traditional FRB systems (Zadeh-Mamdani and Takagi-Sugeno). They have unique ability to deal with problems containing categorical variables. This is thanks to the fact that only the prototypes are needed to be identified for the *empirical* FRB system either by users (detailed in Sect. 5.3) or by the data-driven approach (detailed in Sect. 5.2), and the system will derive *empirical* membership functions from *data clouds* formed around the prototypes automatically. Moreover, with the proposed *empirical* fuzzy sets and *empirical* FRB systems, one can tackle heterogeneous data and combine categorical (e.g., gender, occupation, number of doors) with continuous and/or discrete variables, as detailed in Sect. 5.5.

The *empirical* fuzzy sets and systems are very suitable for processing streaming data. With an exponential growth in the scale and complexity of the data generated in high volume by sensors, people, society, industry, etc. they are increasingly seen as an untapped resource, which offers new opportunities for extracting aggregated information to inform decision-making in policy and commerce even in real time. The *empirical* approach extends the existing techniques for FRB system identification. Indeed, the traditional methods for designing fuzzy sets were developed in the era when the data were no at such a large scale and were assumed to be mostly available offline, not streaming and possibly stationary. It is practically difficult to design traditional fuzzy models from a huge amount of unlabeled images, big data representing customer choices or preferences, and so on. In contrast, the proposed *empirical* fuzzy sets and systems offer an efficient and data-centered (thus, *empirical*) tool that is clear and intuitive, yet it is not ad hoc, and can facilitate and empower the human experts and users instead of overloading or overwhelming them.

5.7 Examples—Recommender Systems Based on *Empirical* Fuzzy Sets

The convenience of the *empirical* fuzzy sets and FRB systems may significantly enhance the recommendation systems used by retailers. Let us use an example of buying a house. Of course, there are various visible and hidden factors that can influence the customer's decision before buying a particular house, i.e. price, the distances to the city center, schools and main roads, number of rooms, the environment, the safety conditions, the neighborhood, house floor area, etc. Let us, for simplicity, consider only the following four factors/features,

(1) price,
(2) house floor area,
(3) distance to the city center and
(4) distance to the main roads.

If the estate agency wants to have a recommendation system using the traditional FRB systems, they need to build a number of fuzzy sets to categorize the houses based on different features. In this case, the linguistic terms for the features can be (presented in Fig. 5.12):

- price, e.g., "*Economic*", "*Moderate*" and "*Luxury*";
- house floor area, e.g., "*Small*", "*Medium*" and "*Large*";
- distance to the city center, e.g., "*Close*", "*Medium*" and "*Far*";
- distance to the main roads, e.g., "*Close*", "*Medium*" and "*Far*".

Building such fuzzy sets involves a lot of efforts, however, the result is subjective, and the reason is obvious. A big issue is parameterizing the membership functions. Another issue is the type of membership functions as well as the number of linguistic terms used and the ranges of the variables.

Different estate agents as well as different customers may have different perceptions of the features. For example, a middle age customer may think that a

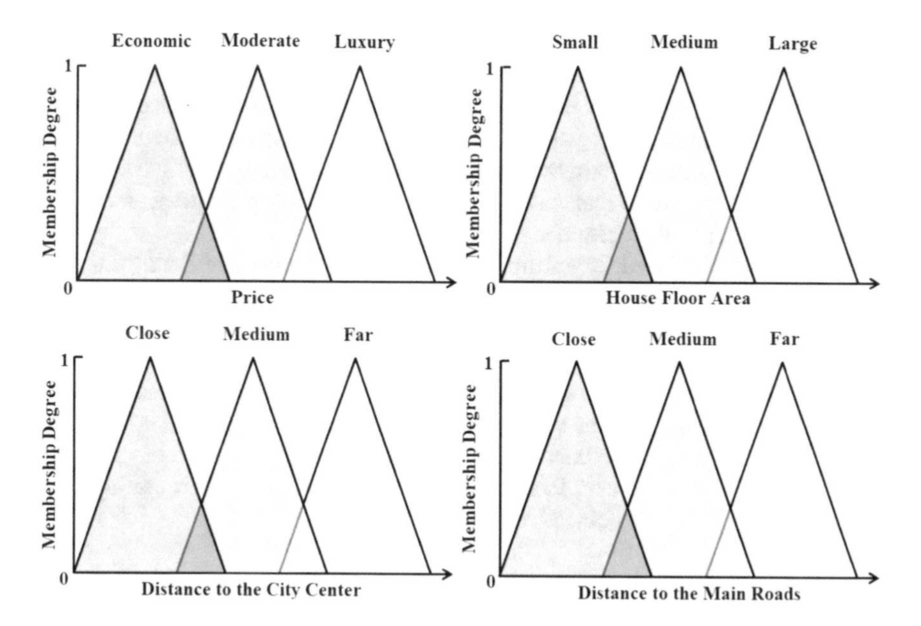

Fig. 5.12 The triangular type membership functions of the traditional FRB system

150 thousand pounds house is affordable. However, a retired customer may think that a house that costs 100 thousand pounds is too expensive. Alternatively, a single customer may think that a 70 m^2 house is "big", but a couple with two children may think that this house is too "small".

Moreover, the preference may not be smoothly monotonic. There is no scientific way to decide the number of linguistic terms to use each time (three or more or less). As a result, the handcrafted FRB systems are difficult to design and use, and this may be the main reason that they are still not widely accepted.

In contrast, when the *empirical* FRB system is used instead, the estate agent only needs to ask customers to select one or more houses they are most satisfied with. These houses can be any real houses in this city regardless whether they are for sale or not. These may also be imaginary, ideal houses as well.

For example, one customer may select three different houses in the city with different features, i.e.

(1) a small house with luxury decoration close to the city center and main roads, denoted by $\boldsymbol{p}_1$;
(2) a medium house with luxury decoration close to the city center, but far away from main roads, denoted by $\boldsymbol{p}_2$;
(3) a large house with economic decoration close to the main roads, but far away from city center, denoted by $\boldsymbol{p}_3$.

Then, the selected houses can be used as prototypes to form the *data clouds* based on the normalized variables from all the available for sale houses in the database. The *empirical* membership functions derived from the *data clouds* formed around the prototypes are visualized in Fig. 5.13.

Based on the degrees of similarity of each available for sale house to the prototypes, the estate agent can easily make a list of recommended houses for each customer. For example, the agent can quickly identify the available for sale houses for which the corresponding membership degrees are above 0.9 ($\mu^{\varepsilon} \geq 0.9$), or the top 10 candidate houses that have the highest membership degrees, and, then, recommend them to the customer.

All of this is achieved by asking each customer a simple question: "Can you, please, tell me the most satisfactory houses in the city you have seen."

As we can see from the above example, there is no need for any parameters or unnecessary efforts, the *empirical* FRB recommendation system only needs the users to give some examples of whatever they think are the best to serve as the prototypes. Then, the system will automatically form the *data clouds* based on the prototypes and calculate the degrees of similarity for all the available products. The recommendation list can, then, be generated automatically in a user-specific, but at the same time, also objective and data-driven manner.

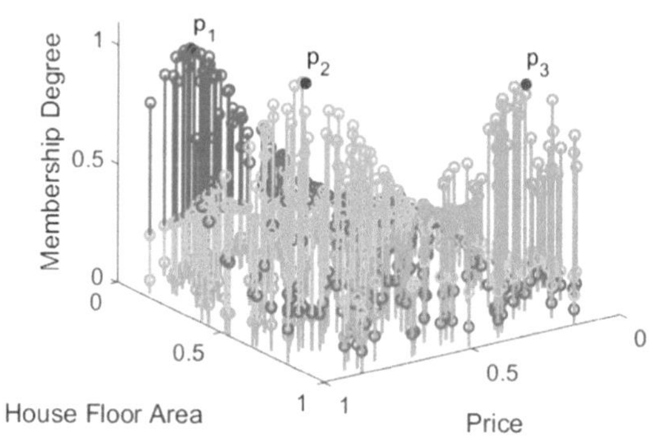

(a) Visualization with features *1)* price and *2)* house floor area

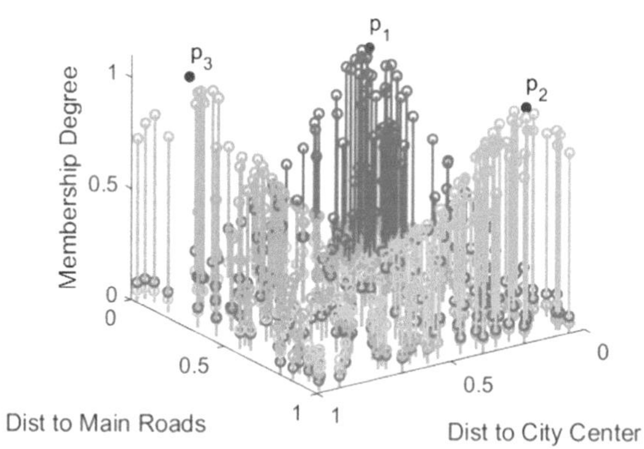

(b) Visualization with features *3)* distance to the city center and *4)* distance to the main roads

Fig. 5.13 Visualization of the *empirical* membership functions based on four normalized attributes of houses (red dots "·" represent the prototypes)

5.8 Conclusions

This chapter introduces the *empirical* fuzzy sets and the FRB systems based on them, named *empirical* FRB systems. Two approaches (subjective and objective) for identifying *empirical* FRB systems are presented in this chapter as well.

Compared with the traditional fuzzy sets and FRB systems, the *empirical* fuzzy sets and FRB systems have the following significant advantages:

(1) they are derived in a transparent, data-driven way without *prior* assumptions *Prior* assumptions;
(2) they effectively combine the data- and human-derived models;
(3) they possess very strong interpretability and high objectiveness;
(4) they facilitate experts or even bypass the involvement of human expertise.

Indeed, the traditional fuzzy sets and FRB systems are prone to the so-called "curse of dimensionality" and handcrafting fuzzy sets for complex, high-dimensional problems may become a tough obstacle due to the exponential growth of the number of fuzzy sets required for the high-dimensional problems.

5.9 Questions

(1) Explain the subjective process of design of *empirical* fuzzy sets and systems. How they differ from the traditional fuzzy sets and fuzzy rule-based systems?
(2) Describe the objective process of design of *empirical* fuzzy sets and systems.
(3) How the *empirical* fuzzy sets and systems deal with categorical variables, such as number of rooms in an apartment, doors of a car, gender, occupation, etc. How would the traditional fuzzy sets and systems deal with such variables? How would a statistical system deal with the same problem?
(4) Give few examples of application areas for the *empirical* fuzzy sets and systems.

5.10 Take Away Items

- *Empirical* fuzzy systems are defined based on prototypes. Their membership functions are extracted from the *empirically* observed data and not assumed a priori/imposed by the human;
- *Empirical* fuzzy sets can be designed using a subjective or an objective approach;
- *Empirical* fuzzy sets and systems have advantages in comparison with the traditional fuzzy sets and systems, including:

  (1) they are derived in a transparent, data-driven way without *prior* assumptions;
  (2) they effectively combine the data- and human-derived models;
  (3) they have very strong interpretability and high objectiveness;
  (4) The involvement of human experts is significantly facilitated or can be bypassed.

- *Empirical* fuzzy sets can deal with categorical variables, such as gender, occupation, number of rooms, number of doors, etc.
- Examples of *empirical* fuzzy sets and systems include, but are not limited to recommender systems, decision support systems, etc.

References

1. E.H. Mamdani, S. Assilian, An experiment in linguistic synthesis with a fuzzy logic controller. Int. J. Man Mach. Stud. **7**(1), 1–13 (1975)
2. T. Takagi, M. Sugeno, Fuzzy identification of systems and its applications to modeling and control. IEEE Trans. Syst. Man. Cybern. **15**(1), 116–132 (1985)
3. R.R. Yager, D.P. Filev, Learning of fuzzy rules by mountain clustering, in *SPIE Conference on Application of Fuzzy Logic Technology*, 1993, pp. 246–254
4. P. Angelov, R. Yager, A new type of simplified fuzzy rule-based system. Int. J. Gen Syst. **41** (2), 163–185 (2011)
5. P. Angelov, *Autonomous Learning Systems: From Data Streams to Knowledge in Real Time* (Wiley, Ltd., 2012)
6. P.P. Angelov, X. Gu, Empirical fuzzy sets. Int. J. Intell. Syst. **33**(2), 362–395 (2017)
7. C.C. Aggarwal, A. Hinneburg, D.A. Keim, On the surprising behavior of distance metrics in high dimensional space, in *International Conference on Database Theory*, 2001, pp. 420–434
8. X. Gu, P.P. Angelov, J. C. Principe, A method for autonomous data partitioning, Inf. Sci. **460–461**, 65–82 (2018)
9. http://www.worldweatheronline.com
10. A. Okabe, B. Boots, K. Sugihara, S.N. Chiu, *Spatial Tessellations: Concepts and Applications of Voronoi Diagrams*, 2nd edn. (Wiley, Chichester, England, 1999)
11. P. Angelov, An approach for fuzzy rule-base adaptation using on-line clustering. Int. J. Approx. Reason. **35**(3), 275–289 (2004)
12. P.P. Angelov, D.P. Filev, An approach to online identification of Takagi-Sugeno fuzzy models, IEEE Trans. Syst. Man, Cybern. Part B Cybern. **34**(1), 484–498 (2004)
13. P. Cortez, A. Silva, Using data mining to predict secondary school student performance, in *5th Annual Future Business Technology Conference*, 2008, pp. 5–12
14. P.P. Angelov, X. Gu, J. Principe, D. Kangin, Empirical data analysis—a new tool for data analytics, in *IEEE International Conference on Systems, Man, and Cybernetics*, 2016, pp. 53–59
15. P. Angelov, X. Gu, D. Kangin, Empirical data analytics. Int. J. Intell. Syst. **32**(12), 1261–1284 (2017)

Chapter 6
Anomaly Detection—*Empirical* Approach

Anomaly detection is one of the very first steps of pre-processing the data (after selecting the type of proximity/distance metric that will be used), which can be performed even before normalization or standardization. It can be intertwined with them, especially with the standardization.

Anomaly detection is an important problem that has been studied and applied to diverse areas and application domains [1]. It is very important on its own (for applications as diverse as insurance and bank data analysis [2, 3]; intruder [4], insider or malware detection [5]; fault detection in technical systems [6]; novelty and object detection in video processing [7], etc. However, in machine learning (and, especially, in autonomous learning systems [8, 9]) anomaly detection is a part of the pre-processing, and the removal or ignoring of anomalous data items is required to avoid them affecting the data-driven models' quality [8]. It is clear that an anomalously high or low value will affect the means, standard deviations, covariances and other statistical characteristics of the dataset or data stream.

Another important role of the anomaly detection, especially, in online data stream and evolving context, is that some of these anomalies serve as triggers to the change (dynamic evolution) of the structure of the models. Indeed, in a dynamically evolving data stream context appearance and formation of a new *data cloud* or cluster starts with a single outlier, which is not associated with any previously existing *data cloud* or cluster. Very similarly to the real life, if more and more data samples appear that are quite similar to this outlier, its role will change from being an outlier to an initiator of a new *data cloud* or cluster.

As the name suggests, the aim of anomaly detection is to discover outliers, rare events [10]. From statistical point of view, this is an analysis of the tail of the distribution. The difficulty comes from the fact that, in reality, the distribution of the data is rarely normal/Gaussian or exactly described by another smooth and well-known function which are usually considered in the theoretical analysis. This has significant effect on the conclusions made regarding the anomaly/outliers.

P. P. Angelov and X. Gu, *Empirical Approach to Machine Learning*, Studies in Computational Intelligence 800,
https://doi.org/10.1007/978-3-030-02384-3_6

In many real situations and applications, i.e. detecting criminal activities, fire alarm, bank fraud detection, machine condition monitoring, human body monitoring, surveillance system for security, etc., the rare values play a key role.

Anomaly detection problem can be seen, in some sense, as an inverse to the clustering: a data point/vector that is with highest *data density*, is the most *typical*, representative one. It is most likely to be a cluster center or prototype [10]; on the contrary, a data point/vector which is far away from all other data samples, with lowest *data density* is potentially an outlier/an anomaly.

Anomalies can be categorized into the following three types [1, 11]:

(1) **Point anomalies**—these are the individual data samples showing significant differences from the rest of the data. This type of anomalies is the most widely studied one [1].
(2) **Contextual anomalies**—these are the data samples being anomalous in a certain context only, but not otherwise. See more details in Sect. 6.1.
(3) **Collective anomalies** are a collection of related data samples anomalous with respect to the entire dataset. The collective anomaly samples may not be anomalies by themselves, but they are anomalous when occur together.

One can also group the anomaly detection approaches into [1]:

(1) **Supervised anomaly detection methods**. Methods of this kind usually train a supervised learning algorithm with a training set consisting of data samples labelled as either "normal" or "anomalous" [12–15]. Supervised anomaly detection methods rely heavily on *prior* knowledge and human expertise, and may suffer from the problem of imbalanced class distributions. Examples of this type of anomaly detection methods include decision trees [14] and one-versus-all support vector machines (SVM) [15]. These techniques require the labels of the observations to be known in advance, which allows the algorithms to learn in a supervised way and generate the desired output after the training. Therefore, supervised approaches are usually more accurate and effective in detecting outliers compared with the statistical methods. However, in real applications, labels are usually unavailable, and it can be very expensive to obtain them.
(2) **Semi-supervised anomaly detection methods**. Semi-supervised methods usually assume that the training set is composed of normal data samples only. Based on this assumption, they learn a model from the training set corresponding to normal behaviors, and use the model to identify anomalies in the validation data [16]. Some semi-supervised methods use inverse logic by assuming that the training set is composed of anomalies only [17]. The main problem of the semi-supervised approaches is that it is practically impossible to get representative collection of all possible normal (or anomalous) data samples in the training set. That is, only *expected* anomalies may be detected, but often in real applications *unexpected* anomalies are that matters most.
(3) **Unsupervised anomaly detection methods**. Unsupervised approaches do not require training samples, and, thus, are more widely used than the two other

types. Unlike the other two types, most of the unsupervised approaches score the data samples based on density or similarity instead of giving labels. Many of the existing unsupervised anomaly detection approaches [18–20], however, require a number of user inputs to be predefined, i.e. thresholds, error tolerances, numbers of nearest neighbors, etc. The selection of these values requires good *prior* domain knowledge and is subjective and problem-specific. The performance of these approaches is highly affected by the choice of these values. Although, some unsupervised approaches do not require such *prior* knowledge explicitly, they usually do this implicitly, by assumptions (e.g. that normal data samples are far more frequent than anomalies, etc.). The most popular unsupervised methods of such type include the so-called "3σ" principle (which follows from the Chebyshev inequality [21]), different density estimation methods, including the kernel density estimation (KDE) and the recursive density estimation (RDE) [22] methods.

These three types of anomaly detection methods differ by the availability of labels in the training sets and can be visualized by the flowcharts given in Fig. 6.1 [23].

Anomaly detection techniques are traditionally based on the statistical analysis [24, 25] of the data sets or streams. They rely on a number of *prior* assumptions about the data generation models (usually assuming Gaussian distribution) and require a certain degree of *prior* knowledge [24] or making ad hoc decisions,

Fig. 6.1 Flowcharts of different types of anomaly detection methods

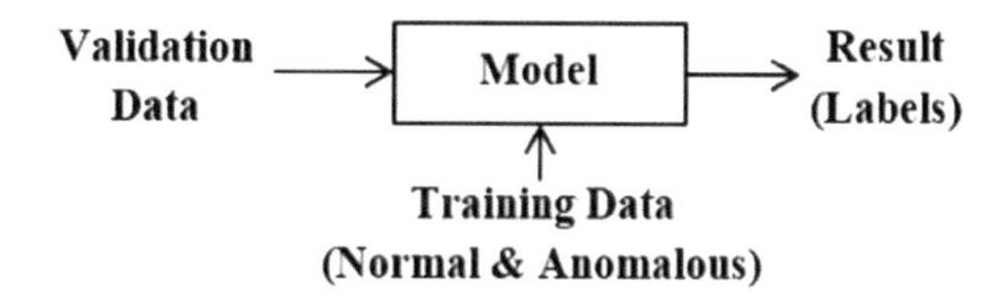

(a) Supervised approaches

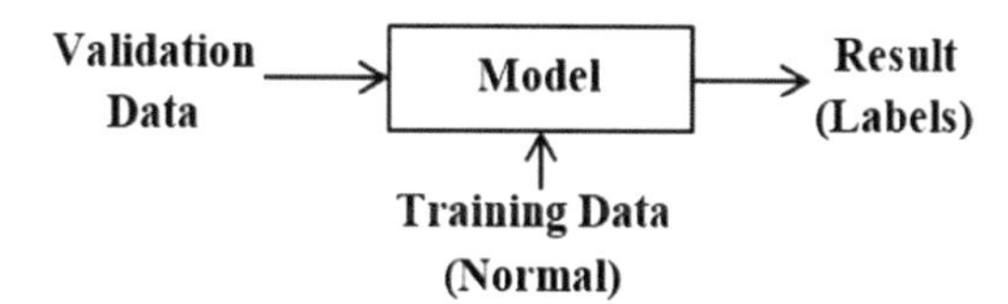

(b) Semi-supervised approaches

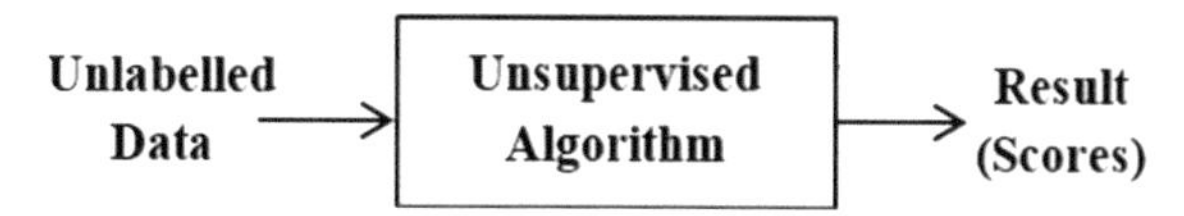

(c) Unsupervised approaches

i.e. regarding the thresholds. However, these *prior* assumptions are only true in the ideal/theoretical situations, i.e. Gaussian, independently and identically distributed (*iid*) data. In reality, however, the *prior* knowledge is more often unavailable.

6.1 Anomalies Are Contextual and Can Be *Global* or *Local*

Anomalies are contextual. A data item/vector can be seen as "normal", but under certain context, it may be regarded as anomalous. For example, a maximum day temperature of 5 °C is quite normal for Lancaster, UK; If it is measured in August, however, it is anomalous (low). The context is related to the time/season.

Anomaly detection is closely related to the definition of "normality", its boundaries (which may be crisp, but may also be fuzzy) as well as to the reference —do we consider all the data or part of it. Even if the method is not data-driven and *empirical*—do we consider the whole range of possible values for the data or only part of it.

A simple example from the everyday life can easily illustrate this. If we are asked to define the "normal" spending pattern, salaries, etc. while we have a good idea about these and we can also get statistical representation relatively easy, "the devil is in the detail". For example, if there are obvious anomalies (extremely high or low values) we must remove these before we calculate the means which will point towards the "normal" values. In addition to this, boundaries are not necessarily crisp/sharp. Such obvious anomalies are *global*, because they are anomalous in regard to all the data or to the whole range of possible values.

Let us imagine we want to consider separately the data we have for the whole country split per counties/districts, cities etc. Now, the "normal" values for London (for example) will obviously not be the same as for Lancaster and so on. The separation does not need to be geographical. It can be a professional split, based on age or education, etc. Obviously, the "normal" spending and salary of a football player or a medical doctor will not be the same as for a librarian, etc. The *local* (per sub-set of data; this subset will obviously form a cluster in the dimensionality of the variables of interest, e.g. salary, spending, etc.) anomalies (and respectively, *normality*) are conceptually close to *contextual* anomalies. The main difference is that the context may be provided by other factors (e.g. season or time) while the separation between *global* and *local* is entirely based on the values in the data space (money in the two illustrative examples provided). For example, for the same person, with the same occupancy, spending a certain amount or getting some salary may be considered differently in terms of whether it is anomalous for the following two cases:

(1) If the spending is made during the holiday or the salary is earned while the person is still under 25, and
(2) If the spending is made during the usual working period or if the salary is when the person is in her/his 40 s or 50 s.

These are examples of *contextual* anomalies. The separation between *global* and *local* is entirely based on the values of money and the data available or possible ranges of values.

Another simple example may concern cars and their size, engine, power, etc. If we consider cars in USA a "normal" car will be much larger than if we consider cars in the city center of Rome, for example. This is again a *contextual* anomaly. However, even only for cars in USA, we may have *global* anomalies which may include Mini Cooper, for example (too small) but also *local* anomalies in different categories/clusters of family cars, SUV, etc.

The aim of an anomaly detection algorithm is to identify the anomalous data samples from the dataset that deviate from the *normal* ones. However, in practice, this basic assumption itself is ambiguous in a wide variety of cases.

Local anomalies for a *2D* case based on the temperature and wind speed readings are depicted in Fig. 6.2. Only part of the data samples used in the previous examples, which represent the data measured during summer and winter in Manchester, UK for the period 2010–2015 (downloadable from [26]) were used in Fig. 6.2 in order to make clear the anomalies.

In this example, four *global* anomalies (circled in red) are far away from the majority of the data. Two data samples (circled in green) are *local* anomalies (in regard to the respective group/cluster, but not *globally*). Finally, one small cluster (circled in magenta), is a group of collective anomaly and can also been seen as a small-size/micro cluster.

Local anomalies and micro clusters are not obvious to detect in many cases compared with the *global* anomalies. Traditional statistical approaches very often ignore the *local* anomalies. Nearly all the available unsupervised anomaly detection algorithms today are for detecting single anomalous instances, namely point anomaly detection [23] instead of micro clusters and *global* anomalies.

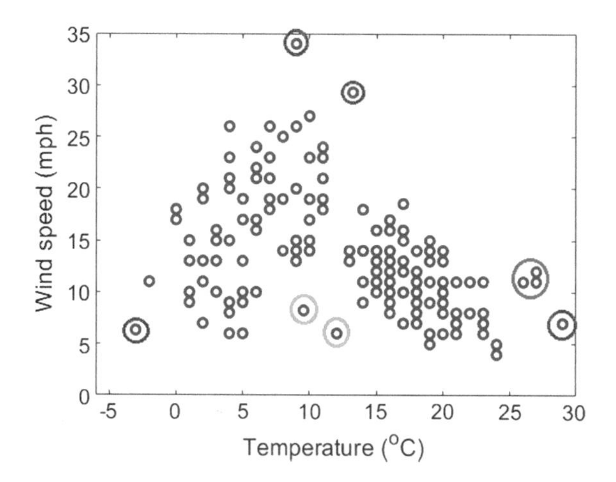

Fig. 6.2 A simple two-dimensional example of *global* (red), *local* (green) and collective (magenta) anomalies of the climate data

6.2 Anomaly Detection Based on Density Estimation

Anomaly detection is inherently linked to the *data density*, D and to the clustering. Indeed, clustering itself is a procedure that groups together data points, which are close together in the data/feature/attribute space. Therefore, clusters naturally appear in areas of the data space which have higher *data density* while, on the contrary, anomalies are data points which have lowest values of *data density* [8].

There is a more complex relation between clustering and anomaly detection which is somewhat like the "chicken and egg" dilemma, but has a more or less clear way of resolution. *Local* anomalies (as described in the previous section and depicted in Fig. 6.2) can only be declared once the clusters are being formed or declared. However, forming clusters before removing (*global*) anomalies will distort the result (cluster boundaries). So, one must detect and remove/ignore anomalies before clustering the data, but at the same time, one can only detect *local* anomalies **after** clusters are being formed. Thus, it is a kind of a "chicken and egg" dilemma.

A partial practical/pragmatic solution includes as a very first step of pre-processing to detect and remove *global* anomalies (which do not depend on the clustering). Then, once these have been removed, one can move to clustering. After that, one can detect and ignore *local* anomalies from the further data analysis, modeling and possible other tasks (classification, prediction, control, etc.).

Anomaly detection (as well as clustering) can be performed based on the *data density*. The traditional approaches go back to the so-called Parzen windows and Kernel Density Estimation [24]. This approach is quite popular but is off line and requires all the data to be available. A much more efficient method for data density estimation, called: Recursive density estimation (RDE) was initially proposed in [27, 28]. The formal name, RDE was firstly used in [28] and further developed in [8], it is also a part of two granted patents [22, 29]. The core of RDE is the *data density*, which, if use Euclidean type of distance, has the form of Cauchy kernel function as it was shown in Chap. 5.

Data density is of paramount importance in model structure design, clustering, classification, anomaly detection and other related problems [8]. RDE, in general, can use Euclidean, Mahalanobis type of distances or cosine similarity. It recursively updates the *data density* by keeping only the key statistical meta-parameters, which include the mean of data, $\boldsymbol{\mu}_K$, the average scalar product, X_K and the current time instance, K, in the memory (time instances $k = 1, 2, \ldots, K$ may also represent the ordered number of data items). It only uses the data sample of the current time instance $\boldsymbol{x}_K$ to update its meta-parameters without keeping the observations in memory, and, thus, is able to detect anomalies from streaming data in a potentially non-stationary environment.

In this section, we use the *data density* expression as described in Sect. 4.4, Eq. (4.32) using Euclidean type of distance estimated at the data sample $\boldsymbol{x}_K$:

$$D_K(\boldsymbol{x}_K) = \frac{1}{1 + \frac{\|\boldsymbol{x}_K - \boldsymbol{\mu}_K\|^2}{X_K - \|\boldsymbol{\mu}_K\|^2}} \tag{6.1}$$

and the recursive expressions of $\boldsymbol{\mu}_K$ and X_K have been given by Eqs. (4.8) and (4.9).

For each newly arrived data sample in the data stream, denoted by $\boldsymbol{x}_K$, $\boldsymbol{\mu}_{K-1}$ and X_{K-1} are firstly updated to $\boldsymbol{\mu}_K$ and X_K, and the *data density* of $\boldsymbol{x}_K$ denoted by $D_K(\boldsymbol{x}_K)$ can be calculated in real time.

Then, the mean of *data densities* at all the previous observations $\{\boldsymbol{x}_1, \boldsymbol{x}_2, \ldots, \boldsymbol{x}_{K-1}, \boldsymbol{x}_K\}$ can be updated as follows [8]:

$$\bar{D}_K \leftarrow \frac{K-1}{K}\bar{D}_{K-1} + \frac{1}{K}D_K(\boldsymbol{x}_K); \quad \bar{D}_1 \leftarrow 1 \tag{6.2}$$

and the standard deviation of the *data densities* is updated as follows:

$$\left(\sigma_K^D\right)^2 \leftarrow \frac{K-1}{K}\left(\sigma_{K-1}^D\right)^2 + \frac{1}{K}\left(\bar{D}_K - D_K(\boldsymbol{x}_K)\right)^2; \quad \left(\sigma_1^D\right)^2 \leftarrow 0 \tag{6.3}$$

One can use the Chebyshev inequality [21] to determine in real time whether $\boldsymbol{x}_{K+1}$ is an anomaly based on Condition 6.1:

Condition 6.1

$$IF\left(D_K(\boldsymbol{x}_K) < \bar{D}_K - n\sigma_K^D\right) \quad THEN(\boldsymbol{x}_K \; is \; an \; anomaly) \tag{6.4}$$

where n corresponds to the well-known "n sigma" principle.

6.3 Autonomous Anomaly Detection

Detecting anomalies traditionally involves human intervention. As a minimum, this concerns making assumptions, defining thresholds, features/attributes, etc. Statistical methods allow moving towards a higher level of autonomy and data-driven mode of operation which is in high demand for real-time and data-rich applications. However, they still require assumptions, thresholds, and features to be determined by a human.

Developing autonomous anomaly detection methods is, therefore, highly desirable. For example, the recently reported autonomous anomaly detection (AAD) method [30] is unsupervised, non-parametric and free from user- and problem-specific assumptions [31, 32]. The AAD approach, firstly, identifies the potential anomalies from the data, and, further, forms *data clouds* (cluster-like groupings of data points, which form Voronoi tessellations [33], thus, having no regular shapes such as hyper-rectangular, circular or ellipsoidal) using the autonomous data partitioning (ADP) algorithm detailed in Chap. 7. Using the data partitioning results, it identifies both the *global* and *local* anomalies from the data. The main procedure of the AAD algorithm is given in the next sub-section. The type of the distance metric used can be different, but for convenience we use the Euclidean type of distance.

6.3.1 Defining Local Areas Under Different Level of Granularity

The concept of "granularity" is introduced in [34]. Usually, a particular problem can be approached at different levels of specificity (detail) depending on the problem complexity, available computing resources, and particular needs. Generally, the higher level of granularity is chosen, the more fine details of the problem, namely, the information of *local* ensemble properties, the learning system is able to grasp. At the same time, the system consumes more computational and memory resources and overfitting may also appear. On the contrary, with low level of granularity, the learning system only obtains the coarse information from the data. Although, the system will be more computationally efficient, its performance may be influenced due to the loss of fine information [35].

Under the Gth level of granularity ($G = 1, 2, 3, \ldots$), the average radius of *local* area of influence, denoted by $R_{K,G}$ around each data sample/prototype is calculated by the following expression in an iterative manner:

$$R_{K,G} = \frac{\sum_{y,z \in \{\boldsymbol{x}\}_K; y \neq z; d^2(y,z) \leq R_{K,G-1}} d^2(\boldsymbol{y}, \boldsymbol{z})}{M_{K,G}}; \quad R_{K,0} = \bar{d}_K^2 \tag{6.5}$$

where $M_{K,G}$ is the number of the pairs of data samples between which the Euclidean square distance is smaller than $R_{K,G-1}$; $R_{K,G-1}$ is the average radius corresponding to $(G-1)$th level of granularity; $\bar{d}_K^2$ is the average distances between any two data samples within $\{\boldsymbol{x}\}_K$:

$$\bar{d}_K^2 = \frac{1}{K^2} \sum_{i=1}^{K} q_K(\boldsymbol{x}_i) \tag{6.6}$$

Compared with the traditional approaches, there are strong advantages in deriving the information of *local* ensemble properties in this way. Firstly, $R_{K,G}$ is guaranteed to be valid all the time. Defining the threshold or hard-coding mathematical principles in advance may suffer from various problems, and the performance of the two approaches is often not guaranteed. For example, in most cases, *prior* knowledge is unavailable, and the hard-coded principles are too sensitive to the nature of the data. In contrast, $R_{K,G}$ is derived from the data directly and is always meaningful. There is no need for *prior* knowledge of data sets/streams, and the level of granularity can be decided based on the requirements of the specific problems. A higher level of granularity allows the AAD approach to detect *local* anomalies based on more fine details, while, a lower level of granularity allows the approach to conduct detection based on coarse details. Thus, users are allowed to have freedom to make decisions, but, at the same time, are not overloaded. Moreover, one can always adapt the system by changing the level of granularity based on the specific needs. Some problems rely heavily on fine details, while others may need generality only.

6.3.2 Identifying Potential Anomalies

In the first stage of the AAD algorithm, the discrete multimod*al typicality* τ^M at $\{\boldsymbol{u}\}_L$ is obtained using Eq. (4.40). By extending τ^M to $\{\boldsymbol{x}\}_K$, one can obtain the multimodal *typicality* at each data sample $\boldsymbol{x}_i$ ($\boldsymbol{x}_i \in \{\boldsymbol{x}\}_K$), denoted by $\{\tau^M(\boldsymbol{x})\}_K$.

The Chebyshev inequality [21] (see Eq. 4.27) describes the probability of a particular data sample $\boldsymbol{x}_i$ ($\boldsymbol{x}_i \in \{\boldsymbol{x}\}_K$) to be n standard deviations, σ_K away from the *global* mean $\boldsymbol{\mu}_K$. For the particular and most popular case of $n = 3$, the probability of $\boldsymbol{x}_i$ to be more than $3\sigma_K$ away from $\boldsymbol{\mu}_K$ is less than $\frac{1}{9}$. In other words, on average, up to one out of nine data samples may be anomalous. Therefore, the AAD approach assumes that $\frac{1}{n^2}$ of the data samples within $\{\boldsymbol{x}\}_K$ are ***potentially*** abnormal (the worst case). Nonetheless, it does not mean that they are actual anomalies.

The AAD approach, firstly, identifies the candidates for *global* anomalies as $\left(\frac{1}{2n^2}\right)$th of the data samples within $\{\boldsymbol{x}\}_K$, which has the smallest τ^M, denoted by $\{\boldsymbol{x}\}_{PA,1}$. Here, n is a small integer corresponding to the n in the "n sigma" rule based on the Chebyshev inequality. In this book, we use $n = 3$ since the "3σ" rule is one of the most popular approaches for identifying *global* anomalies in various applications [10, 36, 37]. However, in traditional approaches, $n = 3$ does directly influence detecting each anomaly. In contrast, in the AAD approach, this is simply the first stage of sub-selection of potential *global* anomalies. We also do not assume Gaussian/normal distribution, but consider it to be arbitrary.

Then, the AAD approach identifies the potential *local* anomalies, denoted by $\{\boldsymbol{x}\}_{PA,2}$. As the next step, the AAD approach identifies the neighboring unique data samples $\{\boldsymbol{x}^*\}_i \subseteq \{\boldsymbol{x}\}_K$ around each unique data sample $\boldsymbol{u}_i$ located in the hypersphere with $\boldsymbol{u}_i$ as the center and $R_{K,G}$ as the radius by Condition 6.2:

Condition 6.2

$$IF\left(d^2(\boldsymbol{u}_i, \boldsymbol{x}_j) \leq R_{K,G}\right) \quad THEN\left(\{\boldsymbol{x}^*\}_i \leftarrow \{\boldsymbol{x}^*\}_i + \boldsymbol{x}_j\right) \tag{6.7}$$

where $j = 1, 2, \ldots, K$; $R_{K,G}$ is a data-derived distance threshold, which defines the radius of the *local* area of influence for each data sample/prototype under the Gth level of granularity, which allows the AAD approach to be more effective in detecting data samples away from the *local* peaks of the *data density* (and *typicality*, respectively). In the AAD algorithm, we used $G = 2$ (the second level of granularity).

The, the *data density* at $\boldsymbol{u}_i$ is calculated *locally* within the hypersphere, a *local* cell of Voronoi tessellation (for the Euclidean type distance) [33], around it as follows:

$$D^*_{K_i}(\boldsymbol{u}_i) = \frac{1}{1 + \frac{\|\boldsymbol{u}_i - \varsigma_i\|^2}{\chi_i - \|\varsigma_i\|^2}} \tag{6.8}$$

where ς_i and χ_i are the respective means of $\{\boldsymbol{x}^*\}_i$ and $\left\{\|\boldsymbol{x}^*\|^2\right\}_i$; $\{\boldsymbol{x}^*\}_i$ is the collection of data samples located in the *local* area around $\boldsymbol{u}_i$; K_i is the number of data samples within $\{\boldsymbol{x}^*\}_i$.

By taking both, the frequency of occurrence of the unique data samples, $\boldsymbol{u}_i$ and the data distribution of the *local* area around $\boldsymbol{u}_i$ into consideration, the *locally* weighted multimodal *typicality* at $\boldsymbol{u}_i$ is given as (also see Sect. 4.6):

$$\tau^{M^*}_K(\boldsymbol{u}_i) = \frac{K_i D^*_{K_i}(\boldsymbol{u}_i)}{\sum_{j=1}^{L} K_j D^*_{K_j}(\boldsymbol{u}_j)} \tag{6.9}$$

By expanding the *locally* weighted multimodal *typicality*, τ^{M^*}, at $\{\boldsymbol{u}\}_L$ to the original dataset $\{\boldsymbol{x}\}_K$ accordingly, the set $\left\{\tau^{M^*}(\boldsymbol{x})\right\}_K$ is obtained. The AAD approach identifies $\{\boldsymbol{x}\}_{PA,2}$ as $\frac{1}{2n^2}$ of the data samples within $\{\boldsymbol{x}\}_K$, which has the smallest τ^{M^*}.

Finally, by combining $\{\boldsymbol{x}\}_{PA,1}$ and $\{\boldsymbol{x}\}_{PA,2}$ together (no more than $\left(\frac{1}{n^2}\right)$th of the data), we obtain the whole set of potential anomalies, $\{\boldsymbol{x}\}_{PA}$ which consists of the upper limit of possible *global* and *local* anomalies according to the Chebyshev inequality.

6.3.3 Forming Data Clouds

In this stage, all the identified potential anomalies $\{\boldsymbol{x}\}_{PA}$ are checked to see whether they are able to form *data clouds* using the ADP algorithm as detailed in Chap. 7. However, we further impose a distance constraint on the final stage of the ADP algorithm after it has identified the most representative prototypes and before it starts to form *data clouds*:

Condition 6.3

$$IF\left(n^* = \underset{p\in\{p\}_{PA}}{\arg\min}(d(\boldsymbol{x}_i,\boldsymbol{p}))\right)AND(d(\boldsymbol{x}_i,\boldsymbol{p}_{n*})\leq\omega_K) \\ THEN(\mathbf{C}_{n^*}\leftarrow\mathbf{C}_{n^*}+\boldsymbol{x}_i) \tag{6.10}$$

where $\boldsymbol{x}_i \in \{\boldsymbol{x}\}_{PA}$; $\{\boldsymbol{p}\}$ is the collection of the most representative prototypes; $\{\mathbf{C}\}_{PA}$ is the *data clouds* formed around $\{\boldsymbol{p}\}_{PA}$; ω_K is the data-derived threshold calculated by Eqs. (7.8)–(7.10). Since we have not introduced the ADP algorithm itself yet, we will return to this again in Chap. 7.

If Condition 6.3 is not met, $\boldsymbol{x}_i$ is identified as an anomaly directly without being assigned to any of the *data clouds* ($\{\boldsymbol{x}\}_A \leftarrow \{\boldsymbol{x}\}_A + \boldsymbol{x}_i$), $\{\boldsymbol{x}\}_A$ denotes the collection of the identified anomalies).

After the *data clouds* are formed from $\{\boldsymbol{x}\}_{PA}$ based on the ADP algorithm, denoted by $\{\mathbf{C}\}_{PA}$, the AAD algorithm enters its last stage.

6.3.4 *Identifying Local Anomalies*

In the final stage, all the formed *data clouds* are checked to see whether they are minor ones (the collection is denoted as $\{\mathbf{C}\}_A$) by Condition 6.4:

Condition 6.4

$$IF\ (S_i < \bar{S}) \quad THEN(\{\mathbf{C}\}_A \leftarrow \{\mathbf{C}\}_A + \mathbf{C}_i) \tag{6.11}$$

where $\mathbf{C}_i \in \{\mathbf{C}\}_{PA}$ and S_i is the corresponding support; $\bar{S}$ is the average support of all the *data clouds* within $\{\mathbf{C}\}_{PA}$. Once $\{\mathbf{C}\}_A$ is identified, we have $\{\boldsymbol{x}\}_A \leftarrow \{\boldsymbol{x}\}_A + \{\mathbf{C}\}_A$.

6.4 Fault Detection

Fault detection (FD) is a specific example of anomaly detection, which forms a large area of application studies and industrial implementations on its own. A technical fault is defined according to Isermann [38] as follows:

> A fault consists of an unpermitted deviation of at least one characteristic property of variable in a system from its acceptable, usual or standard condition.

As we can see, again, like with the anomaly a critical reference is the "normality" (the usual or standard condition). In industrial processes faults are usually also unexpected and concern one or more components of the technical system [39] and can lead to serious losses (stoppages, productivity reduction or even severe accidents) [40].

The two main types of FD methods widely currently used in industry are:

(1) data-driven (or process-history-based) methods for FD, and
(2) model-based methods for FD.

The process-history-based methods are also closely related to statistical methods and signal processing. Their main advantage is that they do not require heavy human involvement and relay primarily on the *empirical* data that are observed and measured from the actual physical process [41]. This can be offline or historical data set or a live, online data stream. Another very important advantage of the data-driven methods is that they can cope with the unpredicted disturbances and possible *data drift* [41].

Model-based FD algorithms start with a model of the plant and are based on the analysis of the deviations from it or residuals. The models can be based on first principles, fuzzy rule-based (FRB) systems, artificial neural networks (ANN), statistical or a hybrid that combines some of these. The downside of such an approach is that having a good quality model is not always possible and such a model may start to deviate from the real data without any fault unless updated (adapted). However, if the update of the model is being done based on the new data that is available. Since, the new data may also contain faults, the problem becomes of the "chicken and egg" type circle. Model-based FD method can be grouped into:

(1) quantitative, and
(2) qualitative.

The quantitative model-based fault detection techniques require accurate mathematical descriptors of the process. Qualitative model-based methods, on the other hand require qualitative knowledge from the human expert.

In both traditional types of FD methods, the assumptions play a very strong role:

(1) in the signal processing type algorithms this concerns the thresholds, types of the distribution of the data (data generation model), randomness of the signal, etc.;
(2) in the model-based algorithms this concerns the model itself.

Many methods has already been developed for data-driven FD, including, but not limited to various forms of support vector machines (SVMs) [42, 43], extreme learning machines [44], time series analysis [45], statistical modelling [46–49], artificial neural networks (ANNs) [50, 51], spectral decomposition [52] methods. The rule-based systems approach [53] sits in between the data-driven and qualitative methods. An important problem that precedes the FD problem itself as well as other learning problems is the feature extraction. For FD, in particular, this may be critically important. Typical approaches for offline feature extraction used in FD context include, but are not limited to kernel principle component analysis (KPCA) or principle component analysis (PCA) [54], online feature selection [8], genetic programming [55].

In this book, we are interested in the data-driven, autonomous FD methods as well as with the link between FD and the more general anomaly detection based on the *data density* and *typicality* as they were introduced in Chap. 4.

The *data density*, *D* (and, respectively, the *typicality*, τ) as described in Chap. 4 can be used to efficiently detect faults. Moreover, the recursive density estimation (RDE) method [8, 22, 29] makes this efficiently in terms of computational power, memory and time allowing real-time FD [39, 56].

A fully autonomous FD based on Condition 6.1 (which itself is based on the *data density* and/or *typicality*) and a fully autonomous fault isolation (FI) can be achieved based on the use of clustering and/or *ALMMo-0* as it was demonstrated in [57].

6.5 Conclusions

In this chapter we describe the new, *empirical* approach to the problem of anomaly detection. It is model-, user- and problem-specific parameter-free and is entirely data driven. It is based on the *data density* and/or on the *typicality* as described earlier, in Chap. 4.

We further make a clear distinction between *global* and *local* or contextual anomalies and propose an autonomous anomaly detection (AAD) method which consists of two stages: in the first stage we detect all potential *global* anomalies and in the second stage we analyze the data pattern by forming *data clouds* and identifying possible *local* anomalies (in regards to these *data clouds*). Finally, the well-known Chebyshev inequality has been simplified by using the standardized eccentricity and a fully autonomous approach and the problem of fault detection (FD) has been outlined.

The latter approach can also be extended to a fully autonomous fault detection and isolation (FDI: detecting and identifying the types of faults) as it is elaborated in our other publications. This is a principally new approach to the very important for practical applications problem of FDI. The actual algorithms are detailed further in Chap. 10 of Part III of the book, and the pseudo-code and MATLAB implementations are given in Appendices B.1 and C.1. Further details on this new approach to anomaly detection and its application to FD as well as to other problems, e.g. human behavior analysis, computer vision, etc., can also be found in our peer reviewed journal and conference publications, e.g. [8, 10, 30, 56–60].

6.6 Questions

(1) What is the difference between the *global* and *local* anomalies?
(2) Explain the contextual anomaly and give an example.
(3) Describe the two stages of the proposed AAD method.
(4) Formulate the Chebyshev inequality through the standardized eccentricity.
(5) How FDI can be solved using the newly proposed AAD method?

6.7 Take Away Items

- Anomaly detection is described from the prism of the new *empirical* approach.
- A clear distinction between *global* and *local* anomalies is made.
- A two stages approach is proposed to autonomous anomaly detection where in the first stage all potential anomalies are detected and in the second stage the data pattern is analysed by forming *data clouds* and identifying possible *local* anomalies (in regards to these *data clouds*).
- The Chebyshev inequality is re-formulated in a much simpler form through the *standardized eccentricity*.
- This new approach is applicable to autonomous fault detection and isolation as well as to other problems.

References

1. V. Chandola, A. Banerjee, V. Kumar, Anomaly detection: a survey. ACM Comput. Surv. **41** (3), p. Article 15 (2009)
2. M. Kirlidog, C. Asuk, A fraud detection approach with data mining in health insurance. Procedia-Social Behav. Sci. **62**, 989–994 (2012)
3. E.W.T. Ngai, Y. Hu, Y.H. Wong, Y. Chen, X. Sun, The application of data mining techniques in financial fraud detection: a classification framework and an academic review of literature. Decis. Support Syst. **50**(3), 559–569 (2011)
4. E.C. Ngai, J. Liu, M.R. Lyu, On the intruder detection for sinkhole attack in wireless sensor networks, in *IEEE International Conference on Communications*, 2006, pp. 3383–3389
5. A. Shabtai, U. Kanonov, Y. Elovici, C. Glezer, Y. Weiss, 'Andromaly': a behavioral malware detection framework for android devices. J. Intell. Inf. Syst. **38**(1), 161–190 (2012)
6. R. Isermann, Model-based fault-detection and diagnosis–status and applications. Ann. Rev. Control **29**(1), 71–85 (2005)
7. L. Itti, P. Baldi, A principled approach to detecting surprising events in video, in *IEEE Computer Society Conference on Computer Vision and Pattern Recognition*, 2005, pp. 631–637
8. P. Angelov, *Autonomous Learning Systems: From Data Streams to Knowledge in Real Time* (Wiley, 2012)
9. P.P. Angelov, X. Gu, J.C. Principe, Autonomous learning multi-model systems from data streams. IEEE Trans. Fuzzy Syst. **26**(4), 2213–2224 (2016)

10. P.P. Angelov, Anomaly detection based on eccentricity analysis, in *2014 IEEE Symposium Series in Computational Intelligence, 'IEEE Symposium on Evolving and Autonomous Learning Systems, EALS, SSCI 2014*, 2014, pp. 1–8
11. A.M. Tripathi, R.D. Baruah, Anomaly detection in data streams based on graph coloring density coefficients, in *IEEE Symposium Series on Computational Intelligence (SSCI)*, 2016, pp. 1–7
12. M.M. Breunig, H.-P. Kriegel, R.T. Ng, J. Sander, LOF: identifying density-based local outliers, in *Proceedings of the 2000 ACM Sigmod International Conference on Management of Data*, 2000, pp. 1–12
13. N. Abe, B. Zadrozny, J. Langford, Outlier detection by active learning, in *IACM International Conference on Knowledge Discovery and Data Mining*, 2006, pp. 504–509
14. S.S. Sivatha Sindhu, S. Geetha, A. Kannan, Decision tree based light weight intrusion detection using a wrapper approach. Expert Syst. Appl. **39**(1), 129–141 (2012)
15. L.M. Manevitz, M. Yousef, One-class SVMs for document classification. J. Mach. Learn. Res. **2**, 139–154 (2002)
16. R. Fujimaki, T. Yairi, K. Machida, An approach to spacecraft anomaly detection problem using kernel feature space, in *The Eleventh ACM SIGKDD International Conference on Knowledge Discovery in Data Mining*, 2005, pp. 401–411
17. D. Dasgupta, F. Nino, A comparison of negative and positive selection algorithms in novel pattern detection, in *IEEE International Conference on Systems, Man, and Cybernetics*, 2000, pp. 125–130
18. V. Hautam, K. Ismo, Outlier detection using k-nearest neighbour graph, in *International Conference on Pattern Recognition*, 2004, pp. 430–433
19. H. Moonesinghe, P. Tan, Outlier detection using random walks, in *Proceedings of the 18th IEEE International Conference on Tools with Artificial Intelligence (ICTAI'06)*, 2006, pp. 532–539
20. M. Salehi, C. Leckie, J.C. Bezdek, T. Vaithianathan, X. Zhang, Fast memory efficient local outlier detection in data streams. IEEE Trans. Knowl. Data Eng. **28**(12), 3246–3260 (2016)
21. J.G. Saw, M.C.K. Yang, T.S.E.C. Mo, Chebyshev inequality with estimated mean and variance. Am. Stat. **38**(2), 130–132 (1984)
22. P. Angelov, Anomalous system state identification, US9390265 B2, 2016
23. M. Goldstein, M. Goldstein, S. Uchida, A comparative evaluation of unsupervised anomaly detection algorithms for multivariate data. PLoS One 1–31 (2016)
24. C.M. Bishop, *Pattern Recognition and Machine Learning* (Springer, New York, 2006)
25. A. Bernieri, G. Betta, C. Liguori, On-line fault detection and diagnosis obtained by implementing neural algorithms on a digital signal processor. IEEE Trans. Instrum. Measur. **45**(5), 894–899 (1996)
26. http://www.worldweatheronline.com
27. P.P. Angelov, D.P. Filev, An approach to online identification of Takagi-Sugeno fuzzy models. IEEE Trans. Syst. Man, Cybern. Part B Cybern. **34**(1), 484–498 (2004)
28. R. Ramezani, P. Angelov, X. Zhou, A fast approach to novelty detection in video streams using recursive density estimation, in *International IEEE Conference Intelligent Systems*, 2008, pp. 14-2–14-7
29. P. Angelov, Machine learning (collaborative systems), 8250004, 2006
30. X. Gu, P. Angelov, Autonomous anomaly detection, in *IEEE Conference on Evolving and Adaptive Intelligent Systems*, 2017, pp. 1–8
31. P. Angelov, X. Gu, D. Kangin, Empirical data analytics. Int. J. Intell. Syst. **32**(12), 1261–1284 (2017)
32. P.P. Angelov, X. Gu, J. Principe, A generalized methodology for data analysis. IEEE Trans. Cybern. **48**(10), 2981–2993 (2018).
33. A. Okabe, B. Boots, K. Sugihara, S.N. Chiu, *Spatial Tessellations: Concepts and Applications of Voronoi Diagrams*, 2nd edn. (Wiley, Chichester, England, 1999)
34. W. Pedrycz, *Granular Computing: Analysis and Design of Intelligent Systems* (CRC Press, 2013)

35. X. Gu, P.P. Angelov, Self-organising fuzzy logic classifier. Inf. Sci. (Ny) **447**, 36–51 (2018)
36. D.E. Denning, An intrusion-detection model. IEEE Trans. Softw. Eng. SE-13(2), 222–232 (1987)
37. C. Thomas, N. Balakrishnan, Improvement in intrusion detection with advances in sensor fusion. IEEE Trans. Inf. Forensics Secur. **4**(3), 542–551 (2009)
38. R. Isermann, Supervision, fault-detection and fault-diagnosis methods-an introduction. Control Eng. Pract. **5**(5), 639–652 (1997)
39. C.G. Bezerra, B.S.J. Costa, L.A. Guedes, P.P. Angelov, An evolving approach to unsupervised and real-time fault detection in industrial processes. Expert Syst. Appl. **63**, 134–144 (2016)
40. V. Venkatasubramanian, R. Rengaswamy, S.N. Kavuri, K. Yin, A review of process fault detection and diagnosis: part III: process history based methods. Comput. Chem. Eng. **27**(3), 327–346 (2003)
41. R.E. Precup, P. Angelov, B.S.J. Costa, M. Sayed-Mouchaweh, An overview on fault diagnosis and nature-inspired optimal control of industrial process applications. Comput. Ind. **74**, 1–16 (2015)
42. S. Mahadevan, S.L. Shah, Fault detection and diagnosis in process data using one-class support vector machines. J. Process Control **19**(10), 1627–1639 (2009)
43. P. Konar, P. Chattopadhyay, Bearing fault detection of induction motor using wavelet and support vector machines (SVMs). Appl. Soft Comput. **11**(6), 4203–4211 (2011)
44. P.K. Wong, Z. Yang, C.M. Vong, J. Zhong, Real-time fault diagnosis for gas turbine generator systems using extreme learning machine. Neurocomputing **128**, 249–257 (2014)
45. J. Hu, K. Dong, Detection and repair faults of sensors in sampled control system, in *IEEE International Conference on Fuzzy Systems and Knowledge Discovery*, 2015, pp. 2343–2347
46. H. Ma, Y. Hu, H. Shi, Fault detection and identification based on the neighborhood standardized local outlier factor method. Ind. Eng. Chem. Res. **52**(6), 2389–2402 (2013)
47. Z. Yan, C.Y. Chen, Y. Yao, C.C. Huang, Robust multivariate statistical process monitoring via stable principal component pursuit. Ind. Eng. Chem. Res. **55**(14), 4011–4021 (2016)
48. V. Chandola, A. Banerjee, V. Kumar, Outlier detection: a survey. ACM Comput. Surv (2017)
49. V. Hodge, J. Austin, A survey of outlier detection methodologies. Artif. Intell. Rev. **22**(2), 85–126 (2004)
50. S.P. King, D.M. King, K. Astley, L. Tarassenko, P. Hayton, S. Utete, The use of novelty detection techniques for monitoring high-integrity plant, in *IEEE International Conference on Control Applications*, 2002, pp. 221–226
51. Y. Li, M.J. Pont, N.B. Jones, Improving the performance of radial basis function classifiers in condition monitoring and fault diagnosis applications where 'unknown' faults may occur. Pattern Recognit. Lett. **23**, 569–577 (2002)
52. R. Fujimaki, T. Yairi, K. Machida, An approach to spacecraft anomaly detection problem using kernel feature space, in *ACM SIGKDD International Conference on Knowledge Discovery in Data Mining*, 2005, pp. 401–410
53. S. Ramezani, A. Memariani, A fuzzy rule based system for fault diagnosis, using oil analysis results. Int. J. Ind. Eng. Prod. Res. **22**, 91–98 (2011)
54. S.W. Choi, C. Lee, J.M. Lee, J.H. Park, I.B. Lee, Fault detection and identification of nonlinear processes based on kernel PCA. Chemom. Intell. Lab. Syst. **75**(1), 55–67 (2005)
55. G. Smits, A. Kordon, K. Vladislavleva, E. Jordaan, M. Kotanchek, Variable selection in industrial datasets using pareto genetic programming, in *Genetic Programming Theory and Practice III* (Springer, Boston, MA, 2006), pp. 79–92
56. B.S.J. Costa, P.P. Angelov, L.A. Guedes, Fully unsupervised fault detection and identification based on recursive density estimation and self-evolving cloud-based classifier. Neurocomputing **150**(Part A), 289–303 (2015)
57. C.G. Bezerra, B.S.J. Costa, L.A. Guedes, P.P. Angelov, An evolving approach to unsupervised and real-time fault detection in industrial processes. Expert Syst. Appl. **63**, 134–144 (2016)

58. C.G. Bezerra, B.S.J. Costa, L.A. Guedes, P.P. Angelov, A comparative study of autonomous learning outlier detection methods applied to fault detection, in *IEEE International Conference on Fuzzy Systems*, 2015, pp. 1–7
59. B.S.J. Costa, P.P. Angelov, L.A. Guedes, Real-time fault detection using recursive density estimation. J. Control. Autom. Electr. Syst. **25**(4), 428–437 (2014)
60. B. Sielly, C.G. Bezerrat, L.A. Guedes, P.P. Angelov, C. Natal, Z. Norte, Online fault detection based on typicality and eccentricity data analytics, in *International Joint Conference on Neural Networks*, 2015, pp. 1–6

Chapter 7
Data Partitioning—*Empirical* Approach

7.1 *Global* Versus *Local*

In probability theory and statistics, the Gaussian function is the most widely used model of the probability distribution. It provides a closed analytical form of a smooth and conveniently differentiable function with solid mathematical basis and is widely used in the natural and social sciences to represent real-valued random variables. A unimodal Gaussian distribution has two parameters (its *global* mean and the standard derivation). For many practical problems such function can be an acceptable approximation of the true model of data generation due to the Central Limit Theorem (CLT) [1].

For example, if we consider the data about the height and weight of people and if there is a huge number of people/observations then the distribution will tend towards a unimodal Gaussian. However, for many problems (e.g. temperature, rainfall, spending data or human behavior) the distribution may be multi-model (with multiple peaks/*local* maxima) and may significantly differ from the normal/unimodal Gaussian.

For example, Fig. 7.1 depicts the probability mass function (PMF), unimodal and multimodal probability density functions (PDF) of the temperature readings measured during summer and winter in Manchester, UK for the period 2010–2015 (downloadable from [2]) where the multimodal PDF is calculated based on the ground truth.

In reality, the actual data distribution is seldom unimodal. It is much more frequently multimodal [3–6]. The *global* mean of the data cannot present the whole distribution fully [7]. Therefore, in practice, the data is being analyzed separately as part of the pre-processing or it is simply assumed or provided as a subjective input, how many *local* areas of the data pattern there are or may be in the data. For streaming data this is, naturally, extremely difficult or even impossible. This number of *local* data pattern sub-areas can come as a result of clustering [8],

P. P. Angelov and X. Gu, *Empirical Approach to Machine Learning*, Studies in Computational Intelligence 800,
https://doi.org/10.1007/978-3-030-02384-3_7

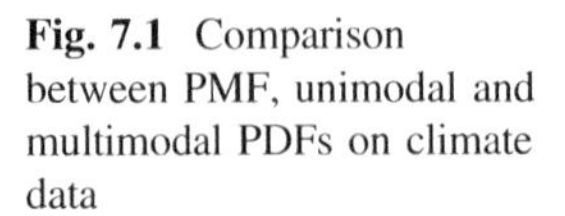

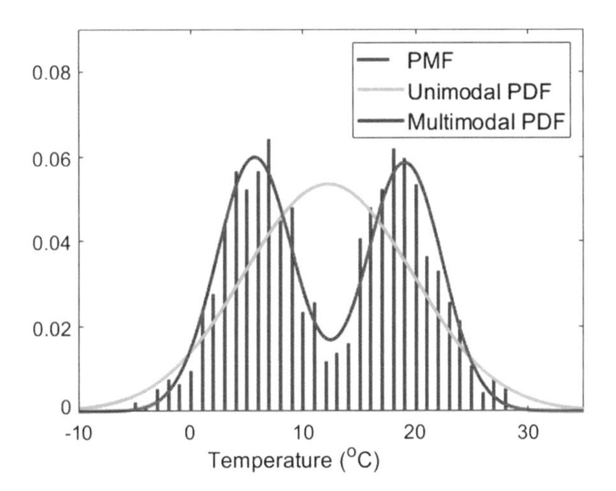

Fig. 7.1 Comparison between PMF, unimodal and multimodal PDFs on climate data

expectation maximization (EM) search algorithms [9–11] or, simply, be provided by a human. Then, one can use so-called mixture Gaussians [12, 13].

A simple *2D* example to visualize the *local* and *global* means and the fact that a unimodal distribution is not likely to be adequate is the well-known benchmark dataset named Aggregation [14] (downloadable from [15]), depicted in Fig. 7.2. There are clearly visible seven *local* groups of data (data samples are represented by the blue dots "·"; the *local* means are represented by larger red dots "·"). The *global* mean (represented by the black asterisk "*") is located in the middle of the data space. The *global* mean is not an actual data point that exists, but an abstraction and is some distance away from any of the actual data points, see Fig. 7.2. In addition, one can see that the clusters' actual shape is not exactly circular or ellipsoidal (even if this is a synthetic dataset).

The problem is that while we as humans can easily see the *2D* data pattern in Fig. 4.2 in real problems with more dimensions and, often, an online mode of

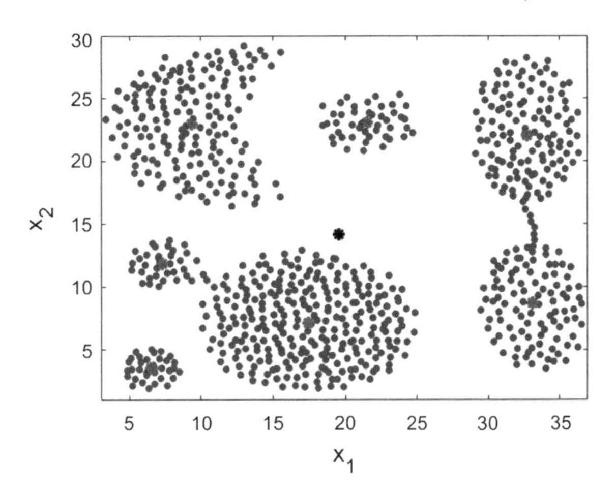

Fig. 7.2 The *global* mean (black asterisk) versus the *local* means (red asterisks)

operation, this is a very difficult problem. Even in offline mode this is a separate problem on its own which is usually done as a part of the pre-processing or in some cases is based on subjective domain expert knowledge. With the multi-modal *typicality* and *data density* (see Sect. 4.6), respectively the peaks are identified in an autonomous manner.

7.2 The Point with the Highest *Data Density/Typicality*

Data density, as described in Sect. 4.4, is a central paradigm in pattern recognition serving as the indictor of the main mode [16–19]. Furthermore, it was proven that the *data density* at the point of the *global* mean (see the black asterisks from Fig. 7.2) is maximum, and equal to 1 [18]. Indeed, with the Euclidean distance, the unimodal *data density* (in both, discrete and continuous forms) is defined through the *global* mean as given in Eqs. (4.32) and (4.45). For the discrete case, the unimodal *data density* is formulated as:

$$D_K(\boldsymbol{x}_i) = \frac{1}{1 + \frac{\|\boldsymbol{x}_i - \boldsymbol{\mu}_K\|^2}{\sigma_K^2}}; \quad \boldsymbol{x}_i \in \{\boldsymbol{x}\}_K \tag{7.1a}$$

and for the continuous case, the unimodal *data density* it is expressed as:

$$D_K^C(\boldsymbol{x}) = \frac{1}{1 + \frac{\|\boldsymbol{x} - \boldsymbol{\mu}_K\|^2}{\sigma_K^2}}; \quad \boldsymbol{x} \in \mathbf{R}^N \tag{7.1b}$$

One may notice from Eqs. (7.1a) and (7.1b) that, $D_K^C(\boldsymbol{\mu}_K) = 1$, while $D_K(\boldsymbol{x}_m) = \max_{\boldsymbol{x}_i \in \{\boldsymbol{x}\}_K} (D_K(\boldsymbol{x}_i))$ may not be exactly equal to 1 unless $\boldsymbol{x}_m = \boldsymbol{\mu}_K$. An example using the Aggregation dataset is depicted in Fig. 7.3.

The unimodal *typicality* given by Eqs. (4.33) and (4.43) (in the discrete and continuous form, respectively) is directly dependent on the unimodal *data density* (in its numerator; divided by the sum/integral which can be seen as a normalizing constant). For the discrete case,

$$\tau_K(\boldsymbol{x}_i) = \frac{D_K(\boldsymbol{x}_i)}{\sum_{j=1}^{K} D_K(\boldsymbol{x}_j)}; \quad \boldsymbol{x}_i \in \{\boldsymbol{x}\}_K \tag{7.2a}$$

and, for the continuous case, the unimodal *typicality* is defined as:

$$\tau_K^C(\boldsymbol{x}) = \frac{D_K^C(\boldsymbol{x})}{\int_{\boldsymbol{x}} D_K^C(\boldsymbol{x}) dx}; \quad \boldsymbol{x} \in \mathbf{R}^N \tag{7.2b}$$

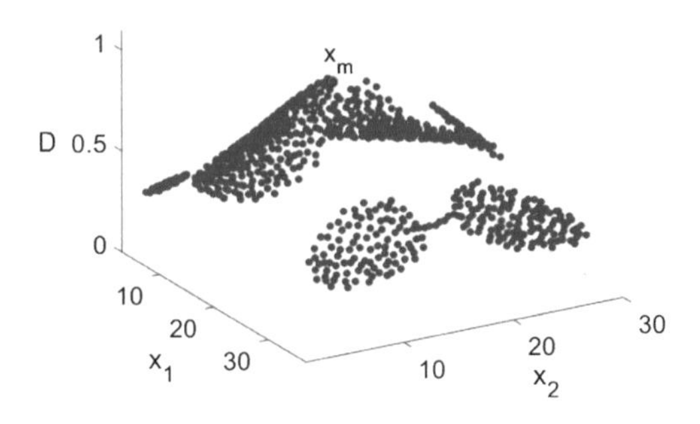

(a) Discrete (the red dot "·" represents $\boldsymbol{x}_m$, where $D_K(\boldsymbol{x}_m) = \max_{\boldsymbol{x}_i \in \{\boldsymbol{x}\}_K} (D_K(\boldsymbol{x}_i)))$

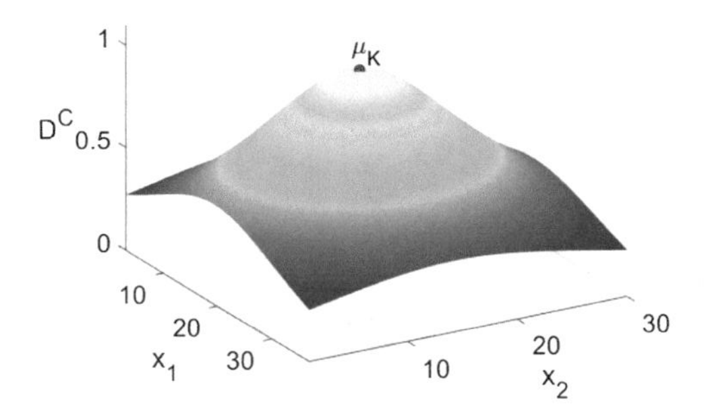

(b) Continuous (the red dot "·" represents the *global* mean, μ_K, where $D_K^C(\mu_K) = 1$)

Fig. 7.3 *Data density*

Similarly, $\tau_K(\boldsymbol{x}_m) = \max_{\boldsymbol{x}_i \in \{\boldsymbol{x}\}_K} (\tau_K(\boldsymbol{x}_i))$ and $\tau_K^C(\boldsymbol{\mu}_K) = \max_{\boldsymbol{x} \in \mathbf{R}^N} \left(\tau_K^C(\boldsymbol{x})\right)$. The value of τ and τ^C is, however $\ll 1$ because the sum (for the discrete case) and the integral (for the continuous case) are equal to 1. An example using the Aggregation dataset is depicted in Fig. 7.4.

Alternatively, by using the multimodal *data density*, one can identify ***all*** *local* modes of the data pattern without any pre-processing techniques. However, for machine learning, some generalization is necessary and beneficial to avoid over-fitting and to get simpler models. Therefore, a kind of "filtering" of most descriptive *local* peaks is preferable.

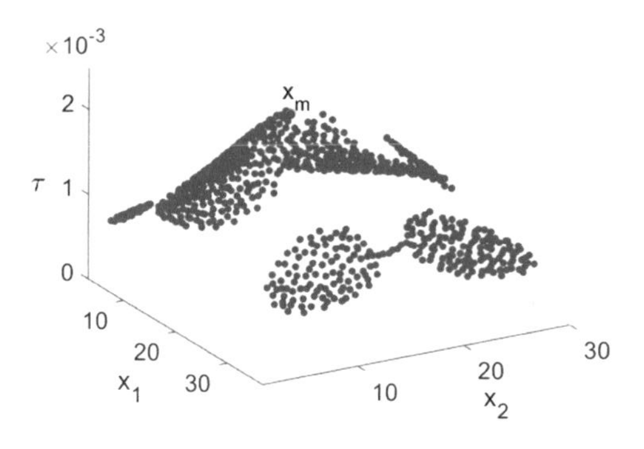

(a) Discrete (the red dot "·" represents $\boldsymbol{x}_m$, where $\tau_K(\boldsymbol{x}_m) = \max_{\boldsymbol{x}_i \in \{\boldsymbol{x}\}_K}(\tau_K(\boldsymbol{x}_i)))$

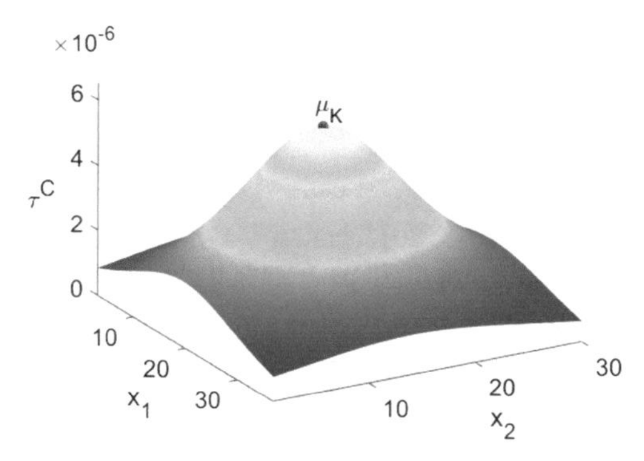

(b) Continuous (the red dot "·" represents the *global* mean, μ_K, where $\tau_K^C(\mu_K) = \max_{\boldsymbol{x} \in \mathbf{R}^N}(\tau_K^C(\boldsymbol{x})))$

Fig. 7.4 *Typicality*

7.3 Data Clouds

The concept of "*data clouds*" was first introduced in [20] and is closely linked with the *AnYa* type FRB systems. *Data clouds* [20] can be seen as a special type of clusters but with some distinctive differences. *Data clouds* are non-parametric, free from external constrains and their shape is not predefined or predetermined by the type of the distance metric used (e.g. in traditional clustering, the shape of clusters derived with the Euclidean distance is always hyper-spherical; clusters formed using Mahalanobis distance are always hyper-ellipsoidal, etc.). Unlike the conventionally defined clusters, *data clouds* directly represent the *local* ensemble properties of the observed data samples. They consist of data samples affiliated with the nearest prototypes forming Voronoi tessellation [21]. There is no need to pre-define any parameters for the *data clouds*; on the contrary, once they are formed around the prototypes, one can extract parameters from the data that form them a posteriori.

Data clouds are used for building the antecedent part of the *AnYa* type as well as for the *empirical* FRB systems. In contrast, traditional membership functions and fuzzy sets used in fuzzy set theory [22–24] often do not represent the real data distributions and, instead, represent some desirable/expected/estimated or subjective preferences [25].

7.4 Autonomous Data Partitioning Method

Autonomous data partitioning (ADP) algorithm which we already referred to earlier will now be described in more detail. It is a recently introduced [26] method. It is fully autonomous *local* mode identification approach for data partitioning. The ADP approach employs the parameter-free *empirical* operators (described in Chap. 4) to disclose the underlying data distribution and ensemble properties of the *empirically* observed data. Based on these operators the ADP method identifies the *local* modes representing the *local* maxima of the *data density* and further partitions the data space into *data clouds* [8, 20] forming Voronoi tessellation [21]. In contrast with the state-of-the-art clustering/data partitioning approaches, the ADP algorithm has the following advantages [26]:

(1) It does not require any user input (*prior* assumptions as well as predefined problem- and user-specific parameters);
(2) It does not impose a model of data generation;
(3) It partitions the data space objectively into non-parametric, shape-free *data clouds*.

ADP algorithm has two versions:

(1) offline, and
(2) evolving.

7.4.1 *Offline ADP Algorithm*

The offline ADP algorithm works with the discrete multimodal *typicality* τ^M (the normalized multimodal *data density*) of the *empirically* observed data samples $\{\boldsymbol{x}\}_K$, and is based on the **ranks** of the observations in terms of their multimodal *typicality* values and *local* ensemble properties. **Ranking** is a nonlinear and discrete operator, and thus, other approaches avoid using it. However, it can provide a different, but very important information about the data distribution, which smooth continuous functions ignore.

The offline version of the ADP is more stable and effective in partitioning static datasets. The main procedure of the offline ADP algorithm consists of three stages as follows.

7.4.1.1 Stage 1: Rank Ordering the Data and Prototype Identification

In this stage, the ADP approach ranks all the unique data samples $\{\boldsymbol{u}\}_L$ in an indexing list, denoted by $\{z\}_L$, based on the values of the multimodal *typicality*. Then, it identifies the prototypes based on the ranking.

Firstly, the multimodal *typicality* [see Eq. (4.40)], $\tau_K^M(\boldsymbol{u}_i) = \frac{D_K^M(\boldsymbol{u}_i)}{\sum_{j=1}^{L} D_K^M(\boldsymbol{u}_j)}$ ($i = 1, 2, \ldots, L$, $\boldsymbol{u}_i \in \{\boldsymbol{u}\}_L$) at all the unique data samples $\{\boldsymbol{u}\}_L$ are calculated, and the unique data sample $\boldsymbol{u}_{j^*}$ with the highest multimodal *typicality* is selected as the first element of $\{z\}_L$:

$$z_1 \leftarrow \boldsymbol{u}_{m^*}; \quad m^* = \underset{\boldsymbol{u}_i \in \{\boldsymbol{u}\}_L}{\arg\max}\left(\tau_K^M(\boldsymbol{u}_i)\right) \tag{7.3}$$

where z_1 is the first element of $\{z\}_L$.

After, z_1 is identified, it is set as the reference sample $\boldsymbol{r} \leftarrow z_1$ and z_1 is removed from $\{\boldsymbol{u}\}_L$.

Then, the ADP algorithm finds out the unique data sample nearest to $\boldsymbol{r}$ as z_{j+1} ($j \leftarrow 1$) from the remaining $\{\boldsymbol{u}\}_L$:

$$z_{j+1} \leftarrow \boldsymbol{u}_{n^*}; \quad n^* = \underset{\boldsymbol{u}_i \in \{\boldsymbol{u}\}_L}{\arg\min}(d(\boldsymbol{u}_i, \boldsymbol{r})) \tag{7.4}$$

and, similarly, z_{j+1} is removed from $\{\boldsymbol{u}\}_L$ and is set as the new reference $\boldsymbol{r} \leftarrow z_{j+1}$.

By repeating the above process ($j \leftarrow j+1$) until $\{\boldsymbol{u}\}_L = \emptyset$, the ranked unique data samples $\{z\}_L$ and their corresponding ranked multimodal *typicality* values, denoted by $\{\tau^M(z)\}_L$ are obtained.

At the end of this stage, all the unique data samples corresponding to the *local* maxima of $\{\tau^M(z)\}_L$ are identified based on Condition 7.1, and they are used as the prototypes for creating Voronoi tessellations in the next stage:

Condition 7.1

$$\begin{aligned} &IF\ \left(\tau_K^M(z_i) > \tau_K^M(z_{i-1})\right)\ AND\ \left(\tau_K^M(z_i) > \tau_K^M(z_{i+1})\right) \\ &THEN\ (z_i\ \textit{is one of the local maxima}) \end{aligned} \tag{7.5}$$

The collection of all the *local* maxima is denoted by $\{z^*\}$.

For the sake of an illustration, let us consider the same real climate data measured in Manchester, UK during the period of 2010–2015 as used in the previous

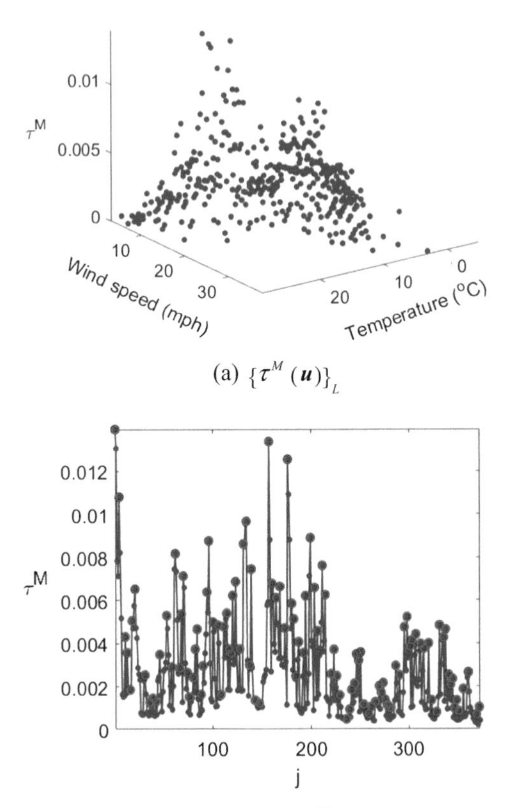

(a) $\{\tau^M(\boldsymbol{u})\}_L$

(b) Identified *local* maxima of $\{\tau^M(z)\}_L$ (red circles "o")

Fig. 7.5 The multimodal *typicality* and the *local* maxima

chapters (downloadable from [2]). The values of multimodal *typicality* of $\{\boldsymbol{x}\}_K$ and the identified *local* maxima of $\{\tau^M(z)\}_L$ are depicted in Fig. 7.5. The prototypes within the data space are depicted in Fig. 7.6.

7.4.1.2 Stage 2: Creating the Voronoi Tessellation

Once the *local* maxima, $\{z^*\}$ are identified, they are used as the focal points/attractors of the *data clouds* representing the *local* modes of the data pattern. All the data samples, $\{\boldsymbol{x}\}_K$ are then assigned to the nearest focal point using the "nearest neighbor" principle, which naturally creates a Voronoi tessellation [21] and forms *data clouds* (the collection of *data clouds* are denoted by $\{\mathbf{C}\}$):

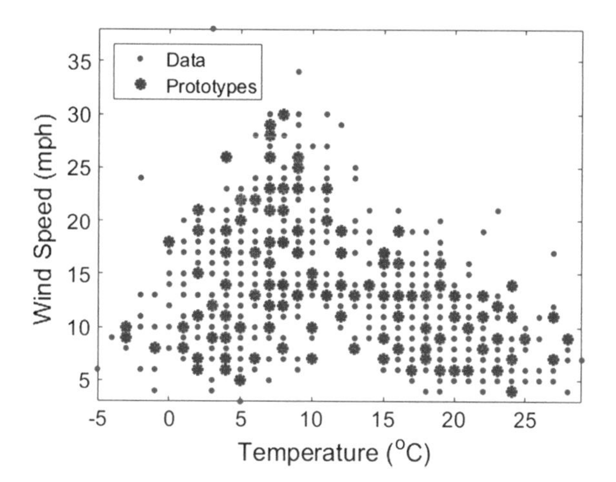

Fig. 7.6 The identified prototypes within the data space

$$\mathbf{C}_{n^*} \leftarrow \mathbf{C}_{n^*} + \boldsymbol{x}_j; \quad n^* = \underset{z_i^* \in \{z^*\}}{\arg\min}\left(d\left(\boldsymbol{x}_j, z_i^*\right)\right) \tag{7.6}$$

After all the *data clouds* are identified (assuming there are *M data clouds* in the data space), one can obtain their corresponding centers as $\boldsymbol{c}_i$ and supports as S_i $(i = 1, 2, \ldots, M)$.

7.4.1.3 Stage 3: Filtering *Local* Modes

As the prototypes identified in the previous stage may contain some less representative ones, in this stage, the initial Voronoi tessellations are filtered to obtain larger and more descriptive *data clouds*. According to the granularity theory [27], this corresponds to moving to a higher/finer level of granularity (see also the previous chapter).

Firstly, the multimodal *typicality* at the *data cloud* centers $\{\boldsymbol{c}\}$ is calculated weighting through the support of the respective *data cloud* ($\boldsymbol{c}_i \in \{\boldsymbol{c}\}$):

$$\tau_K^M(\boldsymbol{c}_i) = \frac{S_i D_K(\boldsymbol{c}_i)}{\sum_{j=1}^{M} S_j D_K(\boldsymbol{c}_j)} \tag{7.7}$$

In order to identify the *local* maxima of multimodal *typicality*, three objectively derived quantifiers of the data pattern are introduced as follows.

The average distance between any pair of the existing *data cloud* centers, namely *local* modes of the *data density*, denoted by γ_K:

$$\gamma_K = \frac{1}{M(M-1)} \sum_{j=1}^{M-1} \sum_{l=j+1}^{M} d(\boldsymbol{c}_j, \boldsymbol{c}_l) \tag{7.8}$$

The average distance between any pair of existing centers, ϖ_K with a distance smaller than γ_K (M_γ is the number of such pairs):

$$\varpi_K = \frac{1}{M_\gamma} \sum_{\substack{\boldsymbol{y},\boldsymbol{z} \in \{\boldsymbol{c}\},\, \boldsymbol{y} \neq \boldsymbol{z} \\ d(\boldsymbol{y},\boldsymbol{z}) \leq \gamma_K}} d(\boldsymbol{y}, \boldsymbol{z}) \tag{7.9}$$

The average distance between any pair of existing *data cloud* centers, ω_K with a distance smaller than ϖ_K (M_ϖ is the number of such pairs):

$$\omega_K = \frac{1}{M_\varpi} \sum_{\substack{\boldsymbol{y},\boldsymbol{z} \in \{\boldsymbol{c}\},\, \boldsymbol{y} \neq \boldsymbol{z} \\ d(\boldsymbol{y},\boldsymbol{z}) \leq \varpi_K}} d(\boldsymbol{y}, \boldsymbol{z}) \tag{7.10}$$

Note that, γ_K, ϖ_K and ω_K are not problem-specific, but are data-derived and parameter-free. The quantifier ω_K can be viewed as an estimation of the distances between the strongly connected *data clouds* representing the *local* ensemble properties of the whole data set. Moreover, instead of relying on a fixed threshold, γ_K, ϖ_K and ω_K derived from the dataset objectively are meaningful all the time. The relationship between γ_K, ϖ_K and ω_K is depicted in Fig. 7.7 using the same climate dataset.

Each center $\boldsymbol{c}_i \in \{\boldsymbol{c}\}$ is compared with the centers, denoted by $\{\boldsymbol{c}^*\}_i$, of the neighboring *data clouds* $\{\mathbf{C}^*\}_i$ ($i = 1, 2, \ldots, M$) in terms of their multimodal *typicality* for *local* maxima identification by the following Condition:

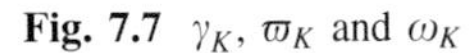

Fig. 7.7 γ_K, ϖ_K and ω_K

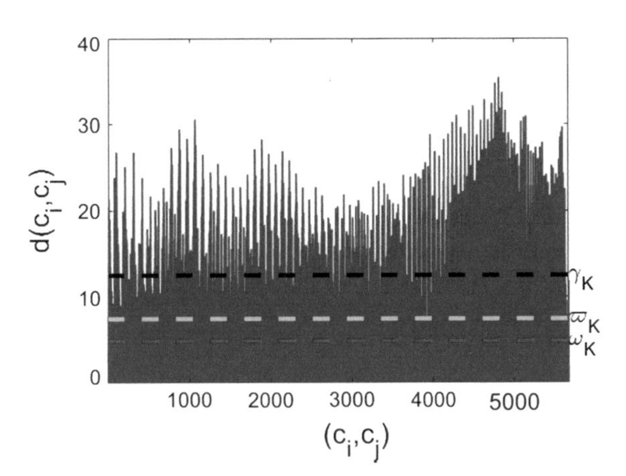

Condition 7.2

$$\begin{array}{c} IF\left(\tau_K^M(\boldsymbol{c}_i)=\max\left(\tau_K^M(\boldsymbol{c}_i),\{\tau^M(\boldsymbol{c}^*)\}_i\right)\right) \\ THEN\,(\boldsymbol{c}_i\,is\,one\,of\,the\,local\,maxima) \end{array} \tag{7.11}$$

where $\{\tau^M(\boldsymbol{c}^*)\}_i$ is the collection of multimodal *typicality* calculated at the centers of neighbouring data clouds $\{\mathbf{C}^*\}_i$, which is identified by Condition 7.3 ($j = 1,2,\ldots,M$ and $i \neq j$):

Condition 7.3

$$IF\left(d\left(\boldsymbol{c}_i,\boldsymbol{c}_j\right)\leq\frac{\omega_K}{2}\right)\quad THEN\left(\mathbf{C}_j\in\{\mathbf{C}^*\}_i\right) \tag{7.12}$$

The criterion of the neighboring range [Eq. (7.12)] is defined in this way because two centers with the distance smaller than γ_K can be considered to be potentially relevant; λ_K is the average distance between the *data cloud* centers of any two potentially relevant *data clouds*. Therefore, if $d\left(\boldsymbol{c}_i,\boldsymbol{c}_j\right)\leq\frac{\omega_K}{2}$, $\boldsymbol{c}_i$ and $\boldsymbol{c}_j$ are strongly influencing each other and, the data samples within the two corresponding *data clouds* are highly related. Therefore, the two *data clouds* are considered as neighbors in the sense of spatial distance. This criterion also guarantees that only small (less important) *data clouds* that significantly overlap with large (more important) ones will be removed during the filtering operation thanks to the multiplicative weights imposed on the multimodal *typicality* by the corresponding supports of $\{\mathbf{C}\}$.

After the filtering operation, the *data cloud* centers with *local* maximum values of multimodal *typicality*, denoted by $\{\boldsymbol{c}^{**}\}$, are obtained. Then, $\{\boldsymbol{c}^{**}\}$ are used as prototypes ($\{z^*\}\leftarrow\{\boldsymbol{c}^{**}\}$) for creating Voronoi tessellations as described in stage 2 and are filtered in stage 3 again.

Aligned with the granularity theory [27], each time the filtering operation is repeated, the partitioning achieves a higher/finer level of specificity (detail).

Stages 2 and 3 are repeated until all the distances between the centers of existing *data clouds* exceed $\frac{\omega_K}{2}$. Finally, we obtain the remaining centers with the *local* maxima of τ^M re-denoted by $\{\boldsymbol{p}\}$ and use them as the focal points to form *data clouds* by creating Voronoi tessellations, Eq. (7.6). After the *data clouds* are formed, the corresponding centers, standard deviations, supports, members and other parameters of the formed *data clouds* can be extracted *post factum*. The identified prototypes representing the *local* maxima of the multimodal *typicality* after each filtering round are depicted in Fig. 7.8. The finally formed *data clouds* are depicted in Fig. 7.9.

Fig. 7.8 The prototypes after each filtering round

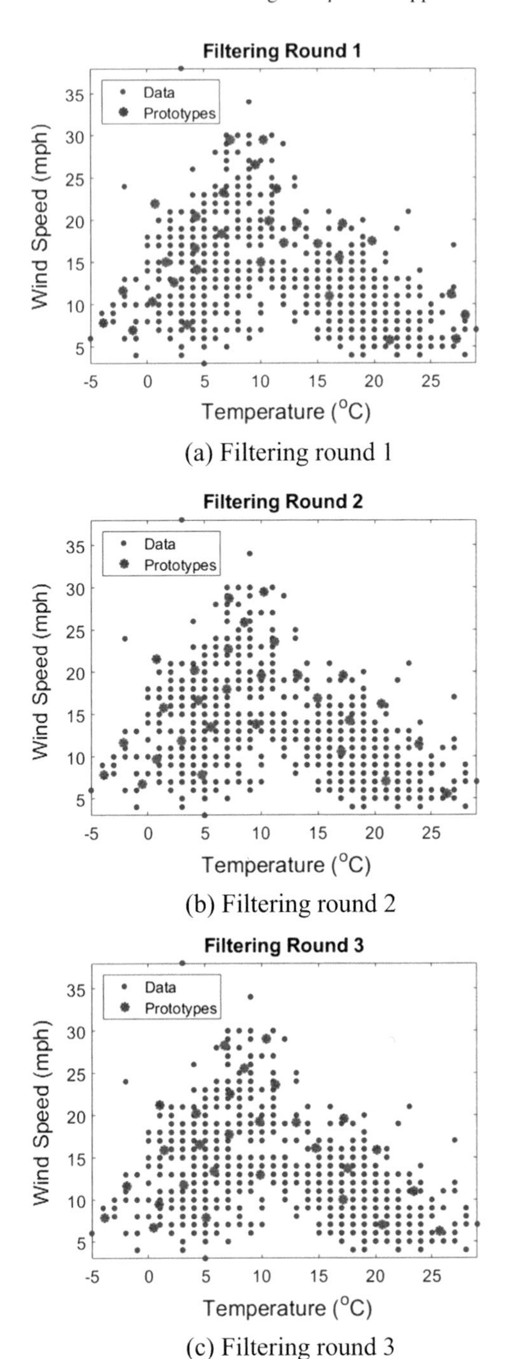

(a) Filtering round 1

(b) Filtering round 2

(c) Filtering round 3

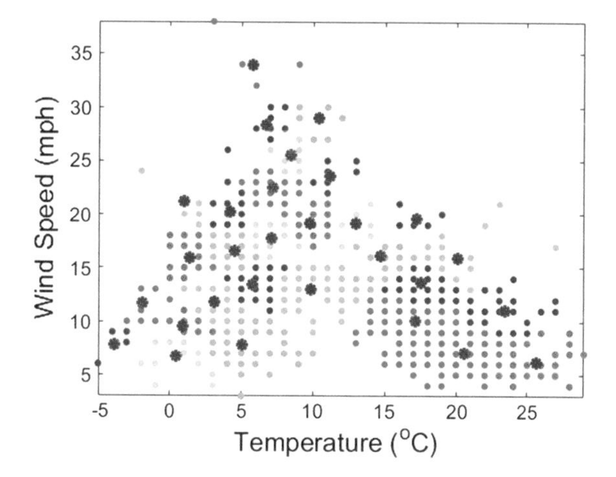

Fig. 7.9 The *data clouds* formed around the identified prototypes (red asterisks "*")

7.4.2 Evolving ADP Algorithm

The evolving ADP algorithm is for streaming data processing and it works with the *data density*, *D*. This algorithm is able to start "from scratch", and a hybrid between the evolving and the offline versions can also be conducted, whereby the offline ADP is applied initially to a relatively small part of the data followed by evolving ADP over the remaining data.

The main procedure of the evolving algorithm consists of three stages and is described as follows. Since the evolving version is using the recursive calculation expression of the *data density*, we present the algorithm with the Euclidean distance for simpler derivation; however, alterative types of distance/dissimilarity measure can be used as well.

7.4.2.1 Stage 1: Initialization

This stage is only needed if the algorithm starts "from scratch".

The firstly observed data sample of the data stream, $\boldsymbol{x}_1$ is selected as the first prototype. The *global* meta-parameters of the ADP algorithm are initialized by $\boldsymbol{x}_1$ as follows.

$$K \leftarrow 1 \tag{7.13a}$$

$$M \leftarrow 1 \tag{7.13b}$$

$$\boldsymbol{\mu}_K \leftarrow \boldsymbol{x}_1 \tag{7.13c}$$

$$X_K \leftarrow \|\boldsymbol{x}_1\|^2 \tag{7.13d}$$

and the meta-parameters of the first *data cloud*, denoted by $\mathbf{C}_1$, are initialized as follows:

$$\mathbf{C}_1 \leftarrow \{\boldsymbol{x}_1\} \tag{7.14a}$$

$$\boldsymbol{c}_{K,1} \leftarrow \boldsymbol{x}_1 \tag{7.14b}$$

$$S_{K,1} \leftarrow 1 \tag{7.14c}$$

After the initialization of the system structure and meta-parameters (both, *global* and *local*), the ADP algorithm starts to self-evolve its structure and update the parameters based on the arriving data samples.

7.4.2.2 Stage 2: System Structure and Meta-parameters Update

For each newly arriving data sample ($K \leftarrow K+1$), denoted as $\boldsymbol{x}_K$, the *global* meta-parameters $\boldsymbol{\mu}_K$ and X_K are updated firstly using Eqs. (4.8) and (4.9). Then, the *data densities* at $\boldsymbol{x}_K$ and the centers of all the existing *data clouds*, namely, $D_K(\boldsymbol{x}_K)$ and $D_K(\boldsymbol{c}_{K-1,i})$ ($i = 1, 2, \ldots, M$) are calculated.

Condition 7.4 is checked to decide whether $\boldsymbol{x}_K$ is able to become a new prototype and forms a *data cloud* around itself (this condition was first proposed in [28–31], in which D was defined differently):

Condition 7.4

$$\begin{gathered} IF \left(D_K(\boldsymbol{x}_K) > \max_{i=1,2,\ldots,M} \left(D_K(\boldsymbol{c}_{K-1,i}) \right) \right) \\ OR \left(D_K(\boldsymbol{x}_K) < \min_{i=1,2,\ldots,M} \left(D_K(\boldsymbol{c}_{K-1,i}) \right) \right) \\ THEN \, (\boldsymbol{x}_K \; is \; a \; new \; prototype) \end{gathered} \tag{7.15}$$

If the above condition is met, a new *data cloud* is added with $\boldsymbol{x}_K$ as the prototype:

$$M \leftarrow M+1 \tag{7.16a}$$

$$\mathbf{C}_M \leftarrow \{\boldsymbol{x}_K\} \tag{7.16b}$$

$$\boldsymbol{c}_{K,M} \leftarrow \boldsymbol{x}_K \tag{7.16c}$$

$$S_{K,M} \leftarrow 1 \tag{7.16d}$$

Otherwise, the center of the nearest *data cloud* $\mathbf{C}_{n^*}$ to $\boldsymbol{x}_K$ is found, denoted as $\boldsymbol{c}_{K-1,n^*}$, and Condition 7.5 is checked to see whether $\boldsymbol{x}_K$ is associated with the nearest *data cloud* $\mathbf{C}_{n^*}$:

Condition 7.5

$$IF\ \left(d\left(\boldsymbol{x}_K,\boldsymbol{c}_{K-1,n^*}\right)\leq \tfrac{\gamma_K}{2}\right)\quad THEN\ (\boldsymbol{x}_K\ is\ assigned\ to\ \mathbf{C}_{n^*}) \tag{7.17}$$

As the ADP algorithm is of "one pass" type, it is less memory- and computation-efficient to calculate γ_K at each time a new data sample arrives. Instead, since the average distance between all the data samples $\sqrt{\bar{d}_K}$ is approximately equal to γ_K ($\gamma_K \approx \sqrt{\bar{d}_K}$), γ_K can be replaced by $\sqrt{\bar{d}_K}$ as follows [26, 32]:

$$\gamma_K \approx \sqrt{\bar{d}_K} = \sqrt{2\left(X_K - \|\boldsymbol{\mu}_K\|^2\right)} \tag{7.18}$$

If Condition 7.5 is satisfied, $\boldsymbol{x}_K$ is associated with the nearest *data cloud* $\mathbf{C}_{n^*}$. The meta-parameters of this *data cloud*, namely S_{K-1,n^*} and $\boldsymbol{c}_{K-1,n^*}$ are updated using the following expressions:

$$S_{K,n^*} \leftarrow S_{K-1,n^*} + 1 \tag{7.19a}$$

$$\boldsymbol{c}_{K,n^*} \leftarrow \frac{S_{K-1,n^*}}{S_{K,n^*}}\boldsymbol{c}_{K-1,n^*} + \frac{1}{S_{K,n^*}}\boldsymbol{x}_K \tag{7.19b}$$

Otherwise, $\boldsymbol{x}_K$ is added as a new prototype, and a new *data cloud* is formed with meta-parameters set by expressions (7.16).

The meta-parameters of the prototypes and *data clouds* to which there is no new data sample assigned in the current processing cycle stay the same for the next newly observed data sample.

7.4.2.3 Stage 3: *Data Clouds* Formation

When there are no more data samples, the identified *local* modes (renamed as $\{\boldsymbol{p}\}$) are used to form *data clouds* using Eq. (7.6). The parameters of these *data clouds* can be extracted *post factum*.

7.4.3 Handling Outliers in ADP

After all *data clouds* are formed around the prototypes, one may notice that some *data clouds* are with support equal to *1*. This means that there is no sample associated with these *data clouds* except for the prototypes themselves. Prototypes of this type are considered to be outliers and they are assigned to the nearest normal *data clouds* (the ones for which $S > 1$) with the meta-parameters updated using Eq. (7.19a, b).

7.5 *Local* Optimality of the Methods

In this section, the analysis of *local* optimality of the solutions obtained by the ADP algorithm is conducted.

7.5.1 *Mathematical Formulation of the Problem*

The *local* optimization problems of the ADP algorithm can be formulated by the following mathematical programming problem [33]:

Problem 7.1

$$f(\mathbf{W}, \mathbf{P}) = \sum_{i=1}^{M} \sum_{j=1}^{K} w_{i,j} d(\boldsymbol{x}_j, \boldsymbol{p}_i) \quad (7.20)$$

where P stands for the number of prototypes; $\mathbf{P} = [\boldsymbol{p}_1, \boldsymbol{p}_2, \ldots, \boldsymbol{p}_M] \in \mathbf{R}^{N \times M}$ ($\boldsymbol{p}_i \in \{\boldsymbol{p}\}$); $\mathbf{W} = [w_{i,j}]_{M \times K}$ is a $M \times K$ dimensional real matrix subject to the following constrains ($i = 1, 2, \ldots, M, j = 1, 2, \ldots, K$):

$$w_{i,j} \geq 0 \text{ and } \sum_{i=1}^{M} w_{i,j} = 1 \quad (7.21)$$

and the collection of $\mathbf{W}$ that satisfies Eq. (7.21) is denoted by $\{\mathbf{W}\}_O$.

Problem 7.1 is a nonconvex problem, a *local* minimum point of $f(\mathbf{W}, \mathbf{P})$ does not need to be a *global* minimum [33]. The necessary conditions for *global* optimality of a mathematical programming problem as the one described above is given by the well-known Karush–Kuhn–Tucker condition [34]. It has been proven in [33] that with the square Euclidean distance, partially optimal solutions are always the *locally* optimal solutions as given by Theorem 7.1 as follows.

Theorem 7.1 Consider Problem 7.1 where $d(\boldsymbol{x}_j, \boldsymbol{p}_i) = (\boldsymbol{x}_j - \boldsymbol{p}_i)^T (\boldsymbol{x}_j - \boldsymbol{p}_i)$, a partially optimal solution of Problem 7.1 is a *local* minimum point.

Therefore, based on Theorem 7.1, one can try to find a partially optimal solution of Problem 7.1 in order to obtain a *locally* optimal solution, and the definition of a partially optimal solution is as follows [35]:

Definition 7.1 A point $(\mathbf{W}^*, \mathbf{P}^*)$ is a partially optimal solution for Problem 7.1 on condition that the two inequalities are met:

$$f(\mathbf{W}^*, \mathbf{P}^*) \leq f(\mathbf{W}, \mathbf{P}^*) \quad \text{for all } \mathbf{W} \in \{\mathbf{W}\}_O \tag{7.22a}$$

$$f(\mathbf{W}^*, \mathbf{P}^*) \leq f(\mathbf{W}^*, \mathbf{P}) \quad \text{for all } \mathbf{P} \in \mathbf{R}^{N \times M} \tag{7.22b}$$

Therefore, by solving the following two problems, a partially optimal solution can be obtained [35]:

Problem 7.2 Given $\hat{\mathbf{P}} \in \mathbf{R}^{N \times M}$, minimize $f(\mathbf{W}, \hat{\mathbf{P}})$ subject to $\mathbf{W} \in \{\mathbf{W}\}_O$.

Problem 7.3 Given $\hat{\mathbf{W}} \in \{\mathbf{W}\}_O$, minimize $f(\hat{\mathbf{W}}, \mathbf{P})$ subject to $\mathbf{P} \in \mathbf{R}^{N \times M}$.

and $(\mathbf{W}^*, \mathbf{P}^*)$ is a partially optimal solution of Problem 7.1 if $\mathbf{W}^*$ solves Problem 7.2 with $\hat{\mathbf{P}} = \mathbf{P}^*$, and $\mathbf{P}^*$ solves Problem 7.3 with $\hat{\mathbf{W}} = \mathbf{W}^*$.

7.5.2 Local *Optimality Analysis of the Data Partitioning Methods*

As it was given in the previous subsection, if $(\mathbf{W}, \mathbf{P})$ is a partially optimal solution of Problem 7.1, the data partitioning result, namely, the M *data clouds* around the prototypes $\{\boldsymbol{p}\}$, is *locally* optimal.

For the prototypes $(\mathbf{P})$ obtained by the ADP algorithm, the Voronoi tessellations $(\mathbf{W})$ created around them is the minimum solution of Problem 7.2 by definition [see Eq. (7.6)]. However, there is no guarantee that the solution $(\mathbf{W}, \mathbf{P})$ is the minimum solution of Problem 7.3 simply because the prototype (focal point) $\boldsymbol{p}_i$ is not necessarily equal to the center, denoted by $\boldsymbol{\mu}_i$, of the *data cloud* $\mathbf{C}_i$. The other greedy steps in the algorithmic procedure (i.e. multiple peaks, filtering) might shift $\{\boldsymbol{p}\}$ away from the centers of *data clouds* formed around them.

Therefore, it is obvious to conclude the data partitioning results obtained by the ADP algorithm as described in Sect. 7.4 are not *locally* optimized. Nonetheless, one can still attain the *locally* optimal solution by further applying an iterative process on the obtained partitioning results, for example, using the well-known K-means clustering algorithm [36], which consists of the following steps.

Step 1. Set the initial prototypes, $\mathbf{P}$ obtained by the ADP algorithm as described in Sect. 7.4 as $\mathbf{P}^t$ ($t = 0$, which indicates the current number of iterations);

Step 2. Solve Problem 7.2 by setting $\hat{\mathbf{P}} \leftarrow \mathbf{P}^t$ and obtain $\mathbf{W}^t = \left[w_{i,j}^t\right]_{M \times K}$ as the optimal solution, which is expressed by the following expression ($j = 1, 2, \ldots, K$):

$$\begin{cases} w_{i,j}^t = 1 & i = \underset{l=1,2,\ldots,M}{\arg\min}\left(d\left(\boldsymbol{x}_j, \boldsymbol{p}_l\right)\right) \\ w_{i,j}^t = 0 & i \in \text{else} \end{cases} \tag{7.23}$$

Step 3. Solve Problem 7.3 by setting $\hat{\mathbf{W}} \leftarrow \mathbf{W}^t$ and find the new prototypes denoted by $\mathbf{P}^{t+1}$;

The solution of Problem 7.3 is not as obvious as Problem 7.2, however, with the given $\mathbf{W}^t$, Problem 7.3 is equivalent to the problem of finding $\mathbf{P}^{t+1} \in \mathbf{R}^{N \times M}$, which satisfies the following mathematical programming problem:

$$f_1\left(\mathbf{P}^{t+1}\right) = \min_{\mathbf{Z} \in \mathbf{R}^{N \times M}} (f_1(\mathbf{Z})) \tag{7.24}$$

where $f_1(\mathbf{Z}) = f(\mathbf{W}^t, \mathbf{Z})$, and $f_1(\mathbf{Z})$ can be reformulated as:

$$\begin{aligned} f_1(\mathbf{Z}) = & \sum_{\forall j \in \left\{h | w_{1,h}^t = 1\right\}} d\left(\boldsymbol{x}_j, z_1\right) + \sum_{\forall j \in \left\{h | w_{2,h}^t = 1\right\}} d\left(\boldsymbol{x}_j, z_2\right) + \cdots + \sum_{\forall j \in \left\{h | w_{M,h}^t = 1\right\}} d\left(\boldsymbol{x}_j, z_M\right) \\ & = f_2(z_1) + f_2(z_2) + \cdots + f_2(z_M) \end{aligned} \tag{7.25}$$

where $f_2(z_i) = \sum_{\forall j \in \left\{h | w_{i,h}^t = 1\right\}} d\left(\boldsymbol{x}_j, z_i\right)$, $i = 1, 2, \ldots, M$.

Equation (7.25) further simplify the problem of minimizing $f_1(\mathbf{Z})$ as a problem of finding $\boldsymbol{p}_i^{t+1} \in \mathbf{R}^N$ $(i = 1, 2, \ldots, M)$ that serves as the minimum solution of $f_2(z_i)$:

$$f_2\left(\boldsymbol{p}_i^{t+1}\right) = \min_{z_i \in \mathbf{R}^N} (f_2(z_i)) \tag{7.26}$$

For the square Euclidean distance used $\left(d\left(\boldsymbol{x}_j, \boldsymbol{p}_i^{t+1}\right) = \left(\boldsymbol{x}_j - \boldsymbol{p}_i^{t+1}\right)^T \left(\boldsymbol{x}_j - \boldsymbol{p}_i^{t+1}\right) = \sum_{l=1}^{N} \left(x_{j,l} - p_{i,l}^{t+1}\right)^2\right)$, $f_2(z_i)$ is a convex function and is differentiable for $z_i \in \mathbf{R}^N$. Therefore, the *locally* optimal solution of $f_2(z_i)$ corresponds to the point where $f_2(z_i)$ has the minimum value. The partial derivative of $f_2(z_i)$ at each dimension is denoted by:

$$\frac{\partial f_2(z_i)}{\partial z_{i,l}} = \sum_{\forall j \in \left\{h | w_{i,h}^t = 1\right\}} \frac{\partial d\left(\boldsymbol{x}_j, z_i\right)}{\partial z_{i,l}} \tag{7.27}$$

Since $d\left(\boldsymbol{x}_j, z_i\right) = \left(\boldsymbol{x}_j - z_i\right)^T \left(\boldsymbol{x}_j - z_i\right)$, we have

$$\frac{\partial f_2(z_i)}{\partial z_{i,l}} = 2 \sum_{\forall j \in \left\{h | w_{i,h}^t = 1\right\}} \left(z_{i,l} - x_{j,l}\right) \tag{7.28}$$

According to Fermat's Theorem, if $z_i = \boldsymbol{p}_i^{t+1} = \left[p_{i,1}^{t+1}, p_{i,2}^{t+1}, \ldots, p_{i,N}^{t+1}\right]^T$ then $\frac{\partial f_2(z_i)}{\partial z_{i,l}} = 0$ $(l = 1, 2, \ldots, N)$, and the following equation is obtained:

$$\sum_{\forall j \in \left\{h | w_{i,h}^t = 1\right\}} \left(p_{i,l}^{t+1} - x_{j,l}\right) = p_{i,l}^{t+1} \cdot \sum_{j=1}^{K} w_{i,j}^t - \sum_{\forall j \in \left\{h | w_{i,h}^t = 1\right\}} x_{j,l} \tag{7.29}$$

As it is demonstrated by Eq. (7.29), $\boldsymbol{p}_i^{t+1}$ is the mean of the data samples that associate with the *data cloud* formed around $\boldsymbol{p}_i^t$, namely:

$$\boldsymbol{p}_i^{t+1} = \frac{1}{\sum_{j=1}^{K} w_{i,j}^t} \sum_{\forall j \in \left\{h | w_{i,h}^t = 1\right\}} \boldsymbol{x}_j; \quad i = 1, 2, \ldots, M \tag{7.30}$$

Step 4. Solve Problem 7.2 by setting $\mathbf{P}^{t+1}$ to $\hat{\mathbf{P}}$ and obtain $\mathbf{W}^{t+1}$ as the optimal solution.

Step 5. If $f\left(\mathbf{W}^{t+1}, \mathbf{P}^{t+1}\right) = f\left(\mathbf{W}^t, \mathbf{P}^t\right)$, the optimum is reached, the iterations stop and $\left(\mathbf{W}^*, \mathbf{P}^*\right) \leftarrow \left(\mathbf{W}^{t+1}, \mathbf{P}^{t+1}\right)$. Otherwise, $t \leftarrow t+1$ and the iterative process goes back to step 3.

The above iterative process guarantees a partially optimal solution of the Problem 7.1 as stated in Theorem 7.2 as follows [33].

Theorem 7.2 $(\mathbf{W}, \mathbf{P})$ converges to a partially optimal solution $\left(\mathbf{W}^*, \mathbf{P}^*\right)$ in a finite number of iterations.

In short, by involving an iterative process to minimize the objective function in a similar way as described in [33], the ADP algorithm can always converge to the *locally* optimal partitions in a small number of iterations.

Let us continue the example given in Fig. 7.9, by applying the iterative optimization process, one can finally achieve the *locally* optimized prototypes shown in Fig. 7.10a, where we also compare their positions with the original ones, and the value of $f(\mathbf{W}, \mathbf{P})$ calculated after each iteration is given in Fig. 7.10b.

7.6 Importance of the Proposed Methods

In this chapter, the non-parametric data partitioning algorithm, which we call ADP, is presented. When compared with the traditional clustering approaches, it has the following specific features:

Fig. 7.10 The *local* optimization process of the data partitioning result

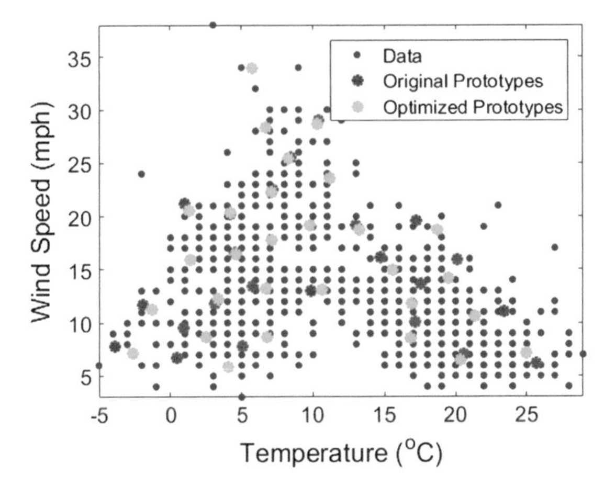

(a) *Locally* optimized prototypes

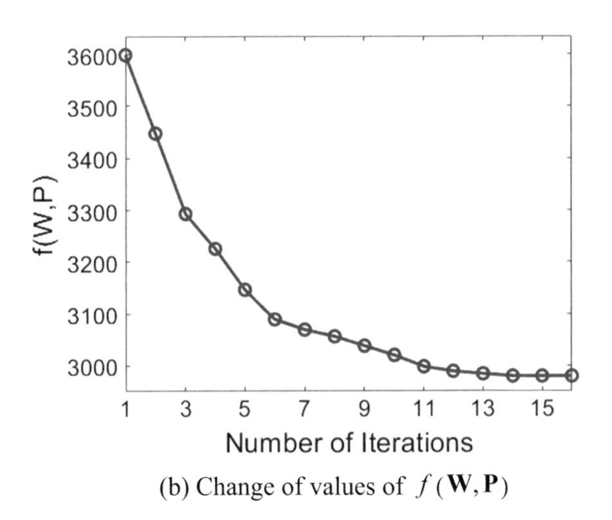

(b) Change of values of $f(\mathbf{W}, \mathbf{P})$

(1) It is autonomous, self-organizing and entirely data-driven;
(2) It is free from user- and problem-specific parameters;
(3) It is based on the ensemble properties and mutual distribution of the *empirically* observed data.

To be more specific, for the ADP algorithm, it involves a fundamentally different data processing approach based on **rank** operators. **Rank** operators are normally avoided in clustering because they are non-linear operators, which do not have a continuous derivative (are not smooth), while most clustering algorithms prefer the linear mean operator. The specificity of the **rank** operator plays a central role in the creation of more parsimonious partitions, specifically when augmented with *local* mode definitions that are parameter-free. For the offline version, the ADP algorithm identifies prototypes from the data samples based on their **ranks** in terms of the *data densities* and mutual distances, and use the prototypes to aggregate data samples around them creating Voronoi tessellations [21]. For the evolving version of the ADP algorithm, it has a more flexible evolving structure compared with other online approaches due to its prototype-based nature. In addition, it replaces the pre-defined threshold, which is commonly used in other online approaches, with a dynamically changing threshold derived from the data. Therefore, the ADP algorithm is able to obtain a more stable, effective and objective partitioning compared with other approaches.

For any data partitioning/clustering algorithm that identifies the underlying data pattern through *local* partitions, the *local* optimality of the solution is of great importance to the validity and effectiveness of the overall approach. The *local* optimization algorithm described in Sect. 7.5.2 is of paramount importance since it is able to guarantee the *local* optimality of the partitioning results obtained by the ADP algorithm as well as other approaches with a similar operating mechanism in few iterations. Moreover, its application is not limited to the data partitioning/clustering only. The algorithm is also applicable to the autonomous learning multi-modal systems with a prototype-based antecedent part as well (this will be presented in Chap. 8). Alternative data partitioning approaches are also possible to use, for example, the self-organizing direction-aware data partitioning approach introduced in [37].

7.7 Conclusions

In this chapter, we introduced a new approach to partition the data autonomously *into data clouds* which form a Voronoi tessellation. This can be seen as clustering, but it has some specific differences mainly in the shape of the *data clouds* gets as well as in the specific way they were formed. The object of both, clustering and data partitioning, is transforming the large amount of raw data into a much smaller (manageable) number of more representative aggregations, which can have semantic meaning.

The proposed new ADP algorithm has two forms/types:

(1) offline, and
(2) online and evolving.

In addition, we formulate and propose an algorithm to guarantee the *local* optimality of the structure that was derived.

As a result of these proposed methods, one can start with raw data and end up with a *locally* optimal structure of *data clouds* represented by their focal points/prototypes, which are nothing else but the peaks in terms of *data density* and *typicality* (the points with *locally* maximum values of the *data density* and *typicality*). This structure is then ready to be used for analysis, building a multi-model classifiers, predictors, controllers or for fault isolation methods.

On this basis, in the following chapters we move to address and solve these problems in a new, *empirical* way (driven by actual observed data and not by pre-defined restrictive assumptions and imposed model structures).

7.8 Questions

(1) How does ADP differ from the traditional clustering algorithms?
(2) How do *data clouds* differ from the traditional clusters?
(3) How does the offline version of the ADP differ from the online version?

7.9 Take Away Items

- A new data partitioning approach which forms *data clouds* is proposed; it resembles clustering, but differs from the traditional clustering;
- The proposed ADP method is based on *data density* and *typicality,* and is autonomous and not restricted by assumptions;
- The Voronoi tessellation formed by the ADP method results in the *data clouds* with irregular shapes; the prototypes/focal points of the *data clouds* are the *local* peaks of the *data density/typicality*;
- The new ADP method has two forms (offline and online);
- The *data clouds* can be used for forming the structure of more complex models (classifiers, predictors, controllers) or used directly for data analysis or fault isolation.

References

1. G.A. Brosamler, An almost everywhere central limit theorem. Math. Proc. Cambridge Philos. Soc. **104**(3), 561–574 (1988)
2. http://www.worldweatheronline.com
3. S.Y. Shatskikha, Multivariate Cauchy distributions as *locally* Gaussian distributions. J. Math. Sci. **78**(1), 102–108 (1996)
4. C. Lee, Fast simulated annealing with a multivariate Cauchy distribution and the configuration's initial temperature. J. Korean Phys. Soc. **66**(10), 1457–1466 (2015)
5. S. Nadarajah, S. Kotz, Probability integrals of the multivariate t distribution. Can. Appl. Math. Q. **13**(1), 53–84 (2005)

6. A. Corduneanu, C.M. Bishop, in *Variational Bayesian Model Selection for Mixture Distributions*, Proceedings of Eighth International Conference on Artificial Intelligent Statistics (2001), pp. 27–34
7. E. Tu, L. Cao, J. Yang, N. Kasabov, A novel graph-based k-means for nonlinear manifold clustering and representative selection. Neurocomputing **143**, 109–122 (2014)
8. P. Angelov, *Autonomous Learning Systems: From Data Streams to Knowledge in Real Time* (Wiley, New York, 2012)
9. M. Aitkin, D.B. Rubin, Estimation and hypothesis testing in finite mixture models. J. R. Stat. Soc. Ser. B (Methodol.) **47**(1), 67–75 (1985)
10. C.E. Lawrence, A.A. Reilly, An expectation maximization (EM) algorithm for the identification and characterization of common sites in unaligned biopolymer sequences. Proteins Struct. Funct. Bioinforma. **7**(1), 41–51 (1990)
11. J.A. Bilmes, A gentle tutorial of the EM algorithm and its application to parameter estimation for gaussian mixture and hidden markov models. Int. Comput. Sci. Inst. **4**(510), 126 (1998)
12. D.A. Reynolds, T.F. Quatieri, R.B. Dunn, Speaker verification using adapted Gaussian mixture models. Digit. Signal Process. **10**(1), 19–41 (2000)
13. C.E. Rasmussen, The infinite Gaussian mixture model. Adv. Neural. Inf. Process. Syst. **12** (11), 554–560 (2000)
14. A. Gionis, H. Mannila, P. Tsaparas, Clustering aggregation. ACM Trans. Knowl. Discov. Data **1**(1), 1–30 (2007)
15. http://cs.joensuu.fi/sipu/datasets/
16. P. Angelov, X. Gu, D. Kangin, Empirical data analytics. Int. J. Intell. Syst. **32**(12), 1261–1284 (2017)
17. P.P. Angelov, X. Gu, J. Principe, D. Kangin, in *Empirical* Data Analysis—A New Tool for *Data Analytics*, IEEE International Conference on Systems, Man, and Cybernetics (2016), pp. 53–59
18. P. Angelov, Fuzzily connected multimodel systems evolving autonomously from data streams. IEEE Trans. Syst. Man, Cybern. Part B Cybern. **41**(4), 898–910 (2011)
19. P. Angelov, R. Yager, Density-based averaging—a new operator for data fusion. Inf. Sci. (Ny) **222**, 163–174 (2013)
20. P. Angelov, R. Yager, A new type of simplified fuzzy rule-based system. Int. J. Gen Syst. **41** (2), 163–185 (2011)
21. A. Okabe, B. Boots, K. Sugihara, S.N. Chiu, *Spatial Tessellations: Concepts and Applications of Voronoi Diagrams*, 2nd edn. (Wiley, Chichester, 1999)
22. L.A. Zadeh, Outline of a new approach to the analysis of complex systems and decision processes. IEEE Trans. Syst. Man Cybern. **1**, 28–44 (1973)
23. E.H. Mamdani, S. Assilian, An experiment in linguistic synthesis with a fuzzy logic controller. Int. J. Man Mach. Stud. **7**(1), 1–13 (1975)
24. T. Takagi, M. Sugeno, Fuzzy identification of systems and its applications to modeling and control. IEEE Trans. Syst. Man. Cybern. **15**(1), 116–132 (1985)
25. P.P. Angelov, X. Gu, J.C. Principe, Autonomous learning multi-model systems from data streams. IEEE Trans. Fuzzy Syst. **26**(4), 2213–2224 (2018)
26. X. Gu, P.P. Angelov, J.C. Principe, A method for autonomous data partitioning. Inf. Sci. (Ny) **460–461**, 65–82 (2018)
27. W. Pedrycz, *Granular Computing: Analysis and Design of Intelligent Systems* (CRC Press, Boca Raton, 2013)
28. P.P. Angelov, D.P. Filev, An approach to online identification of Takagi-Sugeno fuzzy models. IEEE Trans. Syst. Man Cybern. Part B Cybern. **34**(1), 484–498 (2004)
29. P. Angelov, An approach for fuzzy rule-base adaptation using on-line clustering. Int. J. Approx. Reason. **35**(3), 275–289 (2004)
30. P.P. Angelov, D.P. Filev, N.K. Kasabov, *Evolving Intelligent Systems: Methodology and Applications* (2010)
31. P. Angelov, D. Filev, in *On-line Design of Takagi-Sugeno Models*, in International Fuzzy Systems Association World Congress (Springer, Berlin, 2003), pp. 576–584

32. X. Gu, P.P. Angelov, Self-organising fuzzy logic classifier. Inf. Sci. (Ny) **447**, 36–51 (2018)
33. S.Z. Selim, M.A. Ismail, K-means-type algorithms: a generalized convergence theorem and characterization of *local* optimality. IEEE Trans. Pattern Anal. Mach. Intell. **PAMI-6**(1), 81–87 (1984)
34. H.W. Kuhn, A Tucker, in *Nonlinear Programming*, Proceedings of the Second Symposium on Mathematical Statistics and Probability (1951), pp. 481–492
35. R.E. Wendell, A.P. Hurter Jr., Minimization of a non-separable objective function subject to disjoint constraints. Oper. Res. **24**(4), 643–657 (1976)
36. J.B. MacQueen, Some methods for classification and analysis of multivariate observations. 5th Berkeley Symp. Math. Stat. Probab. **1**(233), 281–297 (1967)
37. X. Gu, P. Angelov, D. Kangin, J. Principe, Self-organised direction aware data partitioning algorithm. Inf. Sci. (Ny) **423**, 80–95 (2018)

Chapter 8
Autonomous Learning Multi-model Systems

8.1 Introduction to the *ALMMo* Systems Concept

The concept of multi-model systems is not new on its own [1–3] and it has been applied to machine learning exploiting the centuries' old principle of "divide and rule" [1–3]. Examples of multi-model systems include, but, are not limited to, fuzzy rule-based (FRB) systems [2, 3] and artificial neural networks (ANN) [1]. However, the decisions about the structure, number of *local* models and their design and update were done offline and with heavy involvement of the human expert [1, 2] until the concept of autonomous learning systems [3] was brought up to the scene. The *empirical* approach described in this book can be seen as a further development of the concept of autonomous learning systems which was outlined in [3]. The autonomous data partitioning (ADP) concept described in the previous chapter divides the data space into *local* sub-spaces. This forms the basis of *ALMMo* [4, 5] —the autonomously learning multi-model systems.

Autonomous learning multi-model (*ALMMo*) system is an effective tool for handling complex problems by decomposing a complex problem into a set of simpler ones and combining these afterwards. Multi-model systems have demonstrated their capability in various real applications and have been widely implemented for various purposes, i.e. control, classification, prediction, etc. [1–3, 6–13]. *ALMMo* systems we consider can be seen as a set of *AnYa* type FRB systems [14] but they can also be seen as neuro-fuzzy systems (ANNs with a specific structure and functioning which have high level of transparency and interpretability). *ALMMo* systems can be of either:

(1) zero-order, *ALMMo-0*
(2) first-order, *ALMMo-1*, or
(3) higher-order.

The structure of the antecedent (IF) part of all three types is composed of *data clouds* defined around prototypes (actual data samples/points) forming Voronoi

P. P. Angelov and X. Gu, *Empirical Approach to Machine Learning*, Studies in Computational Intelligence 800,
https://doi.org/10.1007/978-3-030-02384-3_8

tessellation [15]. The design of the antecedent (IF) part of an *ALMMo* system concerns only the identification of prototypes (focal points of the *data clouds*). Prototypes (according to the ADP method described earlier), represent the *local* modes of the data pattern in terms of *data density* and *typicality*.

The consequent (THEN) part of the zero-order *ALMMo* is trivial since it comprises of values representing a label of a class or a singleton. The consequent (THEN) part of the *ALMMo-1* (first-order) system aims to identify the values of the parameters (coefficients of the linear models) that minimize the error between the model prediction and the true value. This is a typical least squares (LS) problem of error minimization [16]. There are different approaches to solve this problem, such as least mean squares (LMS) method [17] or the recursive least squares (RLS) method [18]. However, they would provide the answer only for a single linear model while the *ALMMo-1* structure (since it is an *AnYa* type FRB system) does a fuzzy mixture of locally valid linear models. This required a new approach to be developed for this. To address this, the lead author introduced the fuzzily weighted recursive least squares (FWRLS) method [19].

As the name suggests *ALMMo* system is able to self-organize and self-evolve its multi-model architecture from the streaming data in a non-iterative manner. The specific characteristics that distinguish *ALMMo* from the existing methods and schemes include:

(1) The system employs the non-parametric *empirical* quantities as described in Chap. 4 to disclose the underlying data pattern;
(2) The system structure consists of *data clouds* (the shape of which is not a pre-defined geometric figure) and of self-updating *local* models identified in a data-driven way;
(3) It can, in a natural way, deal with heterogeneous data combining categorical with continuous data [14].

The *ALMMo* system touches the foundation of the complex learning systems for streaming data processing, and it can be extended for different problems and applications including, but not limited to, classification, prediction, control, image processing, etc.

8.2 Zero-Order *ALMMo* Systems

The zero-order autonomous learning multi-model (*ALMMo-0*) system [5] is introduced on the basis of the zero-order *AnYa* type fuzzy rule-based (FRB) systems [3, 14] with a multi-model architecture similar to [20], but identified using the autonomous data partitioning (ADP) method as described in the previous chapter. There is no need to train any parameters due to the feedforward nature of the system structure. Therefore, *ALMMo-0* is very suitable for unsupervised and semi-supervised problems.

In the case of a classification problem, *ALMMo-0* is able to automatically identify the prototypes from the *empirically* observed data and to form *data clouds* per class. Then, sub-systems corresponding to different classes are built using *AnYa* type of fuzzy IF…THEN rules. In this section, we present the *ALMMo-0* system for classification purpose, but *ALMMo-0* can also be applied to, prediction, control, etc. In these problems the outputs represent singleton values and the defuzzification produces the overall output (class label).

8.2.1 Architecture

The architecture of *ALMMo-0* is based on the zero-order *AnYa* type fuzzy rules [3, 14]. An illustrative diagram of its overall structure is depicted in Fig. 8.1, where a zoom-in structure of a zero-order *AnYa* type fuzzy rule is also given.

Again, although *ALMMo-0* can be used for prediction, control and other problems, classification is the problem that most naturally corresponds to *ALMMo-0*. In general, the number of fuzzy rules, M has to be $\geq C$. However, for classifiers, the border case ($M = C$) is also possible/acceptable.

During the training stage (see Fig. 8.1a), in a classification problem due to the multi-model architecture of the *ALMMo-0* system, each sub-system is trained with the data samples of the corresponding class only.

For a data set or stream composed of C different classes, there will be C sub-systems trained independently (one per class). These sub-systems can be updated or removed without influencing the others. Each sub-system contains one fuzzy rule of *AnYa* type formulated around the prototypes generalized or learned from the data samples of the corresponding class. Each fuzzy rule can be viewed as a combination of a number of singleton fuzzy rules that are built upon prototypes connected by a disjunction (logical "*OR*") operator ($i = 1, 2, \ldots, C$):

$$\boldsymbol{R}_i\text{: } IF\left(\boldsymbol{x} \sim \boldsymbol{p}_{i,1}\right) OR\left(\boldsymbol{x} \sim \boldsymbol{p}_{i,2}\right) OR\ldots OR\left(\boldsymbol{x} \sim \boldsymbol{p}_{i,M_i}\right) \quad THEN(Class\ i) \tag{8.1}$$

where $\boldsymbol{p}_{i,j}$ is the jth prototype of the ith fuzzy rule; M_i is the number of identified prototypes.

During the validation stage (see Fig. 8.1b), each time a new data sample $\boldsymbol{x}_K$ is coming, it is evaluated by the zero-order *AnYa* type fuzzy rules corresponding to the different classes. Each fuzzy rule will recommend a *local* decision by applying the "winner takes all" principle to select out the most similar to $\boldsymbol{x}_K$ prototype in terms of the fuzzy degree of confidence. Then, the overall decision-making will balance these *local* decisions and produce the estimated label for this data sample.

The learning and validation processes of the *ALMMo-0* system will be detailed in the following two subsections.

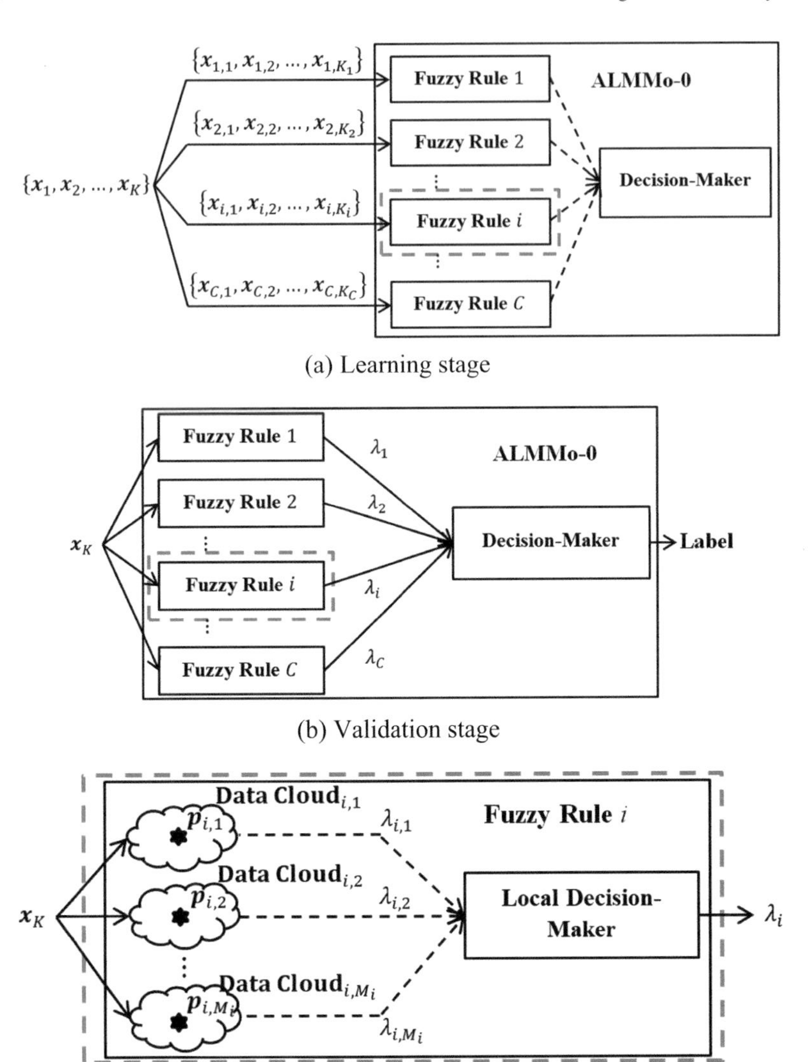

Fig. 8.1 Multi-model architecture of *ALMMo-0*

8.2.2 Learning Process

The *ALMMo-0* system can start learning online "from scratch" from the very beginning of a data stream. It can also be initialized by the available static data in an offline manner and then, continuously update its system structure and meta-parameters based on the newly arrived data samples in a streaming form, which results in a more robust performance. In this subsection, both of these operating modes are presented. As each fuzzy rule is trained in parallel due to the multi-model architecture of the *ALMMo-0*, the learning process is considered to be conducted by the *i*th sub-system ($i = 1, 2, \ldots, C$). The data samples of the *i*th class, are denoted by $\{\boldsymbol{x}\}_{i,K_i} = \{\boldsymbol{x}_{i,1}, \boldsymbol{x}_{i,2}, \ldots, \boldsymbol{x}_{i,K_i}\}\left(\{\boldsymbol{x}\}_{i,K_i} \in \{\boldsymbol{x}\}_K\right)$, and K_i is the number of data samples with $\{\boldsymbol{x}\}_{i,K_i}$. Considering all the classes, we have $\sum_{i=1}^{C} K_i = K$.

During the learning stage of *ALMMo-0* system, each observed data sample, denoted by $\boldsymbol{x}_{i,j}(i = 1, 2, \ldots, C, j = 1, 2, \ldots, K_i)$, will be normalized by its norm:

$$\boldsymbol{x}_{i,j} \leftarrow \frac{\boldsymbol{x}_{i,j}}{\left\|\boldsymbol{x}_{i,j}\right\|} \tag{8.2}$$

This type of normalization can convert the Euclidean distance between any two data samples into a cosine dissimilarity and enhances the ability of the *ALMMo-0* system for high-dimensional data processing [21].

8.2.2.1 System Initialization

The *i*th sub-system is initialized by the first data samples of the *i*th class within the data stream, denoted by $\boldsymbol{x}_{i,1}$, with the *global* meta-parameters set by:

$$K_i \leftarrow 1 \tag{8.3a}$$

$$M_i \leftarrow 1 \tag{8.3b}$$

$$\boldsymbol{\mu}_i \leftarrow \boldsymbol{x}_{i,1} \tag{8.3c}$$

$$X_i \leftarrow 1 \tag{8.3d}$$

and the *local* meta-parameters of the first *data cloud* are set as follows:

$$\mathbf{C}_{i,1} \leftarrow \{\boldsymbol{x}_{i,1}\} \tag{8.4a}$$

$$\boldsymbol{p}_{i,1} \leftarrow \boldsymbol{x}_{i,1} \tag{8.4b}$$

$$S_{i,1} \leftarrow 1 \tag{8.4c}$$

$$r_{i,1} \leftarrow r_o \tag{8.4d}$$

where K_i is the number of data samples observed by the *i*th sub-system (the current time instance); M_i is the number of prototypes within the *i*th sub-system; $\boldsymbol{\mu}_i$ is the *global* mean of the observed data samples of the *i*th class; X_i is the average scalar product, which is always equal to *1* thanks to the normalization (Eq. 8.2). $\mathbf{C}_{i,1}$ denotes the first *data cloud*; $\boldsymbol{p}_{i,1}$ is the first prototype; $S_{i,1}$ is the corresponding support; $r_{i,1}$ is the radius of the area of influence; r_o is a small value to stabilize the initial status of the new-born *data clouds*; more specifically, $r_o = \sqrt{2(1-\cos(30^\circ))}$ is used by default [21].

It has to be stressed that, r_o is not a problem-specific parameter and requires no *prior* knowledge to decide. It is for preventing the newly formed *data clouds* from attracting data samples that are not close enough. It defines a degree of closeness that is interesting and distinguishable. In fact, the value of r_o is adjustable based on the preferences of users to meet a wide variety of problems with specific needs. An alternative approach for deriving r_o based on the mutual distribution of the data and setting the radii of the areas of influence of the existing *data clouds* is given in [22].

The *AnYa* fuzzy rule is then initialized as follows:

$$\boldsymbol{R}_i: \quad IF\left(\boldsymbol{x} \sim \boldsymbol{p}_{i,1}\right) \quad THEN(Class\, i) \tag{8.5}$$

8.2.2.2 System Update

After the sub-system initialization, for each newly arrived data samples $(K_i \leftarrow K_i + 1)$, firstly, the *global* mean $\boldsymbol{\mu}_i$ is updated to $\boldsymbol{\mu}_{i,K_i}$ by $\boldsymbol{x}_{i,K_i}$ using Eq. (4.8). The unimodal *data density* D of the data sample $\boldsymbol{x}_{i,K_i}$ and all the identified prototypes, $\boldsymbol{p}_{i,j} (j = 1, 2, \ldots, M_i)$ are calculated by Eq. (4.32).

Then, Condition 7.4 is checked to see whether $\boldsymbol{x}_{i,K_i}$ becomes a new prototype which will be added to the fuzzy rule [3]. If Condition 7.4 is triggered, a new *data cloud* is being formed around $\boldsymbol{x}_{i,K_i}$ and its parameters are being updated as follows:

$$M_i \leftarrow M_i + 1 \tag{8.6a}$$

$$\mathbf{C}_{i,M_i} \leftarrow \left\{\boldsymbol{x}_{i,K_i}\right\} \tag{8.6b}$$

$$\boldsymbol{p}_{i,M_i} \leftarrow \boldsymbol{x}_{i,K_i} \tag{8.6c}$$

$$S_{i,M_i} \leftarrow 1 \tag{8.6d}$$

$$r_{i,M_i} \leftarrow r_o \tag{8.6e}$$

and a new prototype $\boldsymbol{p}_{i,M_i}$ is added to the fuzzy rule as initialized in Eqs. (8.6a)–(8.6e).

If Condition 7.4 is not satisfied, then the *ALMMo-0* sub-system continues by finding the nearest prototype $\boldsymbol{p}_{i,n^*}$ to $\boldsymbol{x}_{i,K_i}$, which is achieved with Eq. (7.6). Before, $\boldsymbol{x}_{i,K_i}$ is assigned to the nearest *data cloud* $\mathbf{C}_{i,n^*}$, Condition 8.1 is used to check to see whether $\boldsymbol{x}_{i,K_i}$ is close enough or not:

Condition 8.1

$$IF\left(\left\|\boldsymbol{x}_{i,K_i}-\boldsymbol{p}_{i,n^*}\right\|\le r_{i,n^*}\right)\quad THEN\left(\mathbf{C}_{i,n^*}\leftarrow\mathbf{C}_{i,n^*}+\boldsymbol{x}_{i,K_i}\right) \tag{8.7}$$

If Condition 8.1 is satisfied, the meta-parameters of the nearest *data cloud* $\mathbf{C}_{i,n^*}$ are updated as follows:

$$S_{i,n^*}\leftarrow S_{i,n^*}+1 \tag{8.8a}$$

$$\boldsymbol{p}_{i,n^*}\leftarrow\frac{S_{i,n^*}-1}{S_{i,n^*}}\boldsymbol{p}_{i,n^*}+\frac{1}{S_{i,n^*}}\boldsymbol{x}_{i,K_i} \tag{8.8b}$$

$$r_{i,n^*}\leftarrow\sqrt{\frac{1}{2}\left(r_{i,n^*}^2+\left(1-\left\|\boldsymbol{p}_{i,n^*}\right\|^2\right)\right)} \tag{8.8c}$$

and the fuzzy rule is updated accordingly. The radius of the area of influence $\mathbf{C}_{i,n^*}$ is updated in this adaptive way to allow the *data clouds* to converge to the *local* areas where data samples are densely distributed. On the contrary, if Condition 8.1 is not met, a new *data cloud*, $\mathbf{C}_{i,M_i}$ is formed around $\boldsymbol{x}_{i,K_i}$ using Eqs. (8.6a)–(8.6e) and a new prototype $\boldsymbol{p}_{i,M_i}$ is added to the fuzzy rule in the form of Eq. (8.1).

After the system structure and meta-parameters has been updated by $\boldsymbol{x}_{i,K_i}$, the learning algorithm starts a new process cycle for the next newly arrived data samples. The meta-parameters of the *data clouds* that do not receive new members stay the same.

8.2.3 Validation Process

During the validation process, as shown in Fig. 8.1, there is a two-level decision-making process involved for deciding the label of each testing data. This includes the *local* as well as the overall decision-making processes. Both decision-makers follow the "winner takes all" principle. However, alternative principles can be considered as well, i.e. "few winners take all".

Each newly arrived data sample, $\boldsymbol{x}$ is sent to each fuzzy rule $\boldsymbol{R}_i$ to get a firing strength (or score of confidence) denoted by $\lambda_i(i=1,2,\ldots,C)$, which is produced by the *local* decision-maker by selecting out the most similar prototype to $\boldsymbol{x}_K$ within this fuzzy rule [5]:

$$\lambda_i=\max_{j=1,2,\ldots,M_i}\left(\lambda_{i,j}\right) \tag{8.9}$$

where $\lambda_{i,j}=e^{-\left\|\boldsymbol{x}-\boldsymbol{p}_{i,j}\right\|^2}, \boldsymbol{p}_{i,j}\in\{\boldsymbol{p}\}_i$.

However, it has to be stressed that the exponential function to represent λ is just for expanding the small differences in terms of cosine dissimilarities between $\boldsymbol{x}$ and different prototypes $\boldsymbol{p}_{i,j}\left(\boldsymbol{p}_{i,j} \in \{\boldsymbol{p}\}_i\right)$. One can consider alternative functions to replace it, i.e. Cauchy function.

Then, based on the C firing strengths of the C fuzzy rules, respectively (one per rule), the label of $\boldsymbol{x}$ is decided by the overall decision-maker following the "winner takes all" principle:

$$\text{Label} = \underset{i=1,2,\ldots,C}{\arg\max}(\lambda_i) \tag{8.10}$$

8.2.4 Local *Optimality of the* ALMMo-0 *System*

In this subsection, the *local* optimality of the *ALMMo-0* system [5] is studied. Due to the non-parametric nature of both, the consequent (THEN) part and the premise (IF) part (which is prototype-based), the *local* optimality of the *ALMMo-0* system depends solely on the optimal positions of the prototypes (the most representative data samples) in the data space. Therefore, this problem is reduced to finding out a locally optimal partition. From machine learning point of view, this can be considered as locally optimal clustering. A formal mathematical condition for this can be described by mathematical programming problem as described in Sect. 7.5.1 [23].

Considering that the *ALMMo-0* system partitions the data samples of the *i*th class, $\{\boldsymbol{x}\}_{i,K_i}$, into M_i *data clouds*, we reformulate Problem 7.1 for *ALMMo-0* as follows [23]:

Problem 8.1

$$f(\mathbf{W}_i, \mathbf{P}_i) = \sum_{l=1}^{M_i}\sum_{j=1}^{K_i} w_{i,l,j} d\left(\boldsymbol{x}_{i,j}, \boldsymbol{p}_{i,l}\right) \tag{8.11}$$

Problem 8.1 is practically the same as Problem 7.1; the only difference is that the optimal solution of Problem 8.1 only applies to $\{\boldsymbol{x}\}_{i,K_i}$, not the overall dataset $\{\boldsymbol{x}\}_K$. Therefore, we can borrow all the theorems, principles, analysis and conclusions presented in Sect. 7.5 and use them in the Problem 8.1 directly with the only difference we mentioned kept in mind.

Based on Theorem 7.1 we know that the data partitioning result, $\{\mathbf{C}\}_i$, is locally optimal if $(\mathbf{W}_i, \mathbf{P}_i)$ is a partially optimal solution of Problem 8.1. The *ALMMo-0* system that is trained from the static dataset, due to the greedy steps used in the algorithmic procedure (i.e. multiple peaks, filtering), there is no guarantee of the *local* optimality of the partitioning result, which is the same as discussed in Sect. 7.5.2. For the recursive, "one pass" and non-iterative online learning process (either starting "from scratch" or being primed offline), there is also no guarantee that $(\mathbf{W}_i, \mathbf{P}_i)$ can solve both, the Problems 7.2 and 7.3 of the potential *shift* and/or *drift* of the data pattern [24]. Nonetheless, by using the optimization approach presented in Sect. 7.5.2, one can obtain the locally optimal prototypes (the premise, IF part) of the *ALMMo-0* system.

8.3 First-Order *ALMMo* for Classification and Regression

In this section, the autonomous learning multi-model system for streaming data processing, named *ALMMo-1* [4], which is introduced on the basis of the first order *AnYa* type fuzzy rule-based (FRB) systems [3, 14], is presented. Similarly to *ALMMo-0,* its structure is built upon the *data clouds*; all the meta-parameters are extracted from the *empirically* observed data directly with no user- or problem-specific parameters nor *prior* knowledge required and can be recursively updated. Thus, the system is also memory- and computation-efficient. Its system structure is able to evolve online to follow the possible *shifts* and/or *drifts* in the data pattern for the case of streaming data. In this section, we focus on the regression aspect, and present the general architecture, structure identification and identification of the consequent parameters. In the next sections, we use Euclidean distance, however, other types of distance metric and dissimilarity can be considered as well.

8.3.1 Architecture

In *ALMMo*-1, the system structure is composed of a number of first-order *AnYa* type fuzzy rules [14]. Its premise (IF) part design concerns the identification of the prototypes (focal points) of the *data clouds*. Its consequent (THEN) part identification aims to determine the optimal values of the consequent parameters of the *local* linear models [3, 14]. The structure of the *ALMMo-1* system is given in Fig. 8.2 [4, 14].

Each of the linear models in the *ALMMo-1* system (assuming the *i*th one) consists of one first-order *AnYa* type fuzzy rule with the prototype $\boldsymbol{p}_i$ and consequent parameter vector $\boldsymbol{a}_{K,i}$ in the following form:

$$\boldsymbol{R}_i: \quad IF(\boldsymbol{x}_K \sim \boldsymbol{p}_i) \quad THEN\left(y_{K,i} = f_{A,i}\left(\boldsymbol{x}_K, \boldsymbol{a}_{K,i}\right)\right) \tag{8.12}$$

where $\sim$ denotes the similarity, or a fuzzy degree of satisfaction/membership [14]; $\boldsymbol{x}_K = \left[x_{K,1}, x_{K,2}, \ldots, x_{K,N}\right]^T \in \mathbf{R}^N$ is the input vector; $y_{K,i}$ is the output of the *i*th rule; $f_{A,i}(\cdot)$ is the linear output function of the consequent part of $\boldsymbol{R}_i$:

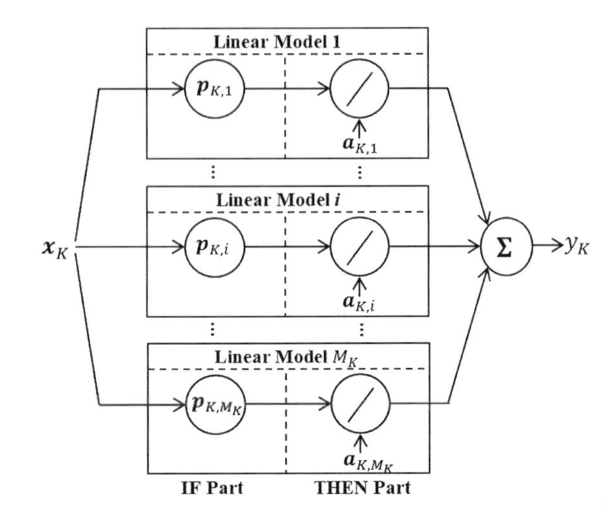

Fig. 8.2 General architecture of *ALMMo-1* system

$$f_{A,i}(\boldsymbol{x}_K, \boldsymbol{a}_{K,i}) = \bar{\boldsymbol{x}}_K^T \boldsymbol{a}_{K,i} \tag{8.13}$$

where $\bar{\boldsymbol{x}}_K^T = [1, \boldsymbol{x}_K^T]$; $\boldsymbol{a}_{K,i} = [a_{K,i,0}, a_{K,i,1}, a_{K,i,2}, \ldots, a_{K,i,N}]^T$.

The overall system output of *ALMMo-1* at the *K*th time instance is mathematically modelled as follows [4, 14, 19]:

$$y_K = f_A(\boldsymbol{x}_K, \boldsymbol{a}_K) = \sum_{i=1}^{M_K} \lambda_{K,i} f_{A,i}(\boldsymbol{x}_K, \boldsymbol{a}_{K,i}) \tag{8.14}$$

where y_K is the overall output; $f_A(\cdot)$ is the nonlinear function that *ALMMo-1* system aims to approximate; $\lambda_{K,i}$ is the firing strength of the ith fuzzy rule; M_k is the number of fuzzy rules.

Different from other autonomous learning systems [14, 19, 20], the firing strength of each rule within the *ALMMo-1* is defined as normalized *data density* calculated per rule locally as follows:

$$\lambda_{K,i} = \frac{D_{K,i}(\boldsymbol{x}_K)}{\sum_{j=1}^{M_K} D_{K,j}(\boldsymbol{x}_K)} \tag{8.15}$$

where

$$D_{K,i}(\boldsymbol{x}_K) = \frac{1}{1 + \frac{S_{K,i}^2 \left\|\boldsymbol{x}_K - \boldsymbol{p}_{K,i}\right\|^2}{(S_{K,i}+1)(S_{K,i}\chi_{K,i} + \|\boldsymbol{x}_K\|^2) - \left\|\boldsymbol{x}_K + S_{K,i}\boldsymbol{p}_{K,i}\right\|^2}}. \tag{8.16}$$

and $\boldsymbol{p}_{K,i}$, $\chi_{K,i}$ and $S_{K,i}$ are the prototype (center), average scalar product and support of the *i*th *data cloud*, $\mathbf{C}_i$; $\boldsymbol{p}_{K,i}$ and $\chi_{K,i}$ can be recursively updated by Eqs. (4.8) and (4.9).

Alternatively, the overall system output of *ALMMo-1* can be reformulated in a more compact form as follows:

$$y_K = f_A(\boldsymbol{x}_K, \boldsymbol{a}_K) = \boldsymbol{X}_K^T \boldsymbol{A}_K \tag{8.17}$$

where $\boldsymbol{X}_K^T = \left[\lambda_{K,1}\bar{\boldsymbol{x}}_K^T, \lambda_{K,2}\bar{\boldsymbol{x}}_K^T, \ldots, \lambda_{K,M_K}\bar{\boldsymbol{x}}_K^T\right]; \boldsymbol{A}_K = \left[\boldsymbol{a}_{K,1}^T, \boldsymbol{a}_{K,2}^T, \ldots, \boldsymbol{a}_{K,M_K}^T\right]^T$.

8.3.2 *Learning Process*

In this section, the learning process of the *ALMMo-1* system is described in detail, which consists of two main stages:

(1) structure identification, and
(2) parameter identification.

8.3.2.1 Structure Identification

A. ***System Initialization***

For the data sample, $\boldsymbol{x}_1$ that first arrives, the *global* meta-parameters of the *ALMMo-1* system are initialized as follows:

$$K \leftarrow 1 \tag{8.18a}$$

$$\boldsymbol{\mu}_K \leftarrow \boldsymbol{x}_1 \tag{8.18b}$$

$$X_K \leftarrow \|\boldsymbol{x}_1\|^2 \tag{8.18c}$$

$$M_K \leftarrow 1 \tag{8.18d}$$

where $\boldsymbol{\mu}_K$ and X_K are the respective *global* means of $\{\boldsymbol{x}\}_K$ and $\{\boldsymbol{x}^T\boldsymbol{x}\}_K$.

The meta-parameters of the first *data cloud*, denoted by $\mathbf{C}_1$, are initialized as follows:

$$\mathbf{C}_1 \leftarrow \{\boldsymbol{x}_1\} \tag{8.19a}$$

$$\boldsymbol{p}_{K,1} \leftarrow \boldsymbol{x}_1 \tag{8.19b}$$

$$\chi_{K,1} \leftarrow \|\boldsymbol{x}_1\|^2 \tag{8.19c}$$

$$S_{K,1} \leftarrow 1 \tag{8.19d}$$

The corresponding fuzzy rule $\boldsymbol{R}_1$ is initialized in the same form as Eq. (8.12) with $\boldsymbol{p}_{K,1}$ as its premise (IF) part; its consequent part initialization will be given in Sect. 8.3.2.2.

B. ***Structure Update***

For each newly observed data sample, $\boldsymbol{x}_K (K \leftarrow K+1)$, $\boldsymbol{\mu}_{K-1}$ and X_{K-1} are updated to $\boldsymbol{\mu}_K$ and X_K by $\boldsymbol{x}_K$ using Eqs. (4.8) and (4.9). The *data densities* at $\boldsymbol{x}_K$ and all the previously identified prototypes, denoted by $\boldsymbol{p}_{K-1,j} (j = 1, 2, \ldots, M_K)$ are calculated using Eq. (4.32).

Then, Condition 7.4 is used to check whether $\boldsymbol{x}_K$ becomes a new prototype. If Condition 7.4 is triggered, a new *data cloud* is being formed around $\boldsymbol{x}_K$. Meanwhile, it is also necessary to check whether the newly formed *data cloud* is overlapping with the existing *data clouds*. Condition 8.2 is used to resolve such situations $(j = 1, 2, \ldots, M_{K-1})$:

Condition 8.2

$$IF\left(D_{K,j}(\boldsymbol{x}_K) \geq \frac{1}{1+n^2}\right) \quad THEN\left(\begin{array}{c}\mathbf{C}_j \; overlaps\; with \\ the\; new\; data\; cloud\end{array}\right) \tag{8.20}$$

where $D_{K,j}(\boldsymbol{x}_K)$ is calculated by Eq. (8.16).

The rationale to consider $D_{K,j}(\boldsymbol{x}_K) \geq \frac{1}{1+n^2}$ comes from the well-known Chebyshev inequality [25] in the form of *data density* (Eq. 4.30). Here $n = 0.5$ is used, which is equivalent to $D_{K,j}(\boldsymbol{x}_K) \geq 0.8$, which defines $\boldsymbol{x}_K$ that is less than $\frac{\sigma}{2}$ away from $\boldsymbol{p}_{K-1,j}$, which means that $\boldsymbol{x}_K$ is strongly associated with $\boldsymbol{p}_{K-1,j}$.

If Condition 7.4 is satisfied and Condition 8.2 is not met, a new *data cloud* with $\boldsymbol{x}_K$ as its prototype is added to the system:

$$M_K \leftarrow M_{K-1} + 1 \tag{8.21a}$$

$$\mathbf{C}_{M_K} \leftarrow \{\boldsymbol{x}_K\} \tag{8.21b}$$

$$\boldsymbol{p}_{K,M_K} \leftarrow \boldsymbol{x}_K \tag{8.21c}$$

$$\chi_{K,M_K} \leftarrow \|\boldsymbol{x}_K\|^2 \tag{8.21d}$$

$$S_{K,M_K} \leftarrow 1 \tag{8.21e}$$

In contrast, if Conditions 7.4 and 8.2 are both satisfied, then the nearest overlapping *data cloud*, denoted by $\mathbf{C}_{n^*}$, is merged with the new one as follows:

$$M_K \leftarrow M_{K-1} \tag{8.22a}$$

$$\mathbf{C}_{n^*} \leftarrow \mathbf{C}_{n^*} + \boldsymbol{x}_K \tag{8.22b}$$

$$\boldsymbol{p}_{K,n^*} \leftarrow \frac{\boldsymbol{p}_{K-1,n^*} + \boldsymbol{x}_K}{2} \tag{8.22c}$$

$$\chi_{K,n^*} \leftarrow \frac{\chi_{K-1,n^*} + \|\boldsymbol{x}_K\|^2}{2} \tag{8.22d}$$

$$S_{K,n} \leftarrow \left\lceil \frac{S_{K-1,n^*} + 1}{2} \right\rceil \tag{8.22e}$$

By merging the overlapping *data cloud* with the newly formed one using Eqs. (8.22a)–(8.22e), the structure of the *ALMMo-1* system will remain the same, which, in turn, improves the computation- and memory-efficiency.

If Condition 7.4 is not satisfied, then the system will find the nearest *data cloud*, $\mathbf{C}_{n^*}$, to $\boldsymbol{x}_K$, and update the corresponding meta-parameters of $\mathbf{C}_{n^*}$ as $(M_K \leftarrow M_{K-1})$:

$$\mathbf{C}_{n^*} \leftarrow \mathbf{C}_{n^*} + \boldsymbol{x}_K \tag{8.23a}$$

$$S_{K,n^*} \leftarrow S_{K-1,n^*} + 1 \tag{8.23b}$$

$$\boldsymbol{p}_{K,n^*} \leftarrow \frac{S_{K-1,n^*}}{S_{K,n^*}} \boldsymbol{p}_{K-1,n^*} + \frac{1}{S_{K,n^*}} \boldsymbol{x}_K \tag{8.23c}$$

$$\chi_{K,n^*} \leftarrow \frac{S_{K-1,n^*}}{S_{K,n^*}} \chi_{K-1,n^*} + \frac{1}{S_{K,n^*}} \|\boldsymbol{x}_K\|^2 \tag{8.23d}$$

The meta-parameters of other *data clouds* stay the same for the next processing cycle. The prototypes $\boldsymbol{p}_{K,i}$ $(i = 1, 2, \ldots, M_K)$ are used for updating the premise part of the rule base of the *ALMMo-1* system.

C. ***Online Quality Monitoring***

Since *ALMMo-1* system is targeting the processing of streaming data, it is very important to monitor the quality of the dynamically evolving structure in real time in order to guarantee the computational- and memory-efficiency of the learning system.

The quality of the fuzzy rules within the rule base of *ALMMo-1* system can be characterized by their *utility* [3]. In *ALMMo*-1, *utility,* $\eta_{K,i}$ of the *data cloud* $\mathbf{C}_i$ is the accumulated sum of firing strength of the corresponding fuzzy rule, which is equivalent to its contribution to the overall outputs during the life of the rule (from the time instance at which $\mathbf{C}_i$ was generated till the current time instance). It is the measure of importance of the respective fuzzy rule compared to other rules:

$$\eta_{K,i} = \frac{1}{K - I_i} \Lambda_{K,i}; \quad \eta_{I_i,i} = 1; \quad i = 1, 2, \ldots, M_K \tag{8.24}$$

where I_i is the time instance at which $\mathbf{C}_i$ is initialized; $\Lambda_{K,i}$ is the accumulated firing strength of the ith fuzzy rule, expressed as:

$$\Lambda_{K,i} = \sum_{l=I_i}^{K} \lambda_{l,i} \tag{8.25}$$

where $\lambda_{l,i}$ is the at the lth time instance, which is calculated by Eq. (8.15).

By removing the *data clouds* and their corresponding fuzzy rules with low *utility* [3, 14] based on Condition 8.3, the rule base can be simplified and the system can be updated:

Condition 8.3

$$IF\left(\eta_{K,i} < \eta_o\right) \quad THEN(\mathbf{C}_i \textit{ and } \boldsymbol{R}_i \textit{ are removed}) \tag{8.26}$$

where η_o is a small tolerance constant ($\eta_o \ll 1$).

If $\mathbf{C}_i$ satisfies Condition 8.3, the corresponding fuzzy rule, $\boldsymbol{R}_i$ will be removed from the rule base and its corresponding consequent parameters are deleted as well.

8.3.2.2 Parameter Identification

In this subsection, the parameter identification process of the *ALMMo-1* system is described. The consequent parameters can be learned either *globally* or locally, same as in the case of the evolving Takagi-Sugeno (eTS) systems [19]. In this subsection, we will present the two approaches, namely *global* learning and *local* learning, separately. The online input selection mechanism will also be presented.

A. ***Local Consequents Parameters Learning Approach***

The first *data cloud* of the system, $\mathbf{C}_1$ is initialized by the first data sample, $\boldsymbol{x}_1$, and the corresponding consequent parameters of the first fuzzy rule, namely, the consequent parameter vector $\boldsymbol{a}_{1,1}$ and the covariance matrix $\boldsymbol{\Theta}_{1,1}$, within the rule base are set up as:

$$\boldsymbol{a}_{1,1} \leftarrow \mathbf{0}_{(N+1)\times 1} \tag{8.27a}$$

$$\boldsymbol{\Theta}_{1,1} \leftarrow \Omega_o \mathbf{I}_{(N+1)\times(N+1)} \tag{8.27b}$$

where $\mathbf{0}_{(N+1)\times 1}$ denotes a $(N+1) \times 1$ dimensional zero vector; $\mathbf{I}_{(N+1)\times(N+1)}$ is a $(N+1) \times (N+1)$ dimensional identity matrix; Ω_o is a constant for initializing the covariance matrix.

When a new fuzzy rule is added by the newly arrived data sample $\boldsymbol{x}_K$ during the structure identification stage, the corresponding consequent parameters are added as follows:

$$\boldsymbol{a}_{K-1,M_K} \leftarrow \frac{1}{M_{K-1}} \sum_{j=1}^{M_{K-1}} \boldsymbol{a}_{K-1,j} \tag{8.28a}$$

$$\boldsymbol{\Theta}_{K-1,M_K} \leftarrow \Omega_o \mathbf{I}_{(M+1)\times(M+1)} \tag{8.28b}$$

If an existing fuzzy rule (denoted as the n^*th rule) is replaced by a new one when both Conditions 7.4 and 8.2 are satisfied, the new rule will inherit the consequent parameters of the old one, namely, $\boldsymbol{a}_{K-1,n^*}$ and $\boldsymbol{\Theta}_{K-1,n^*}$ stay the same [19].

After the structure of both, the antecedent and consequent parts of the *ALMMo-1* system is revised, the fuzzily weight recursive least square (FWRLS) [19] approach is used to update the consequent parameters $\left(\boldsymbol{a}_{K-1,i} \text{ and } \boldsymbol{\Theta}_{K-1,i}, i = 1, 2, \ldots, M_K\right)$ of each fuzzy rule locally as follows:

$$\boldsymbol{\Theta}_{K,i} \leftarrow \boldsymbol{\Theta}_{K-1,i} - \frac{\lambda_{K,i}\boldsymbol{\Theta}_{K-1,i}\bar{\boldsymbol{x}}_K\bar{\boldsymbol{x}}_K^T\boldsymbol{\Theta}_{K-1,i}}{1 + \lambda_{K,i}\bar{\boldsymbol{x}}_K^T\boldsymbol{\Theta}_{K-1,i}\bar{\boldsymbol{x}}_K} \tag{8.29a}$$

$$\boldsymbol{a}_{K,i} \leftarrow \boldsymbol{a}_{K-1,i} + \lambda_{K,i}\boldsymbol{\Theta}_{K,i}\bar{\boldsymbol{x}}_K\left(y_K - \bar{\boldsymbol{x}}_K^T\boldsymbol{a}_{K-1,i}\right) \tag{8.29b}$$

where y_K is the reference output of the system.

B. ***Global Consequents Parameters Learning Approach***

For the *global* learning approach, the consequent parameters of the *ALMMo-1* system are initialized corresponding to the first *data cloud* $\mathbf{C}_1$ as follows:

$$\boldsymbol{a}_{1,1} \leftarrow \mathbf{0}_{(N+1)\times 1} \tag{8.30a}$$

$$\boldsymbol{A}_1 \leftarrow \left[\boldsymbol{a}_{1,1}^T\right]^T \tag{8.30b}$$

$$\boldsymbol{\Theta}_1 \leftarrow \Omega_o \mathbf{I}_{(N+1)\times(N+1)} \tag{8.30c}$$

Then, when a new fuzzy rule is added around the data sample, $\boldsymbol{x}_K$ at the Kth time instance during the structure identification stage, the corresponding consequent parameters are added as follows [19]:

$$\boldsymbol{a}_{K-1,M_K} \leftarrow \frac{1}{M_{K-1}} \sum_{j=1}^{M_{K-1}} \boldsymbol{a}_{K-1,j} \tag{8.31a}$$

$$\boldsymbol{A}_{K-1} \leftarrow \left[\boldsymbol{A}_{K-1}^T, \boldsymbol{a}_{K-1,M_K}^T\right]^T \tag{8.31b}$$

$$\boldsymbol{\Theta}_K \leftarrow \begin{bmatrix} \boldsymbol{\Theta}_K & \mathbf{0}_{(M_{K-1}\cdot(N+1))\times(N+1)} \\ \mathbf{0}_{(N+1)\times(M_{K-1}\cdot(N+1))} & \Omega_o\mathbf{I}_{(N+1)\times(N+1)} \end{bmatrix} \tag{8.31c}$$

After the structure of the *ALMMo-1* system is revised, the FWRLS [19] approach is used to update the consequent parameters ($\boldsymbol{A}_{K-1}$ and $\boldsymbol{\Theta}_{K-1}$) of each fuzzy rule in a *global* form as follows:

$$\boldsymbol{A}_K \leftarrow \boldsymbol{A}_{K-1} + \beta_K \boldsymbol{\Theta}_{K-1}\boldsymbol{X}_K\left(y_K - \boldsymbol{X}_K^T\boldsymbol{A}_{K-1}\right) \tag{8.32a}$$

$$\boldsymbol{\Theta}_K \leftarrow \boldsymbol{\Theta}_{K-1} - \beta_K \boldsymbol{\Theta}_{K-1}\boldsymbol{X}_K\boldsymbol{X}_K^T\boldsymbol{\Theta}_{K-1} \tag{8.32b}$$

$$\beta_K \leftarrow \frac{1}{1+\boldsymbol{X}_K^T\boldsymbol{\Theta}_{K-1}\boldsymbol{X}_K} \tag{8.32c}$$

In general, the consequents parameters of the system that are updated locally using the FWRLS algorithm per sub-model (fuzzy rule) are significantly less influenced by the system structure evolution than being *globally* updated, and the calculation is significantly less computationally expensive [3, 19].

C. ***Online Input Selection***

In many practical cases, many of the attributes are inter-related. Therefore, it is of great importance to introduce the online input selection mechanism, which can further simplify the computational and memory resources and improve the overall performance of the system. As a result, Condition 8.4 is introduced to *ALMMo-1* for dealing with this same as in [3].

Condition 8.4

$$\begin{gathered} IF\left(\bar{\alpha}_{K,i,j} < \frac{\varphi_o}{M_K}\sum_{l=1}^{M_K}\bar{\alpha}_{K,l,j}\right) \\ THEN(remove\ the\ jth\ set\ from\ \boldsymbol{R}_i) \end{gathered} \tag{8.33}$$

where $j = 1, 2, \ldots, N; i = 1, 2, \ldots, M_K; \varphi_o$ is a small constant, $\varphi_o \in [0.03, 0.05]$; $\bar{\alpha}_{K,i,j}$ is the normalized accumulated sum of consequent parameter values at the *K*th time instance:

$$\bar{\boldsymbol{\alpha}}_{K,i} \leftarrow \frac{\boldsymbol{\alpha}_{K,i}}{\sum_{j=1}^{N}\alpha_{K,i,j}} \tag{8.34}$$

where $\bar{\boldsymbol{\alpha}}_{K,i} = \left[\bar{\alpha}_{K,i,1}, \bar{\alpha}_{K,i,2}, \ldots, \bar{\alpha}_{K,i,N}\right]^T$; $\boldsymbol{\alpha}_{K,i} = \left[\alpha_{K,i,1}, \alpha_{K,i,2}, \ldots, \alpha_{K,i,N}\right]^T$, which is the accumulated sum of consequent parameter values of $\boldsymbol{R}_i$ recursively calculated by:

$$\boldsymbol{\alpha}_{K,i} \leftarrow \boldsymbol{\alpha}_{K-1,i} + |\boldsymbol{a}_{K,i}|; \quad \boldsymbol{\alpha}_{I_i,i} \leftarrow |\boldsymbol{a}_{I_i,i}| \tag{8.35}$$

If the *j*th set of $\boldsymbol{R}_i$ meets Condition 8.4, it is removed from the fuzzy rule and the dimensionality of the covariance matrix is reduced by removing the corresponding column and row.

8.3.3 Validation Process

Once the system structure and meta-parameters of the *ALMMo-1* system has been updated, the system is ready for the next data sample. When the next data sample $\boldsymbol{x}_K$ $(K \leftarrow K+1)$ comes, the system output is generated as follows:

$$\hat{y}_K = \hat{f}_A(\boldsymbol{x}_K) = \boldsymbol{X}_K^T \boldsymbol{A}_{K-1} \tag{8.36}$$

where $\hat{f}_A(\cdot)$ is the approximation of $f_A(\cdot)$.

After the system performs the prediction, it will update its structure and parameters based on $\boldsymbol{x}_K$ and the system error $(e_K = y_K - \hat{y}_K)$.

8.3.4 Stability Analysis

In this subsection, the theoretical proof of the stability of *ALMMo-1* system is presented. In the analysis below, we consider the *global* consequent parameters learning approach, but the proof can be extended to the *local* learning approach.

According to the universal approximation property of fuzzy systems [26], there exist the optimal parameter vector $\boldsymbol{A}^*$ that approximates the nonlinear dynamic function $f_A(\cdot)$:

$$y_K = \boldsymbol{X}_K^T \boldsymbol{A}^* + e_{A,K} \tag{8.37}$$

where $e_{A,K}$ is the inherent approximation error.

ALMMo-1 system is capable to dynamically evolve their structure. With the increase of the number of fuzzy rules, the inherent approximation error can be reduced arbitrarily. Therefore, it is reasonable to assume that $e_{A,K}$ caused by approximating $f_A(\cdot)$ is bounded with the constant e_0, which is given by:

$$|e_{A,K}| \leq e_0 \tag{8.38}$$

Combining the Eq. (8.36) with Eq. (8.37), the system error becomes:

$$e_K = y_K - \hat{y}_K = \boldsymbol{X}_K^T \tilde{\boldsymbol{A}}_{K-1} + \iota_K \tag{8.39}$$

where $\tilde{\boldsymbol{A}}_{K-1} = \boldsymbol{A}^* - \boldsymbol{A}_{K-1}$.

The following Lemma is useful for the proof derivation [27, 28]:

Lemma 8.1 [27] Define $V(\boldsymbol{s}_K) : \mathbf{R}^N \to \mathbf{R} \geq 0$ as a Lyapunov function for a nonlinear function. If there exists $\mathcal{K}_\infty$ functions, $\delta_1(\cdot), \delta_2(\cdot)$, and $\mathcal{K}$ function $\delta_3(\cdot)$ [29] that for any $\boldsymbol{s}_K \in \mathbf{R}^N, \exists \sigma \in \mathbf{R}$ which satisfies:

$$\delta_1(\boldsymbol{s}_K) \leq V_K \leq \delta_2(\boldsymbol{s}_K) \tag{8.40a}$$

$$V_{K+1} - V_K = \Delta V_K \leq -\delta_3(\|\boldsymbol{s}_K\|) + \delta_3(\sigma) \tag{8.40b}$$

where $V_K = V(\boldsymbol{s}_K); V_{K+1} = V(\boldsymbol{s}_{K+1})$.

Then, the nonlinear system is uniformly stable.

Theorem 8.1 [28] Consider the *ALMMo-1* system described by Eq. (8.17) with the self-evolving structure and parameters update mechanisms described by Sect. 8.3.2, the uniform stability of the system is ensured. The error between the system output $\hat{y}_K$ and the reference output y_K converges to a small neighborhood of zero in which the average identification error satisfies the following inequality:

$$\lim_{K\to\infty} \frac{1}{K} \sum_{i=1}^{K} e_i^2 \leq \left(\frac{e_0}{\zeta}\right)^2 \tag{8.41}$$

where ζ is the lower bound of β_K satisfying $\zeta = \min(\beta_K)$ and e_0 is the upper bound of $e_{A,K}$ (Eq. (8.38)).

Besides, the parameter error $\|\tilde{\boldsymbol{A}}_{K-1}\|$ is bounded satisfying $\|\tilde{\boldsymbol{A}}_{K-1}\| \leq \|\tilde{\boldsymbol{A}}_0\|$.

Proof Consider the following Lyapunov function:

$$V_K = \tilde{\boldsymbol{A}}_{K-1}^T \boldsymbol{\Theta}_{K-1}^{-1} \tilde{\boldsymbol{A}}_{K-1} \tag{8.42}$$

According to the ***Matrix Inversion Lemma*** [30]:

$$\left(\mathbf{W} + \mathbf{YZY}^T\right)^{-1} = \mathbf{W}^{-1} - \frac{\mathbf{W}^{-1}\mathbf{YY}^T\mathbf{W}^{-1}}{\mathbf{Z}^{-1} + \mathbf{Y}^T\mathbf{W}^{-1}\mathbf{Y}} \tag{8.43}$$

By setting $\mathbf{W}^{-1} = \boldsymbol{\Theta}_{K-1}, \mathbf{Y} = \boldsymbol{X}_K$ and $\mathbf{Z}^{-1} = 1$, it follows:

$$\boldsymbol{\Theta}_K^{-1} = \boldsymbol{\Theta}_{K-1}^{-1} + \boldsymbol{X}_K \boldsymbol{X}_K^T \tag{8.44}$$

Combining Eqs. (8.42) and (8.44), it follows that [28]:

$$\begin{aligned}V_{K+1} &= \tilde{\boldsymbol{A}}_K^T\boldsymbol{\Theta}_K^{-1}\tilde{\boldsymbol{A}}_K = \tilde{\boldsymbol{A}}_K^T\left(\boldsymbol{\Theta}_{K-1}^{-1}+\boldsymbol{X}_K\boldsymbol{X}_K^T\right)\tilde{\boldsymbol{A}}_K \\ &= \tilde{\boldsymbol{A}}_K^T\boldsymbol{\Theta}_{K-1}^{-1}\tilde{\boldsymbol{A}}_K + \left(\boldsymbol{X}_K^T\tilde{\boldsymbol{A}}_K\right)^2 \\ &= \left(\tilde{\boldsymbol{A}}_{K-1}-\beta_K\boldsymbol{\Theta}_{K-1}\boldsymbol{X}_Ke_K\right)^T\boldsymbol{\Theta}_{K-1}^{-1}\cdot\left(\tilde{\boldsymbol{A}}_{K-1}-\beta_K\boldsymbol{\Theta}_{K-1}\boldsymbol{X}_Ke_K\right)+\left(\boldsymbol{X}_K^T\tilde{\boldsymbol{A}}_K\right)^2 \\ &= V_K - 2\beta_K\tilde{\boldsymbol{A}}_{K-1}\boldsymbol{X}_Ke_K + \tilde{\boldsymbol{A}}_{K-1}^T\boldsymbol{\Theta}_{K-1}\tilde{\boldsymbol{A}}_{K-1}\beta_K^2e_K^2 + \left(\boldsymbol{X}_K^T\tilde{\boldsymbol{A}}_K\right)^2\end{aligned} \tag{8.45}$$

Additionally, from Eqs. (8.32a) and (8.39), it also follows that:

$$\begin{aligned}\boldsymbol{X}_K^T\tilde{\boldsymbol{A}}_K + e_{A,K} &= \boldsymbol{X}_K^T(\boldsymbol{A}^* - \boldsymbol{A}_K) + e_{A,K} \\ &= \boldsymbol{X}_K^T\boldsymbol{A}^* - \boldsymbol{X}_K^T\boldsymbol{A}_K + e_{A,K} \\ &= \boldsymbol{X}_K^T\boldsymbol{A}^* - \boldsymbol{X}_K^T(\boldsymbol{A}_{K-1}+\beta_K\boldsymbol{\Theta}_{K-1}\boldsymbol{X}_Ke_K) + e_{A,K} \\ &= \boldsymbol{X}_K^T\boldsymbol{A}^* - \boldsymbol{X}_K^T\boldsymbol{A}_{K-1} - \left(\frac{\boldsymbol{X}_K^T\boldsymbol{\Theta}_{K-1}\boldsymbol{X}_Ke_K}{1+\boldsymbol{X}_K^T\boldsymbol{\Theta}_{K-1}\boldsymbol{X}_K}\right) + e_{A,K} \\ &= \beta_Ke_K\end{aligned} \tag{8.46}$$

By substituting Eq. (8.46) into Eq. (8.45), it follows that:

$$V_{K+1} = V_K - \left(\boldsymbol{X}_K^T\tilde{\boldsymbol{A}}_K\right)^2 - \boldsymbol{X}_K^T\boldsymbol{\Theta}_{K-1}\boldsymbol{X}_K\left(\boldsymbol{X}_K^T\tilde{\boldsymbol{A}}_K+e_{A,K}\right)^2 - 2\boldsymbol{X}_K^T\tilde{\boldsymbol{A}}_Ke_{A,K} \tag{8.47}$$

Since, always $\boldsymbol{X}_K^T\boldsymbol{\Theta}_{K-1}\boldsymbol{X}_K\left(\boldsymbol{X}_K^T\tilde{\boldsymbol{A}}_K+e_{A,K}\right)^2 > 0$, Eq. (8.47) can be further simplified as follows:

$$\begin{aligned}V_{K+1} &\le V_K - \left(\boldsymbol{X}_K^T\tilde{\boldsymbol{A}}_K\right)^2 - 2\boldsymbol{X}_K^T\tilde{\boldsymbol{A}}_Ke_{A,K} \\ &\le V_K + e_{A,K}^2 - \left(\boldsymbol{X}_K^T\tilde{\boldsymbol{A}}_K+e_{A,K}\right)^2 \\ &\le V_K + e_{A,K}^2 - \beta_K^2e_K^2\end{aligned} \tag{8.48}$$

Combining with Eq. (8.38), Eq. (8.48) can be expressed as follows:

$$V_{K+1} - V_K = \Delta V_K \le e_o^2 - \beta_K^2e_K^2 \tag{8.49}$$

For the $\mathcal{K}_\infty$ functions, $\delta_1(\cdot)$ and $\delta_2(\cdot)$ [29],

$$\delta_1\left(\tilde{\boldsymbol{A}}_{K-1}\right) = M_{K-1}(N+1)\min\left(\left\|\tilde{\boldsymbol{A}}_{K-1}\right\|^2\right) \tag{8.50a}$$

$$\delta_2\left(\tilde{\boldsymbol{A}}_{K-1}\right) = M_{K-1}(N+1)\max\left(\left\|\tilde{\boldsymbol{A}}_{K-1}\right\|^2\right) \tag{8.50b}$$

the following inequality holds:

$$M_{K-1}(N+1)\min\left(\left\|\tilde{\boldsymbol{A}}_{K-1}\right\|^2\right) \leq V_K \leq M_{K-1}(N+1)\max\left(\left\|\tilde{\boldsymbol{A}}_{K-1}\right\|^2\right) \tag{8.51}$$

From inequalities (8.49) and (8.51) one can conclude that they both satisfy Lemma 8.1. Therefore, the uniform stability of the *ALMMo-1* system is ensured.

By summing up both sides of inequality (8.49) from 1 up to K, we get [28]:

$$\sum_{i=1}^{K} \left(\beta_i^2 e_i^2 - e_o^2\right) \leq V_1 - V_K \tag{8.52}$$

Because $V_K > 0$ (see inequality (8.51)), inequality (8.52) is re-formulated as follows:

$$\frac{1}{K}\sum_{i=1}^{K} \beta_i^2 e_i^2 \leq \frac{1}{K} V_1 + e_o^2 \tag{8.53}$$

When $K \to \infty$, we obtain:

$$\lim_{K\to\infty} \frac{1}{K}\sum_{i=1}^{K} \beta_i^2 e_i^2 \leq e_o^2 \tag{8.54}$$

Furthermore, considering $\varsigma = \min(\beta_K)$, one can get finally inequality (8.41). And since $V_1 \geq V_K$, this indicates that [28]:

$$\tilde{\boldsymbol{A}}_{K-1}^T \boldsymbol{\Theta}_{K-1}^{-1} \tilde{\boldsymbol{A}}_{K-1} \leq \tilde{\boldsymbol{A}}_0^T \boldsymbol{\Theta}_0^{-1} \tilde{\boldsymbol{A}}_0 \tag{8.55}$$

From Eq. (8.44) with the Eq. (8.55), one can easily see that:

$$\lambda_{\min}\left(\boldsymbol{\Theta}_{K-1}^{-1}\right) \geq \lambda_{\min}\left(\boldsymbol{\Theta}_0^{-1}\right) \tag{8.56}$$

where $\lambda_{\min}(\cdot)$ denotes the minimum eigenvalue of the matrix.

Combining with inequality (8.55), inequality (8.56) can be further transformed into [28]:

$$\begin{aligned} &\lambda_{\min}\left(\boldsymbol{\Theta}_0^{-1}\right)\left\|\tilde{\boldsymbol{A}}_{K-1}\right\|^2 \leq \lambda_{\min}\left(\boldsymbol{\Theta}_{K-1}^{-1}\right)\left\|\tilde{\boldsymbol{A}}_{K-1}\right\|^2 \\ &\leq \tilde{\boldsymbol{A}}_{K-1}^T \boldsymbol{\Theta}_{K-1}^{-1} \tilde{\boldsymbol{A}}_{K-1} \leq \tilde{\boldsymbol{A}}_0^T \boldsymbol{\Theta}_0^{-1} \tilde{\boldsymbol{A}}_0 \leq \lambda_{\min}\left(\boldsymbol{\Theta}_0^{-1}\right)\left\|\tilde{\boldsymbol{A}}_0\right\|^2 \end{aligned} \tag{8.57}$$

We can conclude that the following inequality holds [28]:

$$\left\|\tilde{\boldsymbol{A}}_{K-1}\right\| \leq \left\|\tilde{\boldsymbol{A}}_0\right\| \tag{8.58}$$

Remark 8.1 Theorem 8.1 states that the average identification error of the *ALMMo-1* system converges to a small neighborhood of zero. The approximation error is caused by the parameter errors. In an ideal situation, when the parameters of the system converge to the optimal values, the approximation error becomes zero. However, in practice, zero approximation error is hard to be achieved due to the complex nature of the problem, and thus, the approximation error is only expected, in practice, to converge to a very small value around zero.

Assuming that at the Kth time instance, the number of *data clouds* M_K is determined by the autonomous data partitioning (ADP) method. The following theorem [28] demonstrates that the stability and convergence properties will not be affected by adding new rules to the rule base:

Theorem 8.2 [28]: Consider the *ALMMo-1* system described by Eq. (8.13), when a new rule is added ($M_K \leftarrow M_{K-1}+1$), and the consequent parameters $\boldsymbol{A}_K$ and $\boldsymbol{\Theta}_K$ are updated by Eq. (8.29a), the stability of the system is still guaranteed.

Proof Consider the following Lyapunov function:

$$V_K = \tilde{\boldsymbol{A}}_{K-1}^T \boldsymbol{\Theta}_{K-1}^{-1} \tilde{\boldsymbol{A}}_{K-1} \tag{8.59}$$

After a new rule is added to the rule base ($M_K \leftarrow M_{K-1}+1$), Eq. (8.59) is transformed into:

$$\begin{aligned} V_{K+1} &= \tilde{\boldsymbol{A}}_K^T \boldsymbol{\Theta}_K^{-1} \tilde{\boldsymbol{A}}_K \\ &= \begin{bmatrix} \tilde{\boldsymbol{A}}_{K-1} \\ \tilde{\boldsymbol{a}}_{K-1,M_K} \end{bmatrix}^T \begin{bmatrix} \boldsymbol{\Theta}_{K-1} & \boldsymbol{0}_{(M_{K-1}\cdot(N+1))\times(N+1)} \\ \boldsymbol{0}_{(N+1)\times(M_{K-1}\cdot(N+1))} & \Omega_o \mathbf{I}_{(N+1)\times(N+1)} \end{bmatrix}^{-1} \cdot \begin{bmatrix} \boldsymbol{A}_{K-1} \\ \tilde{\boldsymbol{a}}_{K-1,M_K} \end{bmatrix} \\ &= \tilde{\boldsymbol{A}}_{K-1}^T \boldsymbol{\Theta}_{K-1} \tilde{\boldsymbol{A}}_{K-1} + \frac{1}{\Omega_o} \tilde{\boldsymbol{a}}_{K-1,M_K}^T \tilde{\boldsymbol{a}}_{K-1,M_K} = V_K \end{aligned} \tag{8.60}$$

where $\tilde{\boldsymbol{a}}_{K-1,M_K} = \boldsymbol{a}_{K-1,M_K} - \boldsymbol{a}_{M_K}^*$; $\boldsymbol{a}_{M_K}^*$ is the optimal value of $\boldsymbol{a}_{K-1,M_K}$.

Therefore, it is clear that the newly added *data cloud* has **no** influence on the Lyapunov function V_K. Based on Theorem 8.1, the stability of the *ALMMo-1* system is guaranteed.

Remark 8.2 The stability proof is specifically applicable to the *ALMM*o-1 systems presented in this subsection. However, the stability analysis is also applicable to other more general FRB systems satisfying Eq. (8.13) that have structure learning and FWRLS parameter update approach.

8.4 Conclusions

In this chapter we introduce the Autonomous Learning Multi-Model (*ALMMo*) systems. *ALMMo* steps upon the previous research by the lead author into *AnYa* type neuro-fuzzy systems and can be seen as a self-developing, self-evolving, stable, locally optimal proven universal approximator. In this chapter we start with the concept, then we go into the zero- and first-order *ALMMo* in more detail. We describe the architecture, followed by the learning methods. We provide the theoretical proof (using Lyapunov theorem) the stability of the first-order *ALMMo* systems. We further provide the theoretical proof of the *local* optimality which satisfies Karush-Kuhn-Tucker conditions).

The *ALMMo* system does not impose any model of data generation with parameters on the data, thus, has the advantages of being non-parametric, non-iterative and assumption-free. Thus, it is able to select the prototypes in an objective manner, which guarantees the performance and efficiency. They self-develop, self-learn and evolve autonomously. The first-order *ALMMo* systems are also stable.

The applications of the *ALMMo* systems are described in more detail in Chap. 12 and the pseudo-codes of the *ALMMo-0* and *ALMMo-1* are given in Appendixes B.3 and B.4, and the MATLAB implementations are described in Appendixes C.3 and C.4, respectively.

8.5 Questions

(1) What is the difference between *ALMMo-0* and *ALMMo*-1? What is the difference from the higher-order *ALMMo*?
(2) On which theory is the convergence and stability condition based? Which type of *ALMMo* it applies to?
(3) On which condition the property of *local* optimality is based upon? Which types of *ALMMo* it does apply to?
(4) Which types of machine learning, pattern recognition and control problems each of the particular types of *ALMMo* are applicable to?

8.6 Take Away Items

- *ALMMo* is a form of self-learning, self-organizing system which is nonlinear overall, but linear locally;
- *ALMMo* can be zero-, first- or higher-order with *ALMMo-0* typically (but not necessarily only) performing classification and *ALMMo-1* applicable to both, predictors, classifiers as well as controllers;
- *ALMMo-1* has been theoretically proven to provide and guarantee convergence and stability; it was also theoretically proven that the premise/antecedent/IF part

of the *ALMMo* systems of any order can be optimized locally for the price of very few iterations;

- *ALMMo* builds upon the *AnYa* type neuro-fuzzy systems previously introduced by the lead author.

References

1. K.S.S. Narendra, J. Balakrishnan, M.K.K. Ciliz, Adaptation and learning using multiple models, switching, and tuning. IEEE Control Syst. Mag. **15**(3), 37–51 (1995)
2. T. Takagi, M. Sugeno, Fuzzy identification of systems and its applications to modeling and control. IEEE Trans. Syst. Man. Cybern. **15**(1), 116–132 (1985)
3. P. Angelov, *Autonomous learning systems: from data streams to knowledge in real time* (Wiley, New York, 2012)
4. P.P. Angelov, X. Gu, J.C. Principe, Autonomous learning multi-model systems from data streams. IEEE Trans. Fuzzy Syst. **26**(4), 2213–2224 (2018)
5. P.P. Angelov, X. Gu, Autonomous learning multi-model classifier of 0-order (ALMMo-0), in *IEEE International Conference on Evolving and Autonomous Intelligent Systems* (2017), pp. 1–7
6. X. Gu, P.P. Angelov, A.M. Ali, W.A. Gruver, G. Gaydadjiev, Online evolving fuzzy rule-based prediction model for high frequency trading financial data stream, in *IEEE Conference on Evolving and Adaptive Intelligent Systems (EAIS)* (2016), pp. 169–175
7. P. Angelov, Fuzzily connected multimodel systems evolving autonomously from data streams, IEEE Trans. Syst. Man, Cybern. Part B Cybern. **41**(4), 898–910 (2011)
8. J.J. Macias-Hernandez, P. Angelov, X. Zhou, Soft sensor for predicting crude oil distillation side streams using Takagi Sugeno evolving fuzzy models, in *IEEE International Conference on Systems, Man and Cybernetics. ISIC*, vol. 44, no. 1524 (2007), pp. 3305–3310
9. P. Angelov, P. Sadeghi-Tehran, R. Ramezani, An approach to automatic real-time novelty detection, object identification, and tracking in video streams based on recursive density estimation and evolving Takagi-Sugeno fuzzy systems. Int. J. Intell. Syst. **29**(2), 1–23 (2014)
10. X. Zhou, P. Angelov, Real-time joint landmark recognition and classifier generation by an evolving fuzzy system. IEEE Int. Conf. Fuzzy Syst. **44**(1524), 1205–1212 (2006)
11. X. Zhou, P. Angelov, Autonomous visual self-*local*ization in completely unknown environment using evolving fuzzy rule-based classifier, in *IEEE Symposium on Computational Intelligence in Security and Defense Applications* (2007), pp. 131–138
12. P. Angelov, A fuzzy controller with evolving structure. Inf. Sci. (Ny) **161**(1–2), 21–35 (2004)
13. R.E. Precup, H.I. Filip, M.B. Rədac, E.M. Petriu, S. Preitl, C.A. Dragoş, Online identification of evolving Takagi-Sugeno-Kang fuzzy models for crane systems. Appl. Soft Comput. J. **24**, 1155–1163 (2014)
14. P. Angelov, R. Yager, A new type of simplified fuzzy rule-based system. Int. J. Gen Syst. **41** (2), 163–185 (2011)
15. A. Okabe, B. Boots, K. Sugihara, S.N. Chiu, *Spatial tessellations: concepts and applications of Voronoi diagrams*, 2nd edn. (Wiley, Chichester, 1999)
16. S.M. Stiglerin, Gauss and the invention of least squares. Ann. Stat. **9**(3), 465–474 (1981)
17. S. Haykin, B. Widrow (eds.) *Least-mean-square adaptive filters* (Wiley, New York 2003)
18. S. Haykin, *Adaptive filter theory*. (Prentice-Hall, Inc., 1986)
19. P.P. Angelov, D.P. Filev, An approach to online identification of Takagi-Sugeno fuzzy models, IEEE Trans. Syst. Man, Cybern. Part B Cybern. **34**(1), 484–498 (2004)
20. P. Angelov, X. Zhou, Evolving fuzzy-rule based classifiers from data streams. IEEE Trans. Fuzzy Syst. **16**(6), 1462–1474 (2008)

21. X. Gu, P.P. Angelov, D. Kangin, J.C. Principe, A new type of distance metric and its use for clustering. Evol. Syst. **8**(3), 167–178 (2017)
22. X. Gu, P.P. Angelov, Self-organising fuzzy logic classifier. Inf. Sci. (Ny) **447**, 36–51 (2018)
23. S.Z. Selim, M.A. Ismail, K-means-type algorithms: a generalized convergence theorem and characterization of *local* optimality. IEEE Trans. Pattern Anal. Mach. Intell., PAMI-6, no. 1, pp. 81–87, 1984
24. E. Lughofer, P. Angelov, Handling drifts and shifts in on-line data streams with evolving fuzzy systems. Appl. Soft Comput. **11**(2), 2057–2068 (2011)
25. J.G. Saw, M.C.K. Yang, T.S.E.C. Mo, Chebyshev inequality with estimated mean and variance. Am. Stat. **38**(2), 130–132 (1984)
26. L.X. Wang, J.M. Mendel, Fuzzy basis functions, universal approximation, and orthogonal least-squares learning. IEEE Trans. Neural Netw. **3**(5), 807–814 (1992)
27. J. De Jesus Rubio, P. Angelov, J. Pacheco, Uniformly stable backpropagation algorithm to train a feedforward neural network. IEEE Trans. Neural Netw. **22**(3), 356–366 (2011)
28. H.-J. Rong, P. Angelov, X. Gu, J.-M. Bai, Stability of evolving fuzzy systems based on data clouds. IEEE Trans. Fuzzy Syst. https://doi.org/10.1109/tfuzz.2018.2793258 (2018)
29. W. Yu, X. Li, Fuzzy identification using fuzzy neural networks with stable learning algorithms. IEEE Trans. Fuzzy Syst. **12**(3), 411–420 (2004)
30. N.J. Higham, *Accuracy and stability of numerical algorithms*, 2nd edn. (Siam, Philadelphia, 2002)

Chapter 9
Transparent Deep Rule-Based Classifiers

Traditional fuzzy rule-based (FRB) classifiers were successfully and widely implemented for classification purposes in different applications [1, 2] by offering transparent, interpretable structure, but could not reach the levels of performance achieved by deep learning classifiers. Their design also requires handcrafting membership functions, assumptions to be made and parameters to be selected.

In this chapter, a new type of deep rule-based (DRB) classifier with a multilayer architecture is presented. Combining the computer vision techniques, the DRB classifier is designed for image classification with a massively parallel set of zero-order fuzzy rules [2, 3] as the learning engine. With its prototype-based nature, the DRB classifiers generate from the data a transparent and human understandable FRB system structures in a highly efficient manner and offer extremely high classification accuracy at the same time. Their training process is autonomous, non-iterative, non-parametric, online and can start "from scratch".

More importantly, the DRB classifier can start classification from the very first image of each class in the same way as humans do. The DRB classifiers can also work in a semi-supervised mode of learning whereby only a (small) proportion (e.g. 5%) of the data is being labeled initially and all further data is unlabeled; that is, this classifier starts with a small training set and continue in a fully unsupervised mode after that. This mode of operation enhances its ability of handling unlabeled images and allows the classifier to learn new classes actively without human experts' involvement.

Thus, the proposed DRB classifier is able to evolve dynamically its own structure and learn new information (again, very much like humans do). Therefore, the DRB classifier has anthropomorphic characteristics [4]. Thanks to the prototype-based nature of the DRB classifier, the semi-supervised learning process is fully transparent and human-interpretable. These are also anthropomorphic characteristics, typical for humans (people can explain why a face is of a particular person, e.g. because of the nose, ears, hair etc. or because it resembles somebody else (similarly, if we look at an image of a car or a house, etc. we can say "it looks like a sports car" or "it looks like a high rise building" referring to prototypes we

P. P. Angelov and X. Gu, *Empirical Approach to Machine Learning*, Studies in Computational Intelligence 800,
https://doi.org/10.1007/978-3-030-02384-3_9

selected from the data. DRB classifiers can perform classification on out-of-sample images and also support recursive online training on a sample-by-sample basis or a batch-by-batch basis.

In the remainder of this chapter, we will describe the proposed DRB classifier in detail.

9.1 Prototype-Based Classifiers

One of the simplest and widely used methods for conducting classification on unlabeled data samples based on similarities is the nearest neighbor method [5] often also called k-nearest neighbor (kNN), where k represents a predefined integer of the nearest neighbors that will be taken into account (all other data is being ignored). The traditional nearest neighbor method, however, ignores the *global* data space and only considers a pre-defined number (k) of data samples. Furthermore, it is not efficient in the sense that it requires ***all*** the data samples to be stored and accessed which may be prohibitive memory- and time-wise and is computationally inefficient. It can be seen as an extreme example of the so-called prototype-based classifiers [6] as opposed to the vast group of density-based classifiers.

One may consider the popular support vector machine (SVM) [7] classifiers also as a kind of prototype classifiers, because the support vectors (SV) are real data points and, thus,—prototypes. SVM, unlike the simple kNN does not require ***all*** the data to be memorized and accessible and selects the SV based on a form of a model of the data distribution. Indeed, SVM is much more sophisticated (and often, more effective) than the kNN which itself does not go beyond the point-wise information contained in the prototypes to generalize the *local* and, more importantly, the *global* neighborhood. The zero-order autonomous learning multi-model (*ALMMo-0*) systems, presented in Chap. 8 which builds on the earlier eClass0 [2], simpl_eClass0 [8] and TEDAClass-0 [9] are also prototype-based. Other well-known prototype-based classifiers include learning vector quantization (LVQ) [10].

Prototypes play an instrumental role in the prototype-based systems, which decide the performance of the classifiers as well as the effectiveness and validity of the knowledge gained from the data.

The main difference between various types of prototype-based classifiers is in the prototype identification process, which leads to the differences in performance, computational efficiency and system transparency. For example, the proposed DRB classifier has some distant resemblance with the kNN method with $k = 1$ if the so-called "winner takes all" approach is used. However, it is much more sophisticated and powerful, because:

(1) It is using fuzzy sets which allow to determine the degree of similarity (a real value in the range of $[0, 1]$) to each prototype;

(2) The prototypes are determined as the peaks of the *typicality*, respectively *data density* distribution and are, thus the most likely, the most typical, the most representative data points/vectors;
(3) The prototypes are small number (much smaller than the total number of data points, $M \ll K$ due to the autonomous data partitioning (ADP) algorithm used to extract them from the data.

9.2 DRB Classifier Concept

In real-world applications, especially for medical and biological ones, people prefer to keep a number of the most representative samples (prototypes) instead of keeping all the records [11]. Therefore, it is of paramount importance for a learning algorithm to be able to select the most representative data samples out of the data.

The DRB prototype-based classifier [12] is composed of a small set of prototypes, which are the most *typical* and representative data samples within the data space representing the *local* modes of the *typicality* and *data density* distribution, respectively.

The DRB classifier is a general approach serving as a strong alternative to the current deep convolutional neural networks (DCNN) [13–15]. It employs some of the standard image pre-processing and transformation techniques (i.e. normalization, rotation, scaling and segmentation) to improve the generalization ability and the well-known feature descriptors from the field of computer vision (e.g. GIST [16] and histogram of oriented gradients (HOG) [17] or another encoder, e.g. the high-level feature descriptor and encoder represented by the pre-trained VGG-VD-16 [18]) to extract the feature vectors from the training images. In effect, this transforms the real-world problem (images) to a mathematical description within a certain data space defined by the feature vectors:

$$\mathbf{I} \to \boldsymbol{x} \in \mathbf{R}^N \tag{9.1}$$

From the machine learning point of view, these steps equate to pre-processing and feature extraction, see Chap. 2. Moving to the classification itself, the proposed DRB classifier works in the $(N+1)$-dimensional space where the additional $(N+1th)$ dimension is the *typicality* or *data density*, respectively. Thus, a $(N+1)$ dimensional "mountain" of the *typicality* is being formed like the one shown in Fig. 4.16 from the data (images). Each image corresponds to a single N dimensional point and the most typical and representative images will naturally become peaks of the "mountain" when applying ADP method to the data. This can be done in online mode and is non-iterative and non-parametric.

If we want to *locally* optimize the selection of the best (*locally*) prototypes the payoff will be a small number of iterations as described in Sect. 7.5.2. After selecting prototypes (real images which are the most *typical* and representative) transparent and intuitive IF...THEN rules are being formed around them as described in Chap. 8. These rules are of *AnYa* type and form an *ALMMo* system of zero-order, but they have a large number of prototypes and, thus can either be seen as a massively parallel neuro-fuzzy system [13, 15] or by extensive use of disjunction (the logical OR operators) to compress to very compact and easy to understand and interpret *AnYa* type fuzzy rules. It is interesting to notice that the traditional fuzzy sets and systems literature and, more importantly, practice (applications) is almost entirely based on the use of disjunction (logical "*AND*") connective.

9.3 General Architecture of the DRB Classifier

The general architecture of the proposed DRB classifier (Fig. 9.1) is composed of the following layers [12]:

(1) Pre-processing layer (pre-processing blocks);
(2) Feature extraction layer (feature descriptors);
(3) Fuzzy rule-based layer (Massively parallel ensemble of IF...THEN fuzzy rules of *AnYa* type);
(4) Overall decision-maker.

The architecture is simple, entirely data-driven and fully automatic. This contrasts with the "black box" traditional deep learning methods such as DCNN-based approaches which have tens or hundreds of millions of parameters, are hard to interpret and require tens of hours of training using accelerators such as graphics

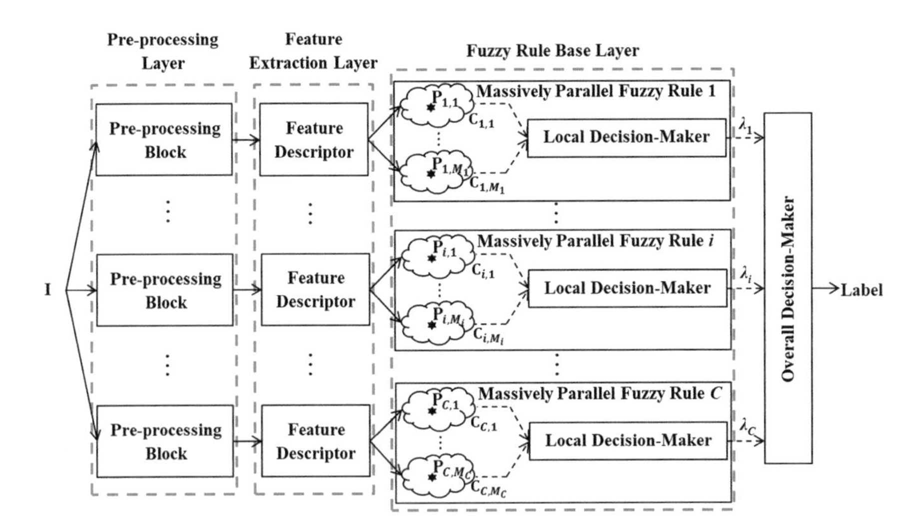

Fig. 9.1 General architecture of the DRB classifier [12]

processing units (GPU) [19, 17]. At the same time, DRB classifiers are able to perform highly accurate classification on various benchmark problems surpassing the state-of-the-art methods, including Bag of Visual Words (BOVW), SVM and traditional deep learning such as DCNN.

A. ***Pre-processing Layer***

The pre-processing blocks within the first layer of the DRB classifier involve only the basic pre-processing techniques; for example, normalization, scaling, rotation and image segmentation. Thus, it is composed of a number of sub-layers serving various purposes. This resembles the historically first DCNNs such as Neocognitron by Fukuda [20]. Details of some of the most widely used image transformation techniques can be found in Sect. 2.2.4.1.

The main purposes of the pre-processing layer within the DRB classifier are:

(1) Improving the generalization ability of the classifier, and
(2) Increasing the efficiency of the feature descriptors in extracting and distilling useful information from the image.

In fact, the sub-structures of the pre-processing block and the usage of the pre-processing techniques may depend on the specific requirements of the problems and users. In this book, we only present the very general architecture, and one may be able to design a more effective pre-processing block with the help of the deeper knowledge of the problem domain.

B. ***Feature Extraction Layer***

For the feature extraction layer, the DRB classifier may employ various feature descriptors that are used in the field of computer vision. Different feature descriptors have different advantages and deficiencies (see Sect. 2.2.4 for more details). In this book, we use two low-level feature descriptors, namely, GIST [16] and HOG [17] as well as one high-level feature descriptor (a pre-trained VGG-VD-16 [18]). However, alternative feature descriptors can also be used. Selecting the most suitable feature descriptor for a particular problem requires *prior* knowledge about the specific problem. Fine-tuning of the high-level feature descriptor to the specific problem can further enhance the performance.

Once the *global* (either low- or high-level) features of the image are extracted and stored, there is no need to repeat the same process again. For example, the number of features when GIST is used is 512 [21] and when VGG-VD-16 is used the number of features vector is 4096 [22]. Thus, $N = 512$ for GIST feature descriptor and $N = 4096$ for VGG-VD-16. Alternative feature descriptors, i.e. HOG [23], scale-invariant feature transform (SIFT) [24], GoogLeNet [25], AlexNet [26], etc., can also be considered for the DRB classifier. One may also combine some of them and obtain an integrated feature vector with higher descriptive ability.

C. ***Fuzzy Rule-Based Layer***

The third and most important layer of the DRB classifier is the massively parallel ensemble of zero-order *AnYa* type fuzzy rules, *ALMMo-0* [27]. The DRB classifier generates autonomously a fully understandable set of fuzzy rules consisting of a (possibly large) number of prototypes after a short "one pass" type training process [13–15] as its "core". The number of prototypes is not pre-defined and depends on the variability of the data (if all the images are quite similar or are drawn form a normal/Gaussian distribution) there will be only one prototype, and so on, following the ADP process described in Chap. 7. The massively parallel fuzzy rules with a single prototype each can be represented in a more compact form by fuzzy rules which have all prototypes per class connected using disjunction operator (logical "*OR*") [15]. The prototypes are identified from the training data directly by a non-iterative, "one pass" process as the most representative prototypes—the *local* peaks of the *typicality* distribution [28].

The learning process of the massively parallel FRB classifier will be described in Sect. 9.4.

D. ***Overall Decision-Maker***

The final layer is the overall decision-maker that decides the winning class label based on the partial suggestion of the per-class fuzzy IF…THEN rules. Each of the per-class fuzzy rules generate a score of confidence (degree of firing) given by the *local* decision-maker of each rule. This layer is only used during the validation stage (not during the training stage, because the DRB classifier is trained per class, and respectively, per rule which itself offers another level of parallelization). The overall decision-maker is applying the "winner takes all" principle same as the *local* (per class) decision makers. Therefore, the DRB classifier actually uses a two-stage decision-making structure during the validation process (see the right half of Fig. 9.1), which will be detailed in Sect. 9.4.2.

In the DRB classifier, a non-parametric rule base consisting of zero-order *AnYa* type [27] fuzzy rules is the main learning "engine":

$$\boldsymbol{R}_i\text{:}\quad IF\ \left(\mathbf{I}\sim\mathbf{P}_{i,1}\right)\ OR\ \left(\mathbf{I}\sim\mathbf{P}_{i,2}\right)\ OR\ldots OR\ \left(\mathbf{I}\sim\mathbf{P}_{i,M_i}\right)\quad THEN\ (Class\ i) \tag{9.2}$$

where $i = 1, 2, \ldots, C$; $\mathbf{P}_{i,j}$ is the jth prototypical images of the ith class; $j = 1, 2, \ldots, M_i$; M_i is the number of prototypes within the fuzzy rule; the feature vectors of the images $\mathbf{I}$ and $\mathbf{P}_{i,j}$ are denoted by $\boldsymbol{x}$ and $\boldsymbol{p}_{i,j}$, respectively.

The rule base makes the DRB classifier interpretable and transparent for human understanding (even to a non-expert) unlike the widely used mainstream deep learning. For example, let us consider the well-known object classification image set called Caltech 101 [29] which has classes such as "butterfly", "dolphin", "helicopter", "lotus", "panda", and "watch". The DRB classifier formed by IF…THEN fuzzy rules of *AnYa* type are depicted in Fig. 9.2.

Fig. 9.2 Illustrative example of *AnYa* type fuzzy rules with Caltech 101 dataset

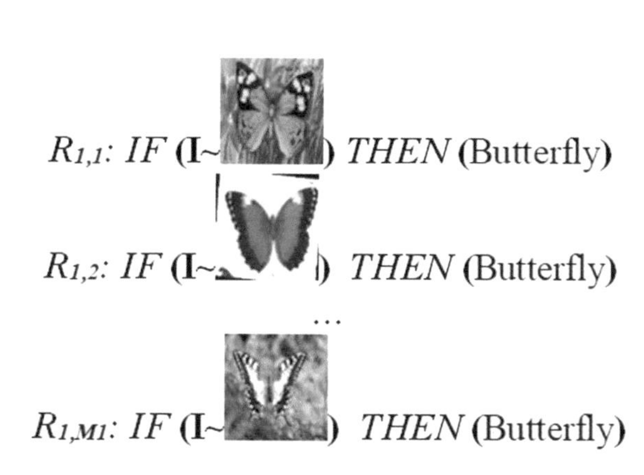

Fig. 9.3 Illustrative example of simpler *AnYa* fuzzy rules

Each *AnYa* type fuzzy rule [27] within the rule base of the DRB system contains a number of prototypes identified from the images of the corresponding class, which are connected by a *local* decision-maker using the "winner takes all" principle. Therefore, each rule can be represented as a series of simpler fuzzy rules with a single prototype connected by a logical "*OR*" operator, see Fig. 9.3 as an example, where the first fuzzy rule in Fig. 9.2, $\boldsymbol{R}_1$ is used to demonstrate the concept. It is obvious that a massive parallelization is possible.

9.4 Functioning of the DRB

As stated in the previous section, the learning process of the DRB classifier is conducted by a fully online learning algorithm, which is, essentially, a modification of the *ALMMo-0* learning procedure as presented in Sect. 8.2.2.1 and is able to start from "from scratch" as well. Through a very efficient, and highly parallelizable learning process, the DRB system automatically identifies prototypical images from the image set/stream and creates Voronoi tessellations [30] around the prototypes per class in the $(N+1)$ dimensional feature space as described in the previous sections. For a training image set/stream consisting of C classes, C independent zero-order FRB systems are trained in parallel (one per class).

In this section, we further describe in more detail the learning and validation processes of the DRB classifier. As the DRB classifier identifies the image prototypes from the very high dimensional feature space, for each image, denoted by $\mathbf{I}$, its corresponding feature vector $\boldsymbol{x}$ is normalized by its norm$\|\boldsymbol{x}\|$, namely, $\boldsymbol{x} \leftarrow \frac{\boldsymbol{x}}{\|\boldsymbol{x}\|}$ as a default before processing (see also Eq. (8.2)). This normalization process effectively enhances the ability of the DRB classifier in handling high dimensional feature vectors [31].

9.4.1 *Learning Process*

During the learning process of the DRB classifier, the prototypes are being identified as the peaks of the *data density* in the $(N+1)$ dimensional feature space derived from the image space [13–15]. Based on these prototypes, the corresponding fuzzy rules are generated. Same as the *ALMMo-0* system, the learning process of the DRB classifier is highly parallelizable (per class and even per rule using a single prototype (before compressing using logical "*OR*"). Due to the fact that the learning algorithm used by the DRB is an extension of the algorithm used by *ALMMo-0* (see Sect. 8.2.2) we call it ***DRB learning algorithm*** and summarize as follows [12]:

A. ***System Initialization***

The ith fuzzy rule is initialized by the first image (denoted by $\mathbf{I}_{i,1}$) of the corresponding class with feature vector denoted by $\boldsymbol{x}_{i,1}$ ($\boldsymbol{x}_{i,1} = \left[x_{i,1,1}, x_{i,1,2}, \ldots, x_{i,1,N}\right]^T$). The meta-parameters of the ith subsystem and the first *data cloud* within the subsystem, denoted by $\mathbf{C}_{i,1}$, are initialized, respectively, as follows:

$$K_i \leftarrow 1 \tag{9.3a}$$

$$M_i \leftarrow 1 \tag{9.3b}$$

$$\boldsymbol{\mu}_i \leftarrow \boldsymbol{x}_{i,1} \tag{9.3c}$$

$$\mathbf{P}_{i,1} \leftarrow \mathbf{I}_{i,1} \tag{9.3d}$$

$$\mathbf{C}_{i,1} \leftarrow \{\mathbf{I}_{i,1}\} \tag{9.3e}$$

$$\boldsymbol{p}_{i,1} \leftarrow \boldsymbol{x}_{i,1} \tag{9.3f}$$

$$S_{i,1} \leftarrow 1 \tag{9.3g}$$

$$r_{i,1} \leftarrow r_o \tag{9.3h}$$

where $\mathbf{P}_{i,1}$ is the first image prototype of the fuzzy rule; $\boldsymbol{\mu}_i$ is the *global* mean of all the feature vectors of the observed images of the ith class; $\boldsymbol{p}_{i,1}$ is the feature prototype corresponding to $\mathbf{P}_{i,1}$, which is, essentially, the mean of the feature vectors of the images within $\mathbf{C}_{i,1}$; $r_o = \sqrt{2(1 - \cos(30^\circ))} \approx 0.5176$ is the degree of similarity (an angle of less than 30° between vectors indicate a *high* degree of similarity; we use this non user- and problem-specific parameter by default and for all problems and images).

With the first image prototype, $\mathbf{P}_{i,1}$, the fuzzy rule is initialized as follows:

$$\boldsymbol{R}_i: \quad IF\ \left(\mathbf{I} \sim \mathbf{P}_{i,1}\right) \quad THEN\ \ (Class\, i) \tag{9.4}$$

B. ***System Recursive Update***

For the newly arrived K_i^{th} ($K_i^{th} \leftarrow K_i^{th} + 1$) training image ($\mathbf{I}_{i,K_i}$) that belongs to the ith class, the *global* mean $\boldsymbol{\mu}_i$ is firstly updated with the feature vector of $\mathbf{I}_{i,K_i}$, denoted by $\boldsymbol{x}_{i,K_i}$, using Eq. (4.8).

We then calculate the *data densities* at the prototypes $\mathbf{P}_{i,j}$ ($j = 1, 2, \ldots, M_i$) and the image $\mathbf{I}_{i,K_i}$ using Eq. (4.32) [32, 33] ($\mathbf{Z} = \mathbf{P}_{i,1}, \mathbf{P}_{i,2}, \ldots, \mathbf{P}_{i,M_i}, \mathbf{I}_{i,K_i}$; $z = \boldsymbol{p}_{i,1}, \boldsymbol{p}_{i,2}, \ldots, \boldsymbol{p}_{i,M_i}, \boldsymbol{x}_{i,K_i}$; z is the corresponding feature vector of $\mathbf{Z}$):

$$D_{K_i}(\mathbf{Z}) = \frac{1}{1 + \frac{\|z - \boldsymbol{\mu}_i\|^2}{1 - \|\boldsymbol{\mu}_i\|^2}} \tag{9.5}$$

Then, Condition 9.1 is checked to see whether $\mathbf{I}_{i,K_i}$ is a new prototype [34], which is a modification of Condition 7.4:

Condition 9.1

$$\begin{aligned} & IF\ \left(D_{K_i}\left(\mathbf{I}_{i,K_i}\right) > \max_{j=1,2,\ldots M_i}\left(D_{K_i}\left(\mathbf{P}_{i,j}\right)\right)\right) \\ & OR\ \left(D_{K_i}\left(\mathbf{I}_{i,K_i}\right) < \min_{j=1,2,\ldots,M_i}\left(D_{K_i}\left(\mathbf{P}_{i,j}\right)\right)\right) \\ & THEN\ \left(\mathbf{I}_{i,K_i}\ is\ a\ new\ image\, prototype\right) \end{aligned} \tag{9.6}$$

Once Condition 9.1 is satisfied, $\mathbf{I}_{i,K_i}$ is set to be a new prototype and it initializes a new *data cloud*:

$$M_i \leftarrow M_i + 1 \tag{9.7a}$$

$$\mathbf{P}_{i,M_i} \leftarrow \mathbf{I}_{i,K_i} \tag{9.7b}$$

$$\mathbf{C}_{i,M_i} \leftarrow \left\{\mathbf{I}_{i,K_i}\right\} \tag{9.7c}$$

$$\boldsymbol{p}_{i,M_i} \leftarrow \boldsymbol{x}_{i,K_i} \tag{9.7d}$$

$$S_{i,M_i} \leftarrow 1 \tag{9.7e}$$

$$r_{i,M_i} \leftarrow r_o \tag{9.7f}$$

If Condition 9.1 is not met, the nearest prototype to $\mathbf{I}_{i,K_i}$, denoted by $\mathbf{P}_{i,n*}$, is identified by the following equation:

$$n* = \underset{j=1,2,\ldots,M_i}{\arg\min}\left(\left\|\boldsymbol{p}_{i,j} - \boldsymbol{x}_{i,K_i}\right\|\right) \tag{9.8}$$

Before $\mathbf{I}_{i,K_i}$ is assigned to $\mathbf{C}_{i,n*}$, Condition 9.2 (a modification of Condition 8.1) is checked to see whether $\mathbf{I}_{i,K_i}$ is located in the area of influence of $\mathbf{P}_{i,n*}$:

Condition 9.2

$$IF\ \left(\left\|\boldsymbol{p}_{i,n*} - \boldsymbol{x}_{i,K_i}\right\| \leq r_{i,n*}\right)\quad THEN\ \left(\mathbf{C}_{i,n*} \leftarrow \mathbf{C}_{i,n*} + \mathbf{I}_{i,K_i}\right) \tag{9.9}$$

If Condition 9.2 is met, it means that $\mathbf{I}_{i,K_i}$ is within the area of influence of the nearest *data cloud*, $\mathbf{C}_{i,n*}$, and it is assigned to the *data cloud* formed around the image prototype $\mathbf{P}_{i,n*}$ with the meta-parameters updated as follows:

$$\boldsymbol{p}_{i,n*} \leftarrow \frac{S_{i,n*}}{S_{i,n*}+1}\boldsymbol{p}_{i,n*} + \frac{1}{S_{i,n*}+1}\boldsymbol{x}_{i,K_i} \tag{9.10a}$$

$$S_{i,n*} \leftarrow S_{i,n*} + 1 \tag{9.10b}$$

$$r_{i,n*} \leftarrow \sqrt{\frac{1}{2}\left(r_{i,n*}\right)^2 + \frac{1}{2}\left(1 - \left\|\boldsymbol{p}_{i,n*}\right\|^2\right)} \tag{9.10c}$$

and the image prototype $\mathbf{P}_{i,n*}$ stays the same.

Otherwise, $\mathbf{I}_{i,K_i}$ becomes a new prototype and a new *data cloud* is initialized by Eqs. (9.7a–9.7f). After $\mathbf{I}_{i,K_i}$ has been processed, the DRB system will update the fuzzy rule accordingly. Then, the next image is read and the system starts a new processing cycle. The algorithm applications and implementations are illustrated in Chap. 13, Appendices B.5 and C.5, respectively.

9.4.2 *Decision Making*

In this subsection, the two-stage decision making mechanism used by the DRB classifier is described in more detail.

9.4.2.1 *Local* Decision-Making

After the system identification procedure (described in Sect. 9.4.1), the DRB system generates C massively parallel fuzzy rules per class. During the validation/usage of the classifier, for each test image $\mathbf{I}$, each one of the C *AnYa* type fuzzy IF…THEN rules will generate a firing strength (or score of confidence),$\lambda_i(\mathbf{I})$ $(i = 1, 2, \ldots, C)$ by their *local* decision-maker based on the feature vector of $\mathbf{I}$, denoted by $\boldsymbol{x}$:

$$\lambda_i(\mathbf{I}) = \max_{j=1,2,\ldots,M_i}\left(\exp\left(-\left\|\boldsymbol{p}_{i,j} - \boldsymbol{x}\right\|^2\right)\right) \tag{9.11}$$

Accordingly, one can get in total C firing strengths from the rule base (one per rule), denoted by $\boldsymbol{\lambda}(\mathbf{I}) = [\lambda_1(\mathbf{I}), \lambda_2(\mathbf{I}), \ldots, \lambda_C(\mathbf{I})]$, which are the inputs passed to the overall decision-maker of the DRB classifier.

9.4.2.2 Overall Decision-Making

For a single DRB classifier, the class label (category) of the testing sample is given by the overall decision-maker, namely, the last layer, following the "winner takes all" principle:

$$\text{Label} = \arg\max_{i=1,2,\ldots,C}(\lambda_i(\mathbf{I})) \tag{9.12}$$

In some applications, i.e. face recognition, remote sensing, object recognition, etc., where *local* information plays a more important role than the *global* information, one may consider to segment (both the training and validation) images into smaller pieces to gain more *local* information. In such cases, the DRB system is trained with segments of training images instead of the full images. In such cases, the overall label of a testing image is given as an integration of all the firing strengths that the DRB sub-systems give to its segments, denoted by $\mathbf{s}_t$ $(t = 1, 2, \ldots, T)$:

$$\text{Label} = \underset{i=1,2,\ldots,C}{\arg\max}\left(\sum_{t=1}^{T}\lambda_i(\mathbf{s}_t)\right) \tag{9.13}$$

where $\mathbf{s}_t$ denotes the tth segment of image $\mathbf{I}$; T is the number of segments.

If a DRB ensemble classifier [1] is used, the label of the testing image is considered as an integration of all the firing strengths that the DRB systems give to the image [13]:

$$\text{Label} = \underset{i=1,2,\ldots,C}{\arg\max}\left(\frac{1}{H}\sum_{n=1}^{H}\lambda_{n,i}(\mathbf{I}) + \max_{n=1,2,\ldots,H}\left(\lambda_{n,i}(\mathbf{I})\right)\right) \tag{9.14}$$

where H is the number of DRB systems in the ensemble.

In contrast with the simple voting mechanism used in many other works, the overall decision made by the DRB ensemble classifier integrates more information to make the judgement by taking both, the overall firing strengths and the maximum firing strengths into consideration.

However, it has to be stressed that, the DRB system is a general approach with simple architecture and fundamental principles. Many different modifications and extensions can be added to the DRB for different purposes. In this book we focus mainly on providing the general principles and demonstrating the concept. More details are provided in journal and conference publications, e.g. [12–15].

9.5 Semi-supervised DRB Classifier

In this section, the DRB classifier as described in the previous sections of this chapter is further extended in the direction of semi-supervised learning strategies for both, offline and online scenarios [12–14]. A strategy for the DRB classifier to actively learn new classes from unlabeled training images is also presented. Thanks to the prototype-based nature of the DRB classifier, the semi-supervised learning process is fully transparent and human-interpretable. It not only can perform classification on out-of-sample images, but also supports recursive online training on a sample-by-sample basis or a chunk-by-chunk basis. Moreover, unlike other semi-supervised approaches, the semi-supervised DRB (SS_DRB) classifier is able to learn new classes actively without human experts' involvement, thus, to self-evolve [34].

Compared with the existing semi-supervised approaches [35–46], the SS_DRB classifier has the following distinctive features because of its prototype-based nature [12–14]:

(1) Its semi-supervised learning process is fully transparent and human-interpretable;
(2) It can be trained online on a sample-by-sample or chunk-by-chunk basis;
(3) It can classify out-of-sample images;
(4) It is able to learn new classes (self-evolve).

In the following subsections, the semi-supervised learning strategies of the DRB classifier in both offline and online scenarios are described. A strategy for the DRB classifier to actively learn new classes from unlabeled training images is also presented.

9.5.1 Semi-supervised Learning from Static Datasets

In an offline scenario, all the unlabeled training images are available as a static dataset and the DRB classifier starts to learn from these images after the learning process based on the labelled ones is finished.

Let us, firstly, define the unlabeled training images as the set $\{\mathbf{V}\}$, and the number of unlabeled training images as K. The main steps of the ***offline SS_DRB learning algorithm*** are as follows:

Step 1. Extract the vector of firing strength to the nearest prototypes for each unlabeled training image, denoted by $\boldsymbol{\lambda}(\mathbf{V}_j) = [\lambda_1(\mathbf{V}_j), \lambda_2(\mathbf{V}_j), \ldots, \lambda_C(\mathbf{V}_j)]$ ($j = 1, 2, \ldots, K$), using Eq. (9.11)
Step 2. Find out all the unlabeled training images satisfying Condition 9.3:

Condition 9.3

$$IF\ \left(\lambda_{\max *}(\mathbf{V}_j) > \Omega_1 \cdot \lambda_{\max **}(\mathbf{V}_j)\right)\quad THEN\ \left(\mathbf{V}_j \in \{\mathbf{V}\}_0\right) \tag{9.15}$$

where $\lambda_{\max *}(\mathbf{V}_j)$ denotes the highest firing strength that $\mathbf{V}_i$ obtains, and $\lambda_{\max **}(\mathbf{V}_j)$ denotes the second highest score; Ω_1 ($\Omega_1 > 1$) is a free parameter; $\{\mathbf{V}\}_0$ denotes the collection of the unlabeled training images that satisfy Condition 9.3. Then, $\{\mathbf{V}\}_0$ is removed from $\{\mathbf{V}\}$.

For the unlabeled training images within $\{\mathbf{V}\}_0$, the DRB classifier is highly confident about the class they belong to and they can be used for updating the structure and meta-parameters of the system. On the other hand, for the remaining images within $\{\mathbf{V}\}$, the DRB classifier is not highly confident about its judgement and, thus, these images cannot be used for updating.

Compared with other pseudo-labelling mechanisms used in other works [16, 47, 48], Condition 9.3 considers not only the mutual distances between the unlabeled images and all the identified prototypes, but also the discernibility of the unlabeled images themselves, which leads to a more accurate pseudo-labelling process.

Step 3. **Rank** the elements within $\{\mathbf{V}\}_0$ in a descending order in terms of the values of $(\lambda_{\max*}(\mathbf{V}) - \lambda_{\max**}(\mathbf{V}))$ $(\mathbf{V} \in \{\mathbf{V}\}_0)$, and denote the ranked set as $\{\mathbf{V}\}_1$

In general, the higher $\lambda_{\max*}(\mathbf{V})$ is, the more similar the image $\mathbf{V}$ is to a particular prototype of the DRB classifier. Meanwhile, the higher the difference $(\lambda_{\max*}(\mathbf{V}) - \lambda_{\max**}(\mathbf{V}))$ is, the less ambiguous the decision made by the DRB classifier is. Since the DRB classifier learns in a sample-by-sample manner from a data stream, by ranking $\{\mathbf{V}\}_0$ in advance, the classifier will, firstly, update itself with images that are the most similar to the previously identified prototypes and have less ambiguity in the decisions about their labels, and later with the less familiar ones, which guarantees a more efficient learning.

Step 4. Update the DRB classifier using the **DRB learning algorithm** presented in Sect. 9.4.1 with the set $\{\mathbf{V}\}_1$. Then the SS_DRB classifier goes back to **step 1** and repeats the whole process, until there are no unlabeled training images within $\{\mathbf{V}\}$ that satisfy Condition 9.3

If the DRB classifier is not designed to learn new classes, once the offline semi-supervised learning process is finished, the labels of all the unlabeled training images will be estimated using Eqs. (9.11) and (9.12) following the “winner takes all” principle. Otherwise, the DRB classifier will, firstly, produce the pseudo labels of the images that can meet Condition 9.3 and, then, learn new classes through the remaining unlabeled images (will self-evolve). The active learning strategy will be detailed in Sect. 9.5.3.

One distinctive feature of the SS_DRB classifier is its robustness to the incorrectly pseudo-labelled images thanks to its prototype-based nature and the “one pass” type (non-iterative) learning process. The incorrectly pseudo-labelled images might influence the SS_DRB classifier in two ways:

(1) they can slightly shift the positions of some of the prototypes, and
(2) they can create new false prototypes by assigning wrong pseudo labels.

However, the SS_DRB classifier has a strong tolerance to a small amount of incorrectly pseudo-labelled images because the errors will not be propagated to the majority of previously existing prototypes in the IF…THEN rule base as detailed in [49].

9.5.2 Semi-supervised Learning from Data Streams

Often in practice, after the learning algorithms have processed the available static data, new data continuously arrives in the form of a data stream (video or time series). The previously introduced semi-supervised approaches [35, 36, 45, 46, 37–44] are limited to offline application due to their operating mechanism. Thanks to the prototype-based nature and the evolving mechanism [2, 34] of the DRB classifier [13–15], online semi-supervised learning can also be conducted.

The online semi-supervised learning of the DRB classifier can be conducted on a sample-by-sample basis or a chunk-by-chunk basis after the supervised training process with the labelled training images finishes. As the online learning process is a modification of the offline learning process as described in Sect. 9.5.1, in this subsection, we only present the main procedure of the semi-supervised learning processes of both types.

The main steps of the ***sample-by-sample online SS_DRB learning algorithm*** are presented firstly as follows:

Step 1. Use Condition 9.3 and the DRB training algorithm to learn from the available unlabeled new image $\mathbf{V}_{K+1}$;
Step 2 (optional). Apply the **active learning algorithm** to actively learn new class from $\mathbf{U}_{K+1}$. This makes this algorithm evolving [34].
Step 3. DRB classifier goes back to **Step 1** and processes the next image $(K \leftarrow K+1)$

The main steps of the ***chunk-by-chunk online SS_DRB learning algorithm*** are presented firstly as follows (initially, $j \leftarrow 1$):

Step 1. Use **offline SS_DRB learning algorithm** to learn from the available chunk of unlabeled images $\{\mathbf{V}\}_H^j$, where H is the chunk size;
Step 2 (optional). Apply the **active learning algorithm** to evolve with new classes from the remaining images in $\{\mathbf{V}\}_H^j$;
Step 3. The DRB classifier goes back to **Step 1** and processes the next chunk $(j \leftarrow j+1)$

It has to be kept in mind that as any other incremental method, the proposed semi-supervised learning is order-dependent. This means that the order of images will influence the results in an online mode.

9.5.3 Active Learning of New Classes, Evolving DRB

In real situations, the labelled training samples may fail to include some classes due to various reasons, i.e. an insufficient amount of *prior* knowledge or change of the data pattern. For example, the pre-trained VGG-VD-16 [50] classifies the three

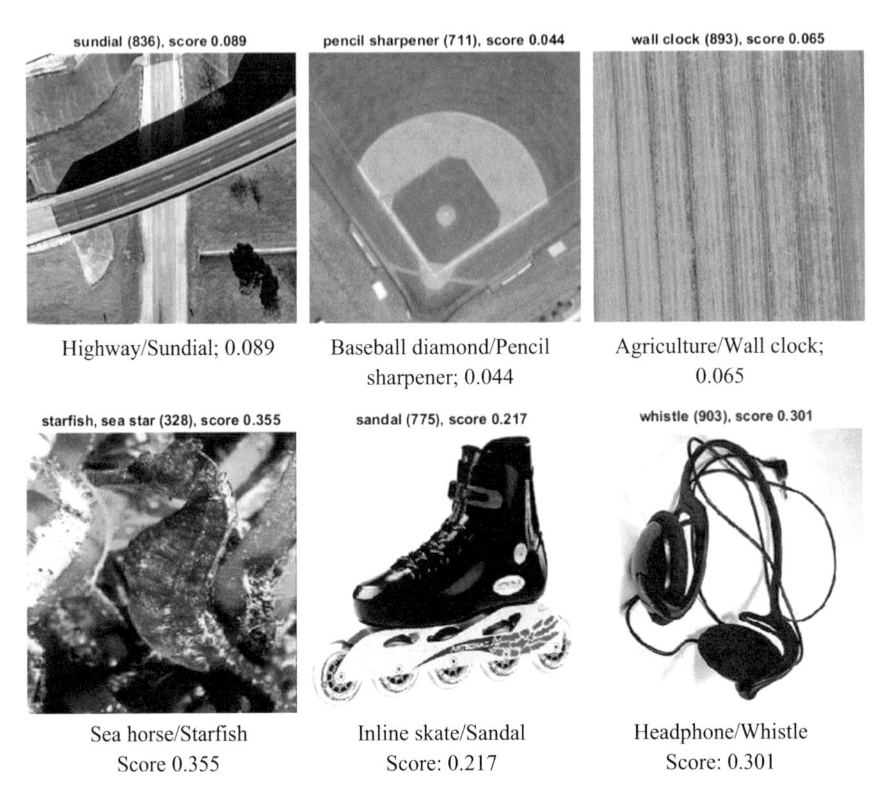

Fig. 9.4 Images misclassified by DCNN [50] (the class label in the left represents the truth; on the right—the output of the DCNN; the value is the score of confidence)

remote sensing images from the well-known UCMerced dataset [51, 52] depicted in Fig. 9.4, top raw (true labels: "highway", "baseball diamond" and "agriculture") wrongly as "sundial", "pencil sharpener" and "wall clock". The images on the bottom raw (taken from the Caltech 101 dataset used earlier) are also misclassified and the classifier was not trained with images of these categories. One can also notice that the degree of confidence is much higher for the second set of misclassifications (0.217 to 0.355) while for the first set of misclassifications it is less than 0.1.

From the example in Fig. 9.4 one can see that it is of paramount importance for a classifier to be able to learn new classes actively, which not only guarantees the effectiveness of the learning process and reduces the requirement for *prior* knowledge, but also enables the human experts to monitor the changes of the data pattern. In addition, this is another anthropomorphic feature (indeed, humans are able to continue to learn during their whole live).

Therefore, the ***active learning algorithm*** for the SS_DRB classifier is introduced as follows:

For an unlabeled training image $\mathbf{V}_j$, if Condition 9.4 is met, it means that the DRB classifier has not seen any similar images before, and therefore, a new class is being added.

Condition 9.4

$$IF\ \left(\lambda_{\max*}(\mathbf{V}_j) \leq \Omega_2\right)\quad THEN\ \left(\mathbf{V}_j \in\ the\ (C+1)th\ class\right) \tag{9.16}$$

where Ω_2 is a free parameter serving as a threshold.

Accordingly, a new fuzzy rule is also added to the rule base with this training image as the first prototype of this new class. In general, the lower Ω_2 is, the more conservatively the SS_DRB classifier behaves when adding new rules to the rule base.

In the offline learning scenario, there may be a number of unlabeled images remaining in $\{\mathbf{V}\}$ after the offline semi-supervised learning process, re-denoted by $\{\mathbf{V}\}_2$. Some of these may satisfy Condition 9.4, and their collection is denoted by $\{\mathbf{V}\}_2$, $\{\mathbf{V}\}_3 \subseteq \{\mathbf{V}\}_2$. As many of the images within $\{\mathbf{V}\}_2$ may actually belong to a few unknown classes, to classify these images, the DRB classifier needs to add a few new fuzzy IF...THEN rules to the existing rule base in an autonomous and proactive manner.

The **active learning algorithm** starts by selecting out the image that has the lowest $\lambda_{\max*}$, denoted by $\mathbf{V}_{\min}$ ($\mathbf{V}_{\min} \in \{\mathbf{V}\}_3$), and adds a new fuzzy rule with $\mathbf{V}_{\min}$ as the corresponding prototype. However, before adding another new fuzzy rule, the DRB classifier repeats the offline semi-supervised learning process on the remaining unselected images within $\{\mathbf{V}\}_2$ to find other unidentified prototypes that are associated with the newly added fuzzy rule. This may solve the potential problem of adding too many rules. After the newly formed fuzzy rule is fully updated, the DRB classifier will start to add the next new rule.

With the active learning strategy, the DRB classifier is able to actively learn from unlabeled training images, gain new knowledge, define new classes and add new rules, correspondingly. Human experts can also examine the new fuzzy rules and give meaningful labels for the new classes by simply checking the prototypes afterwards. This is much less laborious than the usual approach as it only concerns the aggregated prototypical data, not the high volume raw data, and it is more convenient for the human users. However, it is also necessary to stress that identifying new classes and labelling them with human-understandable labels is not compulsory for the DRB classifier to work since in many applications, the classes of the images are predictable based on common knowledge. For example, for handwritten digits recognition problem, there will be images from *10* classes (from "0" to "9"), for characters recognition problem, there will be images from 60 classes (from "a"to "я" and "A"to "Я") for the Cyrillic alphabet [53], 48 for the Greek alphabet (from "α" to "ω" and from "A" to "Ω") and *52* classes (from "a" to "z" and "A" to Z") for the Latin alphabet, etc.

9.6 Distributed Collaborative DRB Classifiers

In real world applications, a distributed collaborative system allows multiple agents to engage in a shared activity, usually, from remote locations. All the agents in the system work together towards a common goal and interact closely with each other to share information. Distributed collaborative systems provide a feasible solution for handling very large scale data by parallelizing into a number of processing units [54]. However, the learning process of the current machine learning algorithms, including the state-of-the-art DCNNs, requires iterative solutions, and, thus, is difficult for parallelization and scaling up. In DCNNs, parallelization can be applied and is boldly claimed, but this concerns image processing (max-pooling), while DRB allows parallelization at a significantly higher degree-up to each individual fuzzy rule and prototype.

Thanks to the prototype-based nature and the non-iterative learning process, DRB classifiers can be distributed in parallel tackling different chunks of the same problem independently, meanwhile, collaboratively achieve the ultimate solution by exchanging the key information only, namely, prototypes as well as other meta-parameters, which is much more efficient than exchanging the whole image set.

In Fig. 9.5, two different types of distributed architecture for the DRB classifiers are presented for collaborative learning from a very large image set. In this figure, the image distributor splits the images into different chunks. The fusion center merges the partial results obtained by different DRB classifiers. The functionality of all the DRB classifiers is exactly the same with only differences in the inputs and outputs [54, 18].

9.6.1 Pipeline Processing

The pipeline architecture (depicted in Fig. 9.5a) is composed of a series of DRB classifiers sequentially learning from the corresponding image chunks $\{\mathbf{I}\}_{K_1}^{1}$,$\{\mathbf{I}\}_{K_2}^{2}$, ... ,$\{\mathbf{I}\}_{K_J}^{J}$ derived from the original image set $\{\mathbf{I}\}_{K}$ [55]. Each DRB classifier receives the meta-parameters (partial result) from the previous predecessor classifier and, then, it updates the meta-parameters with its own image chunk and passes it to the next DRB classifier. The final DRB classifier, namely, the Jth one produces the final output.

This architecture reduces the amount of images being processed by each DRB classifier and, thus, improves the overall processing efficiency and relieves the computational and memory burden. In addition, this architecture preserves the order of processing in the sequential algorithms and can guarantee exactly the same result as if all the images are processed by a single DRB classifier [55].

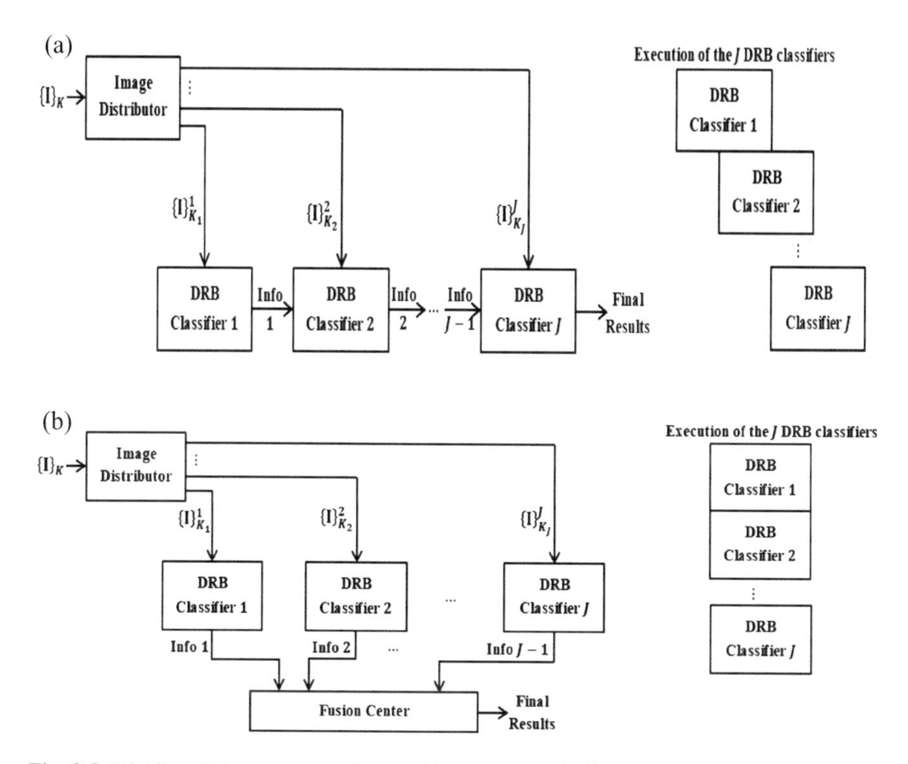

Fig. 9.5 Distributed image processing architectures. **a** Pipeline processing [55]. **b** Parallel processing [34, 56]

9.6.2 Parallel Processing

The parallel architecture (depicted in Fig. 9.5b) allows a full parallelization of the learning process [34, 56]. After the image chunks $\{\mathbf{I}\}^1_{K_1}, \{\mathbf{I}\}^2_{K_2}, \ldots, \{\mathbf{I}\}^J_{K_J}$ are derived from the original image set $\{\mathbf{I}\}_K$, they are sent to individual DRB classifies in parallel. The fusion center receives the partial results obtained by the DRB classifiers and merges them together to produce the final output.

This architecture allows a higher computational and memory efficiency in contrast to the pipeline architecture, but is not able to guarantee the same results as if all the images are processed by a single DRB classifier as it is done by the scheme depicted in Fig. 9.5b [55].

9.7 Conclusions

In this chapter we briefly review the prototype-based classifiers such as kNN, SVM, eClass0, *ALMMo-0* and move to describe the proposed DRB classifier.

Because of its prototype-based nature, the DRB classifier is free from *prior* assumptions about the type of the data distribution, their random or deterministic nature, and there are no requirements to make ad hoc decisions, i.e. regarding the model structure, membership functions, number of layers, etc. Meanwhile, the prototype-based nature allows the DRB classifier to be non-parametric, non-iterative, self-organizing, self-evolving and highly parallel.

The training of the DRB classifier is a modification of the *ALMMo-0* learning algorithm, and, thus, the training process is highly, parallelizable, significantly faster and can start "from scratch". With a prototype-based nature, the DRB classifier is able to identify a number of prototypes from the training images in a fully online, autonomous, non-iterative and non-parametric manner, and, based on this, to build a massively parallel fuzzy rule base for classification. The training and validation processes of the DRB classifier concern only the visual similarity between the identified prototypes and the unlabeled samples, and only involve the very general principles.

The SS_DRB classifier has the following unique features compared with the existing semi-supervised approaches in addition to the unique features of the DRB classifier:

(1) The learning process can be conducted on a sample-by-sample or chunk-by-chunk basis;
(2) It is able to perform out-of-sample classification;
(3) It is able to learn new classes in an active manner and can self-evolve its system structure.
(4) There are no iterations involved in the learning process.

9.8 Questions

(1) Provide examples of prototype-based classification methods. How does the proposed DRB classifier differ from them?
(2) What are the main advantages of the DRB classifier?
(3) How does the DRB classifier relate to the *ALMMo* systems?
(4) Describe how the semi-supervised version of the DRB classifier works. Which are the two types of work of the semi-supervised DRB classifier in regards to the provided data?

9.9 Take Away Items

- The DRB classifier is prototype-based and because of this, it is non-iterative, self-organizing, highly transparent and free from pre-defined parameters as well as *prior* assumptions about the models of data generation and randomness.
- The training of the DRB classifier is highly parallelizable, can start from scratch and is very fast.
- The DRB classifier can be seen as an extension of *ALMMo*.
- It has a semi-supervised version which can work in a sample-by-sample or chunk-by-chunk mode.
- It is able to perform out of sample classification and active learning.

References

1. L. Kuncheva, *Combining Pattern Classifiers: Methods and Algorithms* (Wiley, Hoboken, New Jersey, 2004)
2. P. Angelov, X. Zhou, Evolving fuzzy-rule based classifiers from data streams. IEEE Trans. Fuzzy Syst. **16**(6), 1462–1474 (2008)
3. P.P. Angelov, X. Gu, Autonomous learning multi-model classifier of 0-order (ALMMo-0), in *IEEE International Conference on Evolving and Autonomous Intelligent Systems* (2017), pp. 1–7
4. P.P. Angelov, X. Gu, Towards anthropomorphic machine learning. *IEEE Comput.*, (2018)
5. T. Cover, P. Hart, Nearest neighbor pattern classification. IEEE Trans. Inf. Theory **13**(1), 21–27 (1967)
6. E. Pękalska, R.P.W. Duin, P. Paclík, Prototype selection for dissimilarity-based classifiers. Pattern Recognit. **39**(2), 189–208 (2006)
7. N. Cristianini, J. Shawe-Taylor, *An Introduction to Support Vector Machines and Other Kernel-Based Learning Methods* (Cambridge University Press, Cambridge, 2000)
8. R.D. Baruah, P.P. Angelov, J. Andreu, Simpl_eClass : simplified potential-free evolving fuzzy rule-based classifiers, in *IEEE International Conference on Systems, Man, and Cybernetics (SMC)* (2011), pp. 2249–2254
9. D. Kangin, P. Angelov, J.A. Iglesias, Autonomously evolving classifier TEDAClass. Inf. Sci. (Ny) **366**, 1–11 (2016)
10. T. Kohonen, *Self-organizing Maps* (Springer, Berlin, 1997)
11. P. Perner, Prototype-based classification. Appl. Intell. **28**(3), 238–246 (2008)
12. P.P. Angelov, X. Gu, Deep rule-based classifier with human-level performance and characteristics. Inf. Sci. (Ny) **463–464**, 196–213 (2018)
13. P.P. Angelov, X. Gu, MICE: Multi-layer multi-model images classifier ensemble, in *IEEE International Conference on Cybernetics* (2017), pp. 436–443
14. P. Angelov, X. Gu, A cascade of deep learning fuzzy rule-based image classifier and SVM, in *International Conference on Systems, Man and Cybernetics* (2017), pp. 1–8
15. X. Gu, P. Angelov, C. Zhang, P. Atkinson, A massively parallel deep rule-based ensemble classifier for remote sensing scenes. IEEE Geosci. Remote Sens. Lett. **15**(3), 345–349 (2018)

16. J. Zhang, X. Kong, P.S. Yu, Predicting social links for new users across aligned heterogeneous social networks, in *IEEE International Conference on Data Mining* (2013), pp. 1289–1294
17. D. Ciresan, U. Meier, J. Schmidhuber, Multi-column deep neural networks for image classification, in *Conference on Computer Vision and Pattern Recognition* (2012), pp. 3642–3649
18. P. Angelov, Machine learning (collaborative systems), 8250004 (2006)
19. D.C. Cireşan, U. Meier, L.M. Gambardella, J. Schmidhuber, Convolutional neural network committees for handwritten character classification, in *International Conference on Document Analysis and Recognition*, vol. 10 (2011), pp. 1135–1139
20. K. Fukushima, Neocognitron for handwritten digit recognition. Neurocomputing **51**, 161–180 (2003)
21. A. Oliva, A. Torralba, Modeling the shape of the scene: a holistic representation of the spatial envelope. Int. J. Comput. Vis. **42**(3), 145–175 (2001)
22. G.-S. Xia, J. Hu, F. Hu, B. Shi, X. Bai, Y. Zhong, L. Zhang, AID: a benchmark dataset for performance evaluation of aerial scene classification. IEEE Trans. Geosci. Remote Sens. **55** (7), 3965–3981 (2017)
23. N. Dalal, B. Triggs, Histograms of oriented gradients for human detection, in *IEEE Computer Society Conference on Computer Vision and Pattern Recognition* (2005), pp. 886–893
24. D.G. Lowe, Distinctive image features from scale-invariant keypoints. Int. J. Comput. Vis. **60** (2), 91–110 (2004)
25. C. Szegedy, W. Liu, Y. Jia, P. Sermanet, S. Reed, D. Anguelov, D. Erhan, V. Vanhoucke, A. Rabinovich, C. Hill, A. Arbor, Going deeper with convolutions, in *IEEE Conference on Computer Vision and Pattern Recognition* (2015), pp. 1–9
26. A. Krizhevsky, I. Sutskever, G.E. Hinton, ImageNet classification with deep convolutional neural networks, in *Advances in Neural Information Processing Systems* (2012), pp. 1097–1105
27. P. Angelov, R. Yager, A new type of simplified fuzzy rule-based system. Int. J. Gen Syst **41** (2), 163–185 (2011)
28. P.P. Angelov, X. Gu, J. Principe, A generalized methodology for data analysis. IEEE Trans. Cybern. **48**(10), 2987–2993 (2018).
29. http://www.vision.caltech.edu/Image_Datasets/Caltech101/
30. A. Okabe, B. Boots, K. Sugihara, S.N. Chiu, *Spatial Tessellations: Concepts and Applications of Voronoi Diagrams*, 2nd edn. (Wiley, Chichester, England, 1999)
31. X. Gu, P.P. Angelov, Self-organising fuzzy logic classifier. Inf. Sci. (Ny) **447**, 36–51 (2018)
32. P. Angelov, X. Gu, D. Kangin, Empirical data analytics. Int. J. Intell. Syst. **32**(12), 1261–1284 (2017)
33. P.P. Angelov, X. Gu, J. Principe, D. Kangin, Empirical data analysis—a new tool for data analytics, in *IEEE International Conference on Systems, Man, and Cybernetics* (2016), pp. 53–59
34. P. Angelov, *Autonomous Learning Systems: From Data Streams to Knowledge in Real Time*. Wiley, New York (2012)
35. X. Zhu, Z. Ghahraman, J.D. Lafferty, Semi-supervised learning using gaussian fields and harmonic functions, in *International Conference on Machine Learning* (2003), pp. 912–919
36. D. Zhou, O. Bousquet, T.N. Lal, J. Weston, B. Schölkopf, Learning with *local* and *global* consistency. Adv. Neural. Inform. Process Syst., pp. 321–328 (2004)
37. V. Sindhwani, P. Niyogi, M. Belkin, Beyond the point cloud: from transductive to semi-supervised learning, in *International Conference on Machine Learning*, vol. 1 (2005), pp. 824–831
38. F. Noorbehbahani, A. Fanian, R. Mousavi, H. Hasannejad, An incremental intrusion detection system using a new semi-supervised stream classification method. Int. J. Commun Syst **30**(4), 1–26 (2017)
39. O. Chapelle, A. Zien, Semi-supervised classification by low density separation, in *AISTATS* (2005), pp. 57–64

40. M. Guillaumin, J.J. Verbeek, C. Schmid, Multimodal semi-supervised learning for image classification, in *IEEE Conference on Computer Vision & Pattern Recognition* (2010), pp. 902–909
41. J. Wang, T. Jebara, S.F. Chang, Semi-supervised learning using greedy Max-Cut. J. Mach. Learn. Res. **14**, 771–800 (2013)
42. F. Wang, C. Zhang, H. C. Shen, J. Wang, Semi-supervised classification using linear neighborhood propagation, in *IEEE Conference on Computer Vision & Pattern Recognition* (2006), pp. 160–167
43. S. Xiang, F. Nie, C. Zhang, Semi-supervised classification via *local* spline regression. IEEE Trans. Pattern Anal. Mach. Intell. **32**(11), 2039–2053 (2010)
44. B. Jiang, H. Chen, B. Yuan, X. Yao, Scalable graph-based semi-supervised learning through sparse bayesian model. IEEE Trans. Knowl. Data Eng. (2017). https://doi.org/10.1109/TKDE.2017.2749574
45. J. Thorsten, Transductive inference for text classification using support vector machines. Int. Conf. Mach. Learn. **9**, 200–209 (1999)
46. O. Chapelle, V. Sindhwani, S. Keerthi, Optimization techniques for semi-supervised support vector machines. J. Mach. Learn. Res. **9**, 203–233 (2008)
47. K. Wu, K.-H. Yap, Fuzzy SVM for content-based image retrieval: a pseudo-label support vector machine framework. IEEE Comput. Intell. Mag. **1**(2), 10–16 (2006)
48. D.-H. Lee, Pseudo-label: the simple and efficient semi-supervised learning method for deep neural networks, in *ICML 2013 Workshop: Challenges in Representation Learning* (2013), pp. 1–6
49. X. Gu, P.P. Angelov, Semi-supervised deep rule-based approach for image classification. Appl. Soft Comput. **68**, 53–68 (2018)
50. K. Simonyan, A. Zisserman, Very deep convolutional networks for large-scale image recognition, in *International Conference on Learning Representations* (2015), pp. 1–14
51. Y. Yang, S. Newsam, Bag-of-visual-words and spatial extensions for land-use classification, in *International Conference on Advances in Geographic Information Systems* (2010), pp. 270–279
52. http://weegee.vision.ucmerced.edu/datasets/landuse.html
53. P.T. Daniels, W. Bright (eds.) *The World's Writing Systems*. Oxford University Press on Demand (1996)
54. D. Kangin, P. Angelov, J.A. Iglesias, A. Sanchis, Evolving classifier TEDAClass for big data. Procedia Comput. Sci. **53**(1), 9–18 (2015)
55. P. Angelov, Machine learning (collaborative systems), US 8250004, 2012
56. X. Gu, P.P. Angelov, G. Gutierrez, J. A. Iglesias, A. Sanchis, Parallel computing TEDA for high frequency streaming data clustering, in *INNS Conference on Big Data* (2016), pp. 238–253

Part III
Applications of the Proposed Approach

Chapter 10
Applications of Autonomous Anomaly Detection

Traditional anomaly detection approaches require human intervention, such as, *prior* assumptions, thresholds, features, etc. The autonomous anomaly detection (AAD) method, in contrast, is unsupervised, non-parametric, assumption-free [1].

In this chapter, the implementation and applications of the AAD algorithm is presented. The detailed algorithmic procedure can be found in Sect. 6.3.

10.1 Algorithm Summary

The main procedure of the AAD algorithm involves the following steps:

Step 1. Identify the candidates, $\{\boldsymbol{x}\}_{PA,1}$ for *global* anomalies based on the multimodal *typicality*, τ^{M} of the data;

Step 2. Identify the candidates, $\{\boldsymbol{x}\}_{PA,2}$ for *local* anomalies based on the *locally* weighted multimodal *typicality*, τ^{M*} of the data;

Step 3. Combine the candidates together, $\{\boldsymbol{x}\}_{PA} \leftarrow \{\boldsymbol{x}\}_{PA,1} + \{\boldsymbol{x}\}_{PA,2}$ and apply the autonomous data partitioning (ADP) algorithm to form *data clouds* $\{\mathbf{C}\}_{PA}$ and identify the anomalies $\{\boldsymbol{x}\}_{A}$ from $\{\boldsymbol{x}\}_{PA}$;

Step 4. Identify the *data clouds* formed by anomalies, denoted by $\{\mathbf{C}\}_{A}$ and obtain the anomalies, $\{\boldsymbol{x}\}_{A} \leftarrow \{\boldsymbol{x}\}_{A} + \{\mathbf{C}\}_{A}$ as the final output of the AAD algorithm.

The flowchart of the main procedure of the AAD algorithm is depicted in Fig. 10.1. The pseudo-code of this algorithm is presented in Appendix B.1 and the MATLAB implementation is given in Appendix C.1.

As one can see, the processes A and B, which, respectively, correspond to steps 1 and 2, can be conducted in parallel to speed up the overall anomaly detection process, see the flowchart presented in Fig. 10.2. Nonetheless, in this book, we implement the AAD algorithm without parallelization.

P. P. Angelov and X. Gu, *Empirical Approach to Machine Learning*, Studies in Computational Intelligence 800,
https://doi.org/10.1007/978-3-030-02384-3_10

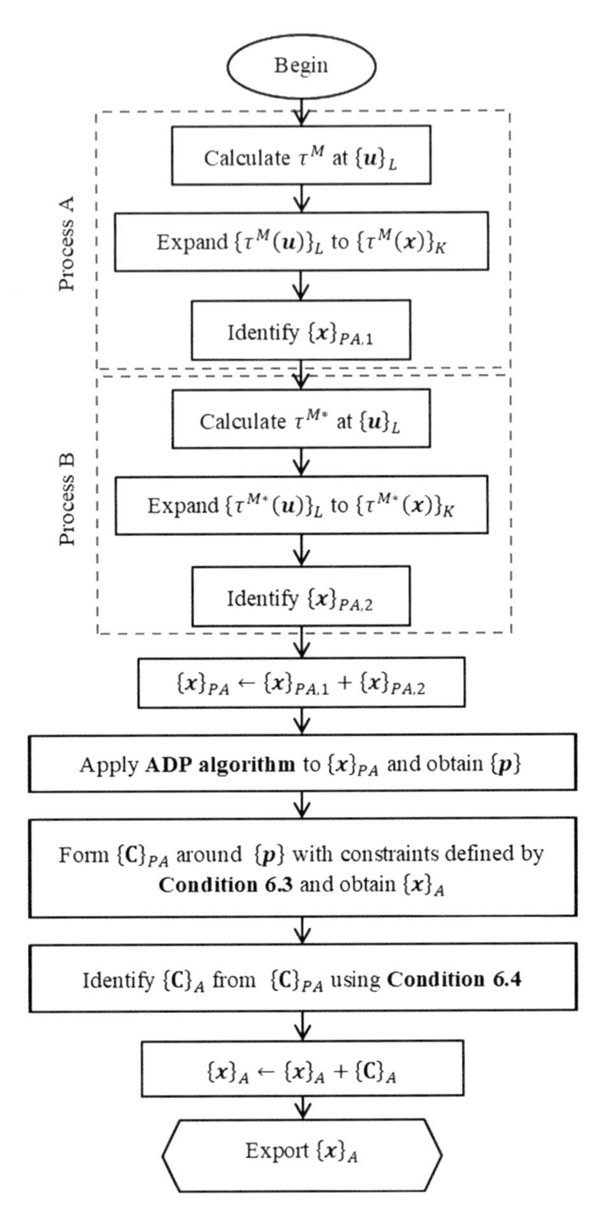

Fig. 10.1 Standard AAD algorithm

10.2 Numerical Examples

In this section, several numerical examples are presented aiming to demonstrate the performance of the AAD algorithm to detect anomalies autonomously.

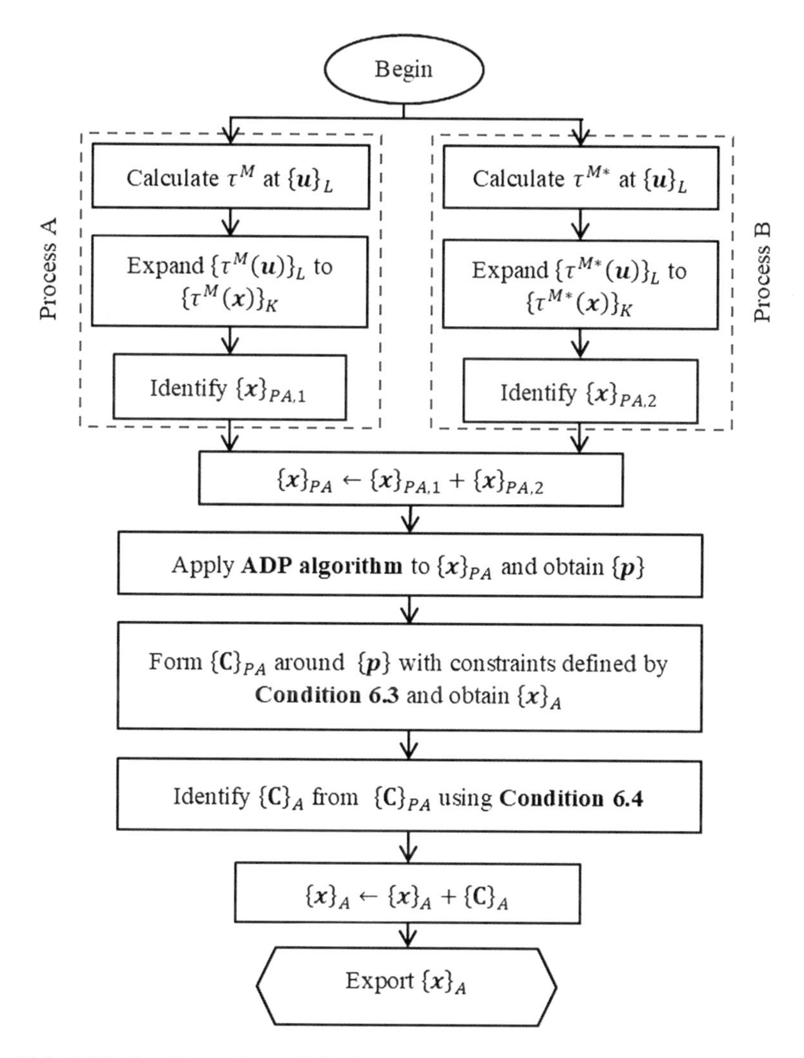

Fig. 10.2 AAD algorithm with parallelization

10.2.1 Datasets for Evaluation

In this section, the following datasets are considered for evaluating the performance of the AAD algorithm on anomaly detection:

(1) Synthetic Gaussian dataset;
(2) User knowledge modelling dataset [2] (downloadable from [3]);
(3) Wine quality dataset [4] (downloadable from [5]);
(4) Wilt dataset [6] (downloadable from [7]).

10.2.1.1 Synthetic Gaussian Dataset

The synthetic Gaussian dataset contains 720 data samples with 2 attributes. There is 1 larger cluster and 2 smaller ones grouping 700 data samples between them. In addition, 4 collective anomalous sets formed by 18 samples as well as 2 single anomalies were identified. The models of the three major clusters extracted from the data $(\boldsymbol{\mu}, \boldsymbol{\Sigma}, S)$ are as follows (in the form of model, $\boldsymbol{x} \sim \mathcal{N}(\boldsymbol{\mu}, \boldsymbol{\Sigma})$ and support, S):

Major cluster 1: $\boldsymbol{x} \sim \mathcal{N}\left([0.00, 3.00], \begin{bmatrix} 0.16 & 0.00 \\ 0.00 & 0.16 \end{bmatrix}\right)$, 300 samples;

Major cluster 2: $\boldsymbol{x} \sim \mathcal{N}\left([2.50, 3.00], \begin{bmatrix} 0.09 & 0.00 \\ 0.00 & 0.09 \end{bmatrix}\right)$, 200 samples;

Major cluster 3: $\boldsymbol{x} \sim \mathcal{N}\left([2.50, 0.00], \begin{bmatrix} 0.09 & 0.00 \\ 0.00 & 0.09 \end{bmatrix}\right)$, 200 samples.

The models of the 4 collective anomalous sets are:

Anomalous set 1: $\boldsymbol{x} \sim \mathcal{N}\left([0.00, 1.00], \begin{bmatrix} 0.09 & 0.00 \\ 0.00 & 0.09 \end{bmatrix}\right)$, 5 samples;

Anomalous set 2: $\boldsymbol{x} \sim \mathcal{N}\left([4.50, 0.00], \begin{bmatrix} 0.09 & 0.00 \\ 0.00 & 0.09 \end{bmatrix}\right)$, 4 samples;

Anomalous set 3: $\boldsymbol{x} \sim \mathcal{N}\left([4.50, 4.00], \begin{bmatrix} 0.04 & 0.00 \\ 0.00 & 0.04 \end{bmatrix}\right)$, 5 samples;

Anomalous set 4: $\boldsymbol{x} \sim \mathcal{N}\left([1.00, -1.00], \begin{bmatrix} 0.04 & 0.00 \\ 0.00 & 0.04 \end{bmatrix}\right)$, 4 samples;

and the two single anomalies are:

Anomalous sample 1: $[2.00, 5.00]$, and;
Anomalous sample 2: $[1.50, 5.00]$.

This dataset is visualized in Fig. 10.3, and the anomalies are circled by red ellipses. It is important to stress that, it is usually very difficult for traditional approaches to detect the collective and single anomalies, which are close to the *global* mean of the dataset.

10.2.1.2 User Knowledge Modelling Dataset

The user knowledge modelling dataset contains 403 samples, and each data sample has five attributes:

(1) the degree of study time for goal object materials;
(2) the degree of repetition number of user for goal object materials;
(3) the degree of study time of user for related objects with goal object;

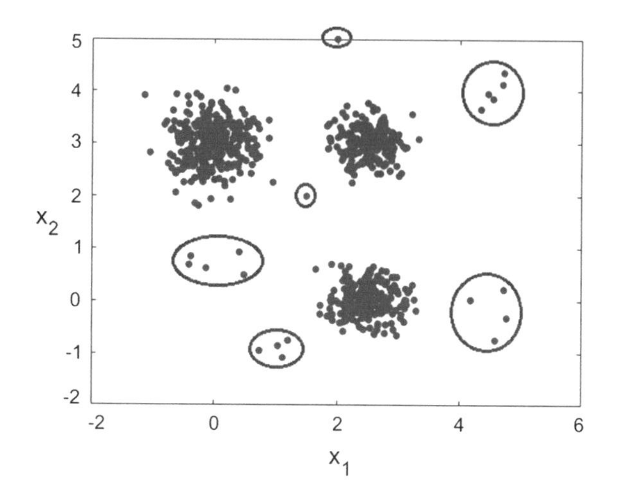

Fig. 10.3 Visualization of the synthetic Gaussian dataset

(4) the exam performance of user for related objects with goal object;
(5) the exam performance of user for goal objects.

and one label.

There are four levels of the user knowledge (four classes),

(1) "*High*": 130 data samples;
(2) "*Middle*": 122 data samples;
(3) "*Low*": 129 data samples;
(4) "*Very Low*":50 data samples.

There are seven anomalies in this dataset:

(1) one anomaly in the class "*Very Low*";
(2) two anomalies in the class "*Low*", and;
(3) four anomalies in the class "*Middle*".

10.2.1.3 Wine Quality Dataset

Wine quality dataset is related to the quality of red and white of Portuguese "Vinho Verde" wines [4]. The dataset consists of two parts, namely:

(1) the red wine dataset, which has 1599 data samples, and
(2) the white wine dataset, which has 4898 data samples.

Each data sample in the red wine and white wine datasets has 11 attributes:

(1) fixed acidity;
(2) volatile acidity;
(3) citric acid;

(4) residual sugar;
(5) chlorides;
(6) free Sulphur dioxide;
(7) total Sulphur dioxide;
(8) density;
(9) pH;
(10) sulphates;
(11) alcohol;

and one label: the score of quality from "3" to "8".

Both datasets are not balanced as there are much more normal wines than the excellent or poor ones.

Within the "red wine" dataset, there are 10 data samples with score "3", 53 data samples with score "4", 681 data samples with score "5", 638 data samples with score "6", 199 data samples with score "7" and 18 data samples with score "8". The numbers of existing anomalies in each class are listed as follows:

(1) Score 3:1 data sample;
(2) Score 4:3 data samples;
(3) Score 5:39 data samples;
(4) Score 6:25 data samples;
(5) Score 7:2 data samples;
(6) Score 8:2 data samples.

In total, there are 72 anomalies.

Within the "white wine" dataset, there are 20 data samples with score "3", 163 data samples with score "4", 1457 data samples with score "5", 2198 data samples with score "6", 880 data samples with score "7" and 175 data samples with score "8". The numbers of existing anomalies in each class are listed as follows:

(1) Score 3:1 data sample;
(2) Score 4:5 data samples;
(3) Score 5:59 data samples;
(4) Score 6:103 data samples;
(5) Score 7:44 data samples;
(6) Score 8:10 data samples.

In total, there are 222 anomalies.

10.2.1.4 Wilt Dataset

Wilt dataset comes from a remote sensing study involving detecting diseased trees in Quickbird imagery [6]. There are two classes in the dataset:

(1) "diseased trees" class, which has 74 samples, and
(2) "other land cover" class, which has 4265 samples.

Each sample has five attributes:

(1) gray-level co-occurrence matrix mean texture;
(2) mean green value;
(3) mean red value;
(4) mean near-infrared value;
(5) standard deviation;

and one label ("other land cover" or diseased trees"). There are 134 anomalies with the label "other land cover" and no anomaly in the "diseased trees" class.

10.2.2 *Performance Evaluation*

In the numerical experiments presented in this book, we use by default, $n = 3$ (corresponding to "3σ" rule) and the second level of granularity, $G = 2$.

10.2.2.1 Numerical Experiments on the Gaussian Dataset

Using the AAD approach, there are 63 potential anomalies $\{\boldsymbol{x}\}_{PA}$ identified from the Gaussian dataset during the first stage (as described in Sect. 6.3.2), which are depicted in Fig. 10.4 marked by the green dots "·".

During the second stage, as described in Sect. 6.3.3, nine *data clouds* are formed from $\{\boldsymbol{x}\}_{PA}$, and also, there are 14 potential anomalies that are isolated from these *data clouds* due to the constraint imposed by condition 6.3. The *data clouds* formed from $\{\boldsymbol{x}\}_{PA}$ are depicted in Fig. 10.5, where the dots in different colors represent different *data clouds*; the dots in black represent the isolated potentials anomalies.

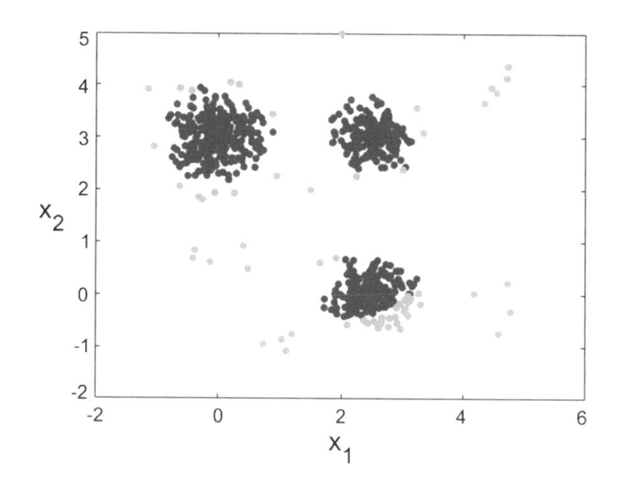

Fig. 10.4 The identified potential anomalies

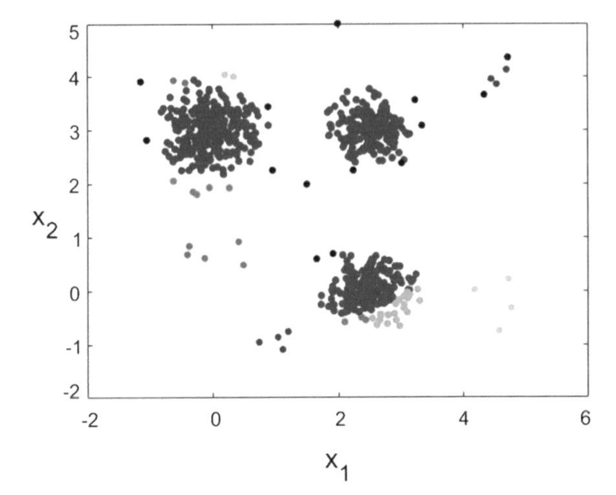

Fig. 10.5 The *data clouds* formed from the potential anomalies

The identified anomalies in the final stage are presented in Fig. 10.6, where the anomalies are presented by red dots. In total, the AAD algorithm identifies 42 anomalies.

Figures 10.4, 10.5 and 10.6 demonstrate that, the AAD algorithm successfully identified all the anomalies. This is because both, the mutual distribution and the ensemble properties of the data samples have been taken into consideration.

For a further evaluation of the AAD algorithm, two well-known traditional approaches are used for comparison:

(1) The well-known "3σ" approach [8–10];
(2) Outlier detection using random walks (ODRW) algorithm [11].

It has to be stressed that the "3σ" approach is based on the *global* mean and *global* standard deviation. The ODRW algorithm requires three parameters to be predefined:

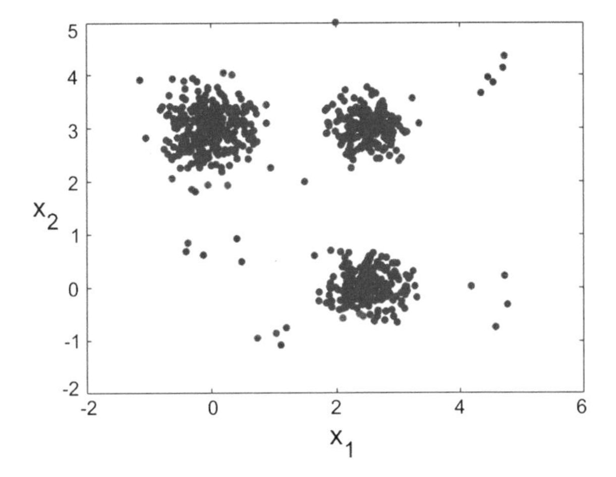

Fig. 10.6 The identified anomalies

(1) error tolerance, ε_o;
(2) similarity threshold, T_o and
(3) number of anomalies, K_A.

In this subsection, $\varepsilon_o = 10^{-6}$ and $T_o = 8$ [11]. To make the results comparable, K_A is set to be the same number of the anomalies as the anomalies identified by the AAD algorithm.

The *global* mean and the standard deviation of the dataset are $\boldsymbol{\mu} = [1.44, 2.13]$ and $\boldsymbol{\sigma} = [1.34, 1.42]$, and the "$3\sigma$" approach failed to detect any anomalies. The result using the ODRW algorithm is shown in Fig. 10.7, where the red dots are the identified anomalies. As one can see, this approach ignored the more than half of the actual anomalies (circled within the orange ellipsoids).

10.2.2.2 Numerical Experiments on the Benchmark Datasets

In this subsection, numerical experiments are conducted for evaluating the performance of the AAD algorithm based on benchmark datasets.

For a better evaluation, the following five measures [11] are used for performance evaluation:

(1) number of identified anomalies (K_A): K_A = True Positive;
(2) precision (Pr): the rate of true anomalies in the detected anomalies, $Pr = \frac{\text{True Positive}}{\text{True Positive} + \text{False Positive}} \times 100\%$;
(3) false alarm rate (Fa): the rate of the true negatives in the identified anomalies, $Fa = \frac{\text{False Positive}}{\text{True Negative} + \text{False Positive}} \times 100\%$;

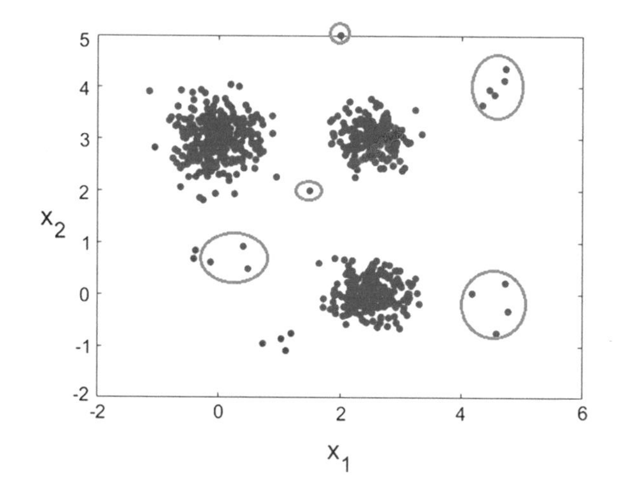

Fig. 10.7 The anomalies identified by the ODRW algorithm

(4) recall rate (Re): the rate of true anomalies the algorithms missed, $Re = \frac{\text{False Negative}}{\text{True Positive} + \text{False Negative}} \times 100\%$.

(5) execution time (t_{exe}): in seconds.

Normally, a good anomaly detection algorithm should be able to produce the results with higher precision rate, lower false alarm and recall rates.

The detection results obtained by the three algorithms on the user knowledge modelling, wine quality and wilt datasets in terms of the five measures are tabulated in Table 10.1.

10.2.3 Discussion and Analysis

From the numerical results presented in this section one can see that the AAD algorithm is able to detect the anomalies with higher precision and lower false alarm rate as well as the recall rate compared with the "3σ" approach and the ODRW algorithm.

The "3σ" approach is the fastest due to its simplicity. However, the performance of the "3σ" approach is decided by the structure of the data because it focuses only on the samples exceeding the *global* 3σ range around the mean. However, when the structure is complex i.e. contains a large number of clusters, or the anomalies are close to the *global* mean, "3σ" approach fails to detect all outliers.

The ODRW algorithm is less efficient in dealing with larger scale datasets. Although its performance is higher than that of the "3σ"approach, it requires the number of anomalies to be predefined in advance, which is very difficult for the users to decide without *prior* knowledge of the problems. Moreover, the other two

Table 10.1 Performance comparison of the anomaly detection algorithms

Dataset	Algorithm	K_A	Pr (%)	Fa (%)	Re (%)	t_{exe} (%)
User knowledge modelling	AAD	21	19.05	4.29	42.86	0.15
	3σ	1	0.00	0.25	100.00	0.03
	ODRW	21	9.52	4.80	71.43	0.04
Red wine quality	AAD	35	54.29	1.05	73.61	0.27
	3σ	141	22.70	7.14	55.56	0.03
	ODRW	35	0.00	2.29	100.00	0.24
White wine quality	AAD	111	47.75	1.24	76.13	1.00
	3σ	396	9.09	7.70	83.78	0.03
	ODRW	111	24.32	1.80	87.84	2.12
Wilt	AAD	108	56.48	1.12	54.48	0.81
	3σ	176	35.23	2.71	53.73	0.03
	ODRW	108	50.93	1.26	58.96	1.60

free parameters required by the ODRW algorithm can also significantly influence the anomaly detection performance.

In contrast, the AAD algorithm can identify the anomalies based on the ensemble properties of the data in a fully unsupervised and autonomous way. It takes not only the mutual distribution of the data within the data space, but also the frequencies of occurrences into consideration. It provides a more objective way for anomaly detection.

More importantly, its performance is not influenced by the structure of the data and is equally effective in detecting collective anomalies as well as individual anomalies.

10.3 Conclusions

In this chapter, the implementation of the AAD algorithm is presented, and numerical examples on benchmark datasets, are presented to demonstrate the performance of the AAD algorithm. From the comparison with the well-known approaches, one can conclude that the AAD algorithm is able to perform highly precise anomaly detection in an unsupervised, objective and efficient manner. The performance of the AAD algorithm is not influenced by the structure of the data and it can handle both collective and individual anomalies effectively.

References

1. X. Gu, P. Angelov, in *Autonomous Anomaly Detection*, in IEEE Conference on Evolving and Adaptive Intelligent Systems (2017), pp. 1–8
2. H.T. Kahraman, S. Sagiroglu, I. Colak, The development of intuitive knowledge classifier and the modeling of domain dependent data. Knowl. Based Syst. **37**, 283–295 (2013)
3. https://archive.ics.uci.edu/ml/datasets/User+Knowledge+Modeling
4. P. Cortez, A. Cerdeira, F. Almeida, T. Matos, J. Reis, Modeling wine preferences by data mining from physicochemical properties. Decis. Support Syst. **47**, 547–553 (2009)
5. https://archive.ics.uci.edu/ml/datasets/Wine+Quality
6. B.A. Johnson, R. Tateishi, N.T. Hoan, A hybrid pansharpening approach and multiscale object-based image analysis for mapping diseased pine and oak trees. Int. J. Remote Sens. **34** (20), 6969–6982 (2013)
7. http://archive.ics.uci.edu/ml/datasets/Wilt
8. D.E. Denning, An intrusion-detection model, *IEEE Trans. Softw. Eng.*, **SE-13**(2), 222–232, 1987
9. P.P. Angelov, in *Anomaly Detection Based on Eccentricity Analysis*, 2014 IEEE Symposium Series in Computational Intelligence, IEEE Symposium on Evolving and Autonomous Learning Systems, EALS, SSCI 2014, 2014, pp. 1–8
10. C. Thomas, N. Balakrishnan, Improvement in intrusion detection with advances in sensor fusion. IEEE Trans. Inf. Forensics Secur. **4**(3), 542–551 (2009)
11. H. Moonesinghe, P. Tan, in *Outlier detection using random walks*, Proceedings of the 18th IEEE International Conference on Tools with Artificial Intelligence (ICTAI'06), 2006, pp. 532–539

Chapter 11
Applications of Autonomous Data Partitioning

As it has been stated in the previous chapters, the mainstream data partitioning/clustering approaches usually rely on *prior* knowledge of the problems and making strong assumptions about data generation models beforehand. As a result, the produced results are often highly subjective. In contrast, the autonomous data partitioning (ADP) algorithm [1] presented in Chap. 7 has the following two distinctive features:

(1) it is entirely data-driven and free from user- and problem- specific parameters;
(2) it does not impose assumptions on the data generation models and, instead, relies entirely on the *empirical* observations, namely the data.

In this chapter, we will describe the implementation and applications of the ADP algorithm. The pseudo-code of this algorithm is presented in Appendix B.2. The MATLAB implementation is given in Appendix C.2.

11.1 Algorithm Summary

11.1.1 Offline Version

The main procedure of the offline version of the ADP algorithm consists of the following steps [1]:

Step 1 **Rank** the unique data samples $\{\boldsymbol{u}\}_L$ in terms of their values of multimodal *typicality* and mutual distances and obtain $\{z\}_L$;

Step 2 Identify local maxima, $\{z^*\}_L$ from $\{z\}_L$ and form *data clouds*, $\{\mathbf{C}\}_L$ using $\{z^*\}_L$ as the prototypes;

P. P. Angelov and X. Gu, *Empirical Approach to Machine Learning*, Studies in Computational Intelligence 800,
https://doi.org/10.1007/978-3-030-02384-3_11

Step 3 Filter the local maxima $\{z^*\}_L$ based on the multimodal *typicality* at $\{z^*\}_L$ and the mutual distances between them, and obtain the most representative local maxima to serve as the identified prototypes $\{\boldsymbol{p}\}$;

Step 4 Form *data clouds* $\{\mathbf{C}\}$ around $\{\boldsymbol{p}\}$ as the final output of the algorithm

The flowchart of the main procedure of the offline ADP algorithm is given in Fig. 11.1.

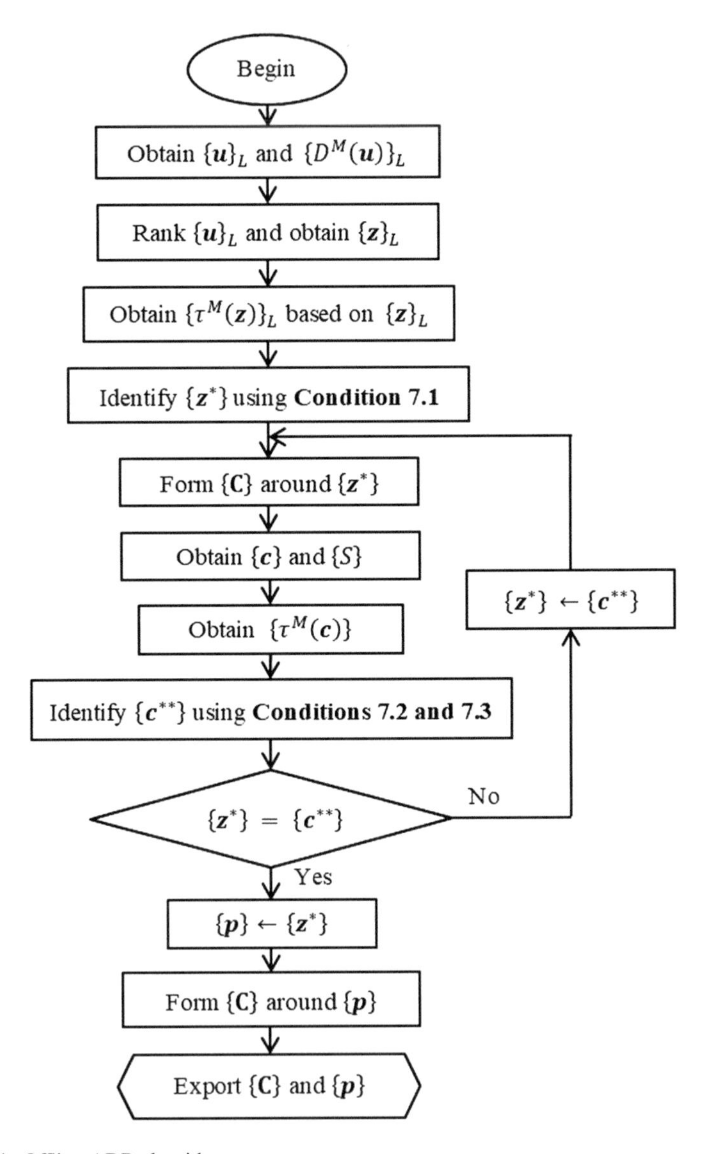

Fig. 11.1 Offline ADP algorithm

11.1.2 Evolving Version

The main procedure of the evolving version of the ADP algorithm has the following steps [1]:

Step 1 Initialize the system with the first data sample $\boldsymbol{x}_1 (K \leftarrow 1)$.

Step 2 For each newly observed data sample, $\boldsymbol{x}_K$ $(K \leftarrow K+1)$, check whether $\boldsymbol{x}_K$ initializes a new *data cloud*. If yes, go to step 3; else, go to step 4;

Step 3 Add a new *data cloud* $\mathbf{C}_M$ $(M \leftarrow M+1)$ to the system with $\boldsymbol{x}_K$ as the center of $\mathbf{C}_M$, denoted by $\boldsymbol{c}_M$, and go to step 5;

Step 4 Use $\boldsymbol{x}_K$ to update meta-parameters of the nearest *data cloud* $\mathbf{C}_{n*}$, and go to step 5;

Step 5 Go to step 2 if new data sample arrives; otherwise, form *data clouds* $\{\mathbf{C}\}$ as the final output of the algorithm with the prototypes $\{\boldsymbol{p}\}\left(\{\boldsymbol{p}\} \leftarrow \{\boldsymbol{c}\}_K\right)$

The flowchart of the main procedure of the evolving ADP algorithm is given in Fig. 11. 2.

11.2 Numerical Examples of Data Partitioning

In this section, the numerical examples are presented for demonstrating the performance of the proposed ADP algorithm.

11.2.1 Datasets for Evaluation

In this section, the following datasets are considered for evaluating the performance of the ADP algorithm:

(1) Banknote authentication dataset [2] (downloadable from [3]);
(2) Cardiotocography dataset [4] (downloadable from [5]);
(3) MAGIC gamma telescope dataset [6] (downloadable from [7]);
(4) Letter recognition dataset [8] (downloadable from [9]).

11.2.1.1 Banknote Authentication Dataset

The banknote authentication dataset was extracted from images that were taken from genuine and forged banknote-like specimens. This dataset is composed of 1372 data samples and each one of them contains four attributes:

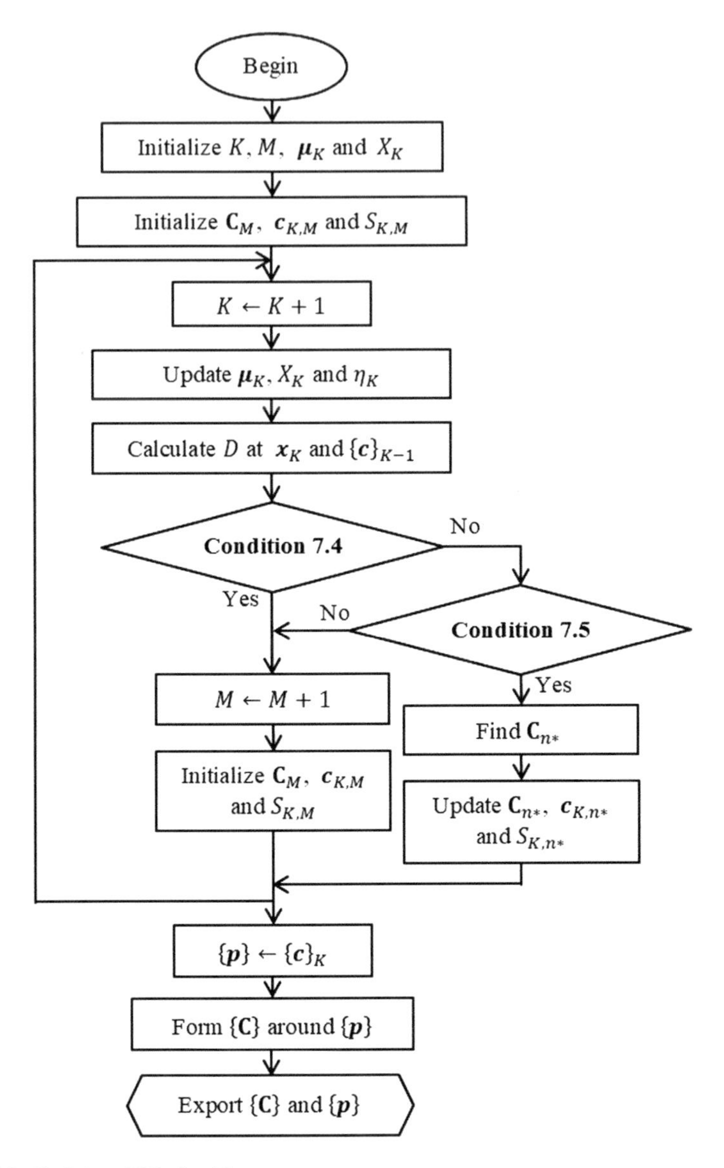

Fig. 11.2 Evolving ADP algorithm

(1) variance of wavelet transformed image;
(2) skewness of wavelet transformed image;
(3) curtosis of wavelet transformed image;
(4) entropy of image.

and one class label ("0" or "1").

11.2.1.2 Cardiotocography Dataset

The cardiotocography dataset contains 2126 fetal cardiotocograms that were automatically processed and the respective diagnostic features were measured. Each data sample contains 21 attributes:

(1) fetal heart rate baseline (beats per minute);
(2) number of accelerations per second;
(3) number of fetal movements per second;
(4) number of uterine contractions per second;
(5) number of light decelerations per second;
(6) number of severe decelerations per second;
(7) number of prolongued decelerations per second;
(8) percentage of time with abnormal short term variability;
(9) mean value of short term variability;
(10) percentage of time with abnormal long term variability;
(11) mean value of long term variability;
(12) width of fetal heart rate histogram;
(13) minimum of fetal heart rate histogram;
(14) maximum of fetal heart rate histogram;
(15) number of histogram peaks;
(16) number of histogram zeros;
(17) histogram mode;
(18) histogram mean;
(19) histogram median;
(20) histogram variance;
(21) histogram tendency;

and one class label ("fetal heart rate pattern class code" from "1" to "10").

11.2.1.3 MAGIC Gamma Telescope Dataset

The MAGIC gamma telescope dataset is generated to simulate registration of high energy gamma particles in a ground-based atmospheric Cherenkov gamma telescope using the imaging technique. This dataset contains 19020 data samples, and each data sample has 10 attributes:

(1) major axis of ellipse;
(2) minor axis of ellipse;
(3) 10-log of sum of content of all pixels;
(4) ratio of sum of two highest pixels over 10-log of sum of content of all pixels;
(5) ratio of highest pixel over 10-log of sum of content of all pixels;
(6) distance from highest pixel to center, projected onto major axis;
(7) third root of third moment along major axis;

(8) third root of third moment along minor axis;
(9) angle of major axis with vector to origin;
(10) distance from origin to center of ellipse;

and one class label ("signal" or "background").

11.2.1.4 Letter Recognition Dataset

The objective of letter recognition dataset is to identify each of the black-and-white rectangular pixel displays as one of the 26 capital letters in the Latin alphabet. This dataset is composed of 20000 data samples, and each of them has 16 attributes:

(1) horizontal position of box;
(2) vertical position of box;
(3) width of box;
(4) high height of box;
(5) total number on pixels;
(6) mean x of on pixels in box;
(7) mean y of on pixels in box;
(8) mean x variance;
(9) mean y variance;
(10) mean x y correlation;
(11) mean of $x \cdot x \cdot y$;
(12) mean of $x \cdot y \cdot y$;
(13) mean edge count left to right;
(14) correlation of x-ege with y;
(15) mean edge count bottom to top;
(16) correlation of y-ege with x;

and one class label (from "1" to "26" corresponding to "A" to "Z").

11.2.2 Performance Evaluation

In this subsection, the performance of the ADP algorithm is evaluated. For clarity, only the partitioning results of the ADP algorithm on banknote authentication dataset are visualized in Fig. 11.3, where dots in different colors denote data samples of different *data clouds*; the black asterisks "*" represent the identified prototypes.

One can see from Fig. 11.3 that the ADP algorithm identified a number of prototypes from the observed data samples and partitioned the datasets into flexible shape *data clouds*. These are formed naturally by attracting data samples around the prototypes forming a Voronoi tessellation [10].

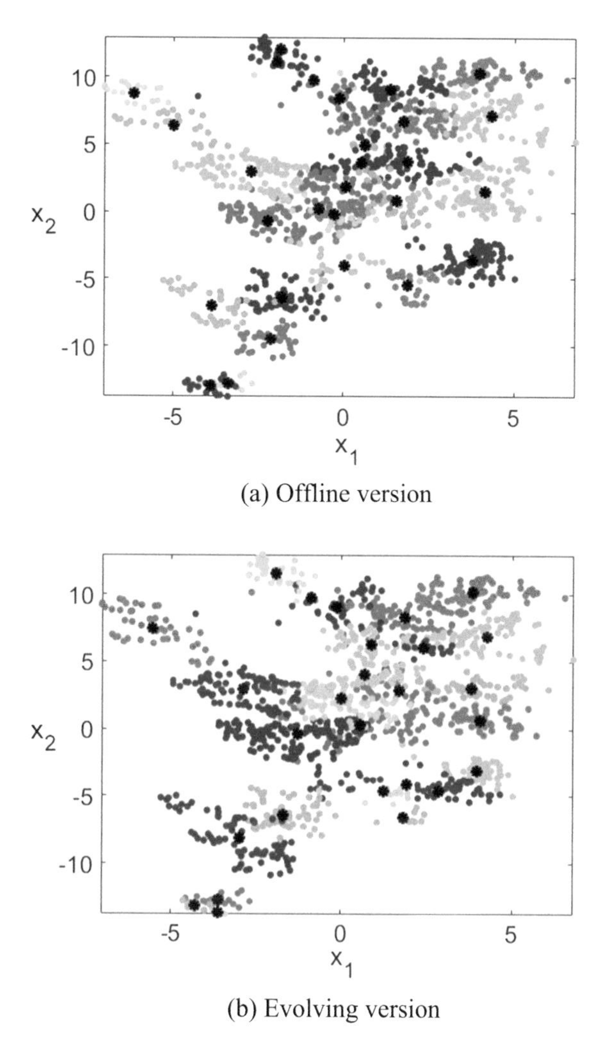

(a) Offline version

(b) Evolving version

Fig. 11.3 Partitioning results for the banknote authentication dataset

However, since there is no clear separation between the data samples of different classes in the high-dimensional, large-scale datasets, it is very hard to directly evaluate the quality of the partitioning results. Therefore, we did not visualize the partitioning results.

The following four indicators can be used for evaluating the performance of the ADP algorithm:

(1) Number of clusters/*data clouds* (M). Ideally, M should be as close as possible to the number of actual classes in the dataset (ground truth). However, this would mean one cluster/*data cloud* per class and is only the best solution if each class has a very simple (hyper-spherical) pattern. However, this is not the case in the vast majority of the real problems.

In most of the cases, data samples from different classes are mixed with each other. The best way to cluster/partition the data with a complex structure is to divide it into smaller partitions (i.e. more than one cluster per class) for a better separation. However, having too many clusters per class is also reducing the generalization ability (leading to overfitting) and the interpretability.

Therefore, we also involve the average number of data samples per cluster/*data cloud* as a quality measure.

(2) Average number of data samples per *data cloud*/cluster (S_a), which is calculated as:

$$S_a = \frac{\sum_{i=1}^{M} S_i}{M} \tag{11.1}$$

where S_i is the support of the *i*th *data cloud*.

In this book, the reasonable value range of S_a is considered as:

$$20 \leq S_a \leq \frac{K}{C} \tag{11.2}$$

If S_a is larger than $\frac{K}{C}$, it means that the number of partitions is smaller the number of actual classes in the dataset, which indicates that the clustering/partitioning algorithm fails to separate the data samples from different classes. If S_a is smaller than 20, it means that there are too many partitions generated by the clustering/partitioning algorithm, which makes the information too trivial for users.

In either case, the clustering/partitioning result is considered as an invalid one.

(3) Purity (Pu) [20], which is calculated based on the result and the ground truth:

$$Pu = \frac{\sum_{i=1}^{M} S_i^D}{K} \tag{11.3}$$

where S_i^D is the number of data samples with the dominant class label in the *i*th cluster. Purity directly indicates separation ability of the clustering algorithm. The higher purity a clustering result has, the stronger separation ability the clustering algorithm exhibits.

(4) Execution time (t_{exe}): the execution time (in seconds) should be as small as possible.

The performance measures of the partitioning results obtained by the ADP algorithm on the four benchmark datasets are tabulated in Table 11.1.

Table 11.1 Performance comparison between different data partitioning/clustering approaches

Dataset	Algorithm	M	S_a	Pu	t_{exe}
Banknote	Offline ADP	28	49.00	0.9811	0.22
	Evolving ADP	27	50.81	0.9803	0.30
	MS	24	57.17	0.9927	0.05
	DBSCAN	48	28.58	0.9402	0.17
	eClustering	*(1)*	*(1372)*	*(0.5554)*	*(0.06)*
	NMM	4	343	0.8462	62.77
	NMIB	20	68.60	0.9913	5.24
Cardiotocography	Offline ADP	90	23.62	0.5691	0.36
	Evolving ADP	48	44.29	0.5118	0.41
	MS	*(1540)*	*(1.38)*	*(0.9661)*	*(0.78)*
	DBSCAN	14	151.86	0.2973	0.45
	eClustering	*(8)*	*(265.75)*	*(0.3500)*	*(0.16)*
	NMM	*4)*	*531.50)*	*0.3015)*	*99.80)*
	NMI	322	6.60	0.6303	27.67
MAGIC gamma telescope	Offline ADP	47	404.68	0.7289	13.71
	Evolving ADP	380	50.05	0.7899	2.65
	MS	*(1469)*	*(12.95)*	*(0.7871)*	*(46.61)*
	DBSCAN	15	1268	0.6247	33.44
	eClustering	6	3170	0.6484	2.61
	NMM	3	6340	0.7345	1560.68
	NMI	*(1578)*	*(12.05)*	*0.7459)*	*(6833.96)*
Letter recognition	Offline ADP	235	85.1	0.6000	15.19
	Evolving ADP	242	82.64	0.5825	2.92
	MS	*(7619)*	*(2.63)*	*(0.9760)*	*(256.07)*
	DBSCAN	51	392.16	0.1584	33.85
	eClustering	*(5)*	*(4000.00)*	*(0.1135)*	*(4.23)*
	NMM	46	434.78	0.4304	6316.60
	NMI	*(14526)*	*(1.38)*	*(0.9975)*	*(6279.86)*

For a better evaluation, we also involve the following five well-known approaches for comparison:

(1) Mean-shift (MS) algorithm [11];
(2) Density-based spatial clustering of applications with noise (DBSCAN) algorithm [12];
(3) eClustering algorithm [13];
(4) Non-parametric mixture model (NMM) algorithm [14];
(5) Non-parametric mode identification (NMI) algorithm [15].

Due to the insufficient *prior* knowledge, the recommended settings of the free parameters from the published literature are used throughout the numerical experiments. The experimental settings of the free parameters of the algorithms are

Table 11.2 Experimental settings of the comparative algorithms

Algorithms	Free parameter(s)	Experimental setting
MS	(1) Bandwidth, ρ (2) Kernel function type	(1) $\rho = 0.5$ [16] (2) Gaussian kernel
DBSCAN	(1) Cluster radius, r (2) Minimum number of data samples within the radius, k	(1) The value of the knee point of the sorted k-dist graph (2) $k = 4$ [12]
eClustering	(1) Initial radius, r (2) Learning parameter, ρ	(1) $r = 0.5$ (2) $\rho = 0.5$ [13]
NMM	(1) *prior* scaling parameter (2) Kappa coefficient, κ	Defined as in [14]
NMI	Grid size	Defined as in [15]

presented in Table 11.2, where the results in italic in brackets are the invalid clustering results obtained by the approaches in the experiments.

The quality measures of the clustering/partitioning results obtained by the comparative algorithms are tabulated in Table 11.1, where the invalid results are presented in bracket.

11.2.3 Discussion and Analysis

Based on the numerical examples and comparisons one can see that:

(1) MS algorithm [11]

MS algorithm is quite efficient when the scale and dimensionality of the dataset is low. However, its computational efficiency decreases quickly in processing large scale and high dimensional datasets. The quality of its clustering results varies dramatically as well. Without *prior* knowledge, it produced invalid clustering results on the high dimensional datasets. This is due to its gradient nature making it highly dependent on the initial guess and being prone to fall into local minima.

(2) DBSCAN algorithm [12]

DBSCAN is a highly efficient online algorithm. However, the quality of its clustering results is very low. One may also notice that, DBSCAN is not effective in handling high-dimensional and large-scale datasets.

(3) eClustering algorithm [13]

eClustering algorithm was one of the most efficient algorithms in the numerical examples. However, it failed to separate the data samples of different classes, especially obvious on the banknote authentication problem.

(4) NMM algorithm [14]

NMM algorithm has a number of pre-defined parameters and coefficients. This algorithm is based on the *prior* assumption that the data follows Gaussian distribution. However, this algorithm failed to give useful clustering results on half of the benchmark problems considered in this chapter. In addition, its computational efficiency is low.

(5) NMI algorithm [15]

NMI algorithm is similar to the NMM algorithm, it has a number of pre-defined parameters, i.e. grid size, interval between two grids, and it assumes that the data has a Gaussian distribution. This algorithm is very accurate on small-scale problems. However, it failed to provide valid results in processing large-scale and high dimensional datasets. Moreover, its computation efficiency is also largely influenced by the size and dimensionality of the data.

(6) ADP algorithm [1]

In contrast, the ADP algorithm is able to produce high quality data partitioning results in an efficient manner. More importantly, the algorithm is free from *prior* assumptions and parameters, which makes it very attractive in real world applications where the *prior* knowledge is very limited. Moreover, the ADP algorithm can operate in different modes and the model can be updated after deployment, which makes it an appealing approach for different problems.

More numerical examples on benchmark problems and comparisons can be found in [1].

11.3 Numerical Examples of Semi-supervised Classification

Most of the current classification approaches require the ground truth (at least part of it for the semi-supervised case) to be available in advance, before training. However, with the ADP algorithm, one can conduct classification with very little supervision by only examining the identified prototypes after the partitioning based on the training data. After the labels of the prototypes are obtained, classification on the validation data can be conducted based on their distances to the prototypes using the "nearest prototype" principle.

In this section, we present the numerical examples to demonstrate the concept.

11.3.1 Datasets for Evaluation

We use the following four benchmark datasets for evaluation.

(1) Banknote authentication dataset [2] (downloadable from [3]);
(2) Tic-Tac-Toe endgame dataset [17] (downloadable from [18]);
(3) Pima Indians diabetes dataset [19] (downloadable from [20]);
(4) Occupancy detection dataset [21] (downloadable from [22]).

Since the details of the banknote authentication dataset have been given in Sect. 11.2.1.1, in this subsection, the details of the other three benchmark datasets will be given only.

11.3.1.1 Tic-Tac-Toe Endgame Dataset

The dataset encodes the complete set of possible board configurations at the end of tic-tac-toe games, where "x" is assumed to have played first. The target concept is "winning for x" (i.e., true when "x" has one of 8 possible ways to create a "three in a row"). This dataset contains 958 data samples, and each sample has 9 attributes:

(1) top-left-square: {"x", "o", "b"};
(2) top-middle-square: {"x", "o", "b"};
(3) top-right-square: {"x", "o", "b"};
(4) middle-left-square: {"x", "o", "b"};
(5) middle-middle-square: {"x", "o", "b"};
(6) middle-right-square: {"x", "o", "b"};
(7) bottom-left-square: {"x", "o", "b"};
(8) bottom-middle-square: {"x", "o", "b"};
(9) bottom-right-square: {"x", "o", "b"};

and one class, namely, "positive" and "negative". In the numerical examples presented in this section regarding this dataset, we convert the categorical data into numerical by encoding "x" as 1, "o" as 5 and "b" as 3.

11.3.1.2 Pima Indians Diabetes Dataset

This dataset is originally from the National Institute of Diabetes and Digestive and Kidney Diseases. The problem is to predict based on diagnostic measurements whether a patient has diabetes. This dataset has 768 data samples, and each one of them has eight attributes:

(1) number of times pregnant;
(2) plasma glucose concentration measured 2 h in an oral glucose tolerance test (mg/dl);
(3) diastolic blood pressure (mm Hg)
(4) triceps skin fold thickness (mm);
(5) 2-h serum insulin (mu U/ml);

(6) body mass index (weight/height in kg/m^2);
(7) diabetes pedigree function;
(8) age (years);

and one class label ("0"—negative or "1"—positive).

11.3.1.3 Occupancy Detection Dataset

The occupancy detection dataset is used for binary classification regarding room occupancy from temperature, humidity, light and CO_2. This dataset is composed of three subsets, one training set, which contains 8143 data samples, and two validation sets, which contain 2665 and 9752 data samples, respectively. Each data sample has six attributes:

(1) time instance, year-month-day hour: minute: second;
(2) temperature, in degree Celsius;
(3) relative humidity, in %;
(4) light, in Lux;
(5) CO_2, in ppm;
(6) humidity ratio, in kg-water-vapor/kg-air;

and one class label ("0" for not occupied, "1" for occupied).

In the numerical examples presented in this section regarding this dataset, we ignored the first attribute, namely, the time instance.

11.3.2 Performance Evaluation

For the banknote authentication, tic-tac-toe endgame and Pima Indians diabetes datasets, we randomly picked out 50% of the data as the training set and use the remaining data for validation. For the occupancy detection dataset, we combine the two validation sets into one.

We use the offline version of the ADP algorithm to identify prototypes from the training set in an unsupervised way and decide their labels based on the true labels of the training samples that are closest to them. Then, we conduct classification on the validation dataset by using the identified prototypes based on the "nearest prototype" principle.

We report in Table 11.3 the overall accuracy of the classification results on the four benchmark datasets after 10 times Monte-Carlo experiments following the same experimental protocol.

We also compare in the same table the ADP algorithm using the same experimental protocol with the well-known classifiers:

Table 11.3 Performance comparison between different classification approaches

Dataset	Algorithm	Overall accuracy (%)
Banknote authentication	ADP	98.25
	SVM	99.56
	KNN	99.88
	NN	99.88
	DT	97.10
Tic-Tac-Toe endgame	ADP	94.89
	SVM	65.51
	KNN	92.69
	NN	92.21
	DT	91.50
Pima Indians diabetes	ADP	66.12
	SVM	64.92
	KNN	72.45
	NN	73.31
	DT	69.22
Occupancy detection	ADP	95.53
	SVM	76.07
	KNN	96.64
	NN	93.69
	DT	93.14

(1) Support vector machine (SVM) classifier with Gaussian kernel [23];
(2) k-nearest neighbor (KNN) classifier with $k = 10$ [24];
(3) Neural network (NN) with three hidden layers of size 20 [25];
(4) Decision tree (DT) classifier [26].

11.3.3 Discussion and Analysis

From Table 11.3 one can see that, the ADP algorithm is able to provide highly accurate classification results with much less supervision compared with other approaches. This is of paramount importance to real applications since the labels of training data are often very expensive to be obtained. In such situations, the supervised classification approaches are not able to provide good performance due to the lack of "ground truth". However, with the ADP approach, one can recognize the underlying patterns of the data at first and analyze the patterns to obtain the labels. This is a much more efficient way of using the human expertise compared with labelling all the data.

11.4 Conclusions

In this chapter, the implementation of the ADP algorithm is presented, and numerical examples on benchmark datasets for data partitioning and classification, are presented to demonstrate the performance of the ADP algorithm.

From the comparisons with the well-known approaches on both tasks presented in this chapter, one can conclude that the ADP algorithm is able to obtain high quality data partitioning results in a highly efficient, objective manner.

Moreover, using the ADP algorithm, one is able to perform classification with very little supervision, which is very important for real-world applications because labelled data is always very expensive to be obtained.

References

1. X. Gu, P.P. Angelov, J.C. Principe, A method for autonomous data partitioning. Inf. Sci. (Ny) **460–461**, 65–82 (2018)
2. V. Lohweg, J.L. Hoffmann, H. Dörksen, R. Hildebrand, E. Gillich, J. Hofmann, J. Schaede, Banknote authentication with mobile devices. Media Watermarking, Security, and Forensics **8665**, 866507 (2013)
3. https://archive.ics.uci.edu/ml/datasets/banknote+authentication
4. D. Ayres-de-Campos, J. Bernardes, A. Garrido, J. Marques-de-Sa, L. Pereira-Leite, SisPorto 2.0: A program for automated analysis of cardiotocograms. J. Matern. Fetal. Med. **9**(5), 311–318 (2000)
5. https://archive.ics.uci.edu/ml/datasets/Cardiotocography
6. R.K. Rock, A. Chilingarian, M. Gaug, F. Hakl, T. Hengstebeck, M. Jiřina, J. Klaschka, E. Kotrč, P. Savický, S. Towers, A. Vaiciulis, W. Wittek, Methods for multidimensional event classification: A case study using images from a Cherenkov gamma-ray telescope. Nucl. Instrum. Methods Phys. Res., Sect. A **516**(2–3), 511–528 (2004)
7. http://archive.ics.uci.edu/ml/datasets/MAGIC+Gamma+Telescope
8. P.W. Frey, D.J. Slate, Letter recognition using Holland-style adaptive classifiers. Mach. Learn. **6**(2), 161–182 (1991)
9. https://archive.ics.uci.edu/ml/datasets/Letter+Recognition
10. A. Okabe, B. Boots, K. Sugihara, S.N. Chiu, *Spatial Tessellations: Concepts and Applications of Voronoi Diagrams*, 2nd edn. (Wiley, Chichester, England, 1999)
11. D. Comaniciu, P. Meer, Mean shift: A robust approach toward feature space analysis. IEEE Trans. Pattern Anal. Mach. Intell. **24**(5), 603–619 (2002)
12. M. Ester, H.P. Kriegel, J. Sander, X. Xu, A density-based algorithm for discovering clusters in large spatial databases with noise. Int. Conf. Knowl. Disc. Data Min. **96**, 226–231 (1996)
13. P.P. Angelov, D.P. Filev, An approach to online identification of Takagi-Sugeno fuzzy models. IEEE Trans. Syst. Man Cybern. Part B Cybern. **34**(1), 484–498 (2004)
14. D.M. Blei, M.I. Filev, Variational inference for Dirichlet process mixtures. Bayesian Anal. **1** (1A), 121–144 (2004)
15. J. Li, S. Ray, B.G. Lindsay, A non-parametric statistical approach to clustering via mode identification. J. Mach. Learn. Res. **8**(8), 1687–1723 (2007)
16. R. Dutta Baruah, P. Angelov, Evolving local means method for clustering of streaming data. IEEE Int. Conf. Fuzzy Syst. **8**, 10–15 (2012)
17. D.W. Aha, Incremental constructive induction: an instance-based approach, in *Machine Learning Proceedings* (1991) pp. 117–121

18. https://archive.ics.uci.edu/ml/datasets/Tic-Tac-Toe+Endgame
19. J.W. Smith, J.E. Everhart, W.C. Dickson, W.C. Knowler, R.S. Johannes, Using the ADAP learning algorithm to forecast the onset of diabetes mellitus, in *Annual Symposium on Computer Application in Medical Care* (1988) pp. 261–265
20. https://archive.ics.uci.edu/ml/datasets/Pima+Indians+Diabetes
21. L.M. Candanedo, V. Feldheim, Accurate occupancy detection of an office room from light, temperature, humidity and CO_2 measurements using statistical learning models. Energy Build. **112**, 28–39 (2016)
22. https://archive.ics.uci.edu/ml/datasets/Occupancy+Detection+
23. N. Cristianini, J. Shawe-Taylor, *An Introduction to Support Vector Machines and Other Kernel-Based Learning Methods* (Cambridge University Press, Cambridge, 2000)
24. P. Cunningham, S.J. Delany, K-nearest neighbour classifiers. Mult. Classif. Syst. **34**, 1–17 (2007)
25. Y. LeCun, Y. Bengio, G. Hinton, Deep learning. Nat. Methods **13**(1), 35–38 (2015)
26. S.R. Safavian, D. Landgrebe, A survey of decsion tree clasifier methodology. IEEE Trans. Syst. Man. Cybern. **21**(3), 660–674 (1990)

Chapter 12
Applications of Autonomous Learning Multi-model Systems

The autonomous learning multi-model (*ALMMo*) system is introduced as a new type of generic autonomous learning systems with multiple *local* simpler models architecture. *ALMMo* system is composed of a number of self-organizing *local* models identified from the *empirical* observations of the data, and has the following specific characteristics that distinguish it from other approaches:

(1) its learning process is computationally efficient, non-iterative and fully data-driven;
(2) it is free from *prior* assumptions and does not impose a parametric data generation model;
(3) it self-develops, self-learns and self-evolves autonomously.

In this chapter, the implementation of both, zero-order and first-order *ALMMo* systems are presented. The pseudo-code of the learning and validation processes of the *ALMMo-0* system is presented in Appendix B.3. The MATLAB implementation is presented in Appendix C.3. The pseudo-code of the learning process of the *ALMMo-1* system is presented in Appendix B.4, and the corresponding MATLAB implementation is presented in Appendix C.4.

12.1 Algorithm Summary

12.1.1 ALMMo-0 *System*

12.1.1.1 Learning Process

As the learning process of the *ALMMo-0* system can be done in parallel for each fuzzy rule, we consider the ith $(i = 1, 2, \ldots, C)$ rule, and only summarize the learning process of an individual fuzzy rule in this section. However, the same

P. P. Angelov and X. Gu, *Empirical Approach to Machine Learning*, Studies in Computational Intelligence 800,
https://doi.org/10.1007/978-3-030-02384-3_12

principle can be applied to all the other fuzzy rules within the rule base of the *ALMMo-0*. The detailed algorithmic procedure can be found in Sect. 8.2.2.

The learning process of the ith fuzzy rule within the *ALMMo-0* system consists of the following steps [1]:

Step 1. Initialize the ith fuzzy rule, $\boldsymbol{R}_i$ with the first data sample $\boldsymbol{x}_{i,1}$ of the ith class $(K_i \leftarrow 1)$.
Step 2. For each newly observed data sample of the ith class, $\boldsymbol{x}_{i,K_i}(K_i \leftarrow K_i + 1)$, check whether $\boldsymbol{x}_{i,K}$ can be a new prototype. If yes, go to step 3, otherwise, go to step 4;
Step 3. Add a new *data cloud*, $\mathbf{C}_{i,M_i}$ $(M_i \leftarrow M_i + 1)$ to the system with $\boldsymbol{x}_{i,K_i}$ as the prototype $\boldsymbol{p}_{i,M_i}$, and go to step 5;
Step 4. Use $\boldsymbol{x}_{i,K_i}$ to update meta parameters of the nearest *data cloud* $\mathbf{C}_{i,n*}$, and go to step 5;
Step 5. Update $\boldsymbol{R}_i$ with the newly updated prototypes, and go to step 2 if the new data sample arrives, otherwise, export $\boldsymbol{R}_i$.

The flowchart of the main procedure of the learning process for the ith fuzzy rule of the *ALMMo-0* system is given in Fig. 12.1.

12.1.1.2 Validation Process

The validation process of the *ALMMo-0* system is summarized as follows:

Step 1. For each newly arrived data sample, $\boldsymbol{x}$, calculate its firing strength λ_i to each fuzzy rule, $\boldsymbol{R}_i$ $(i = 1, 2, \ldots, C)$;
Step 2. Decide the label of $\boldsymbol{x}$ based on the firing strength, λ_i $(i = 1, 2, \ldots, C)$ following the "winner takes all" principle.

The flowchart of the main procedure of the validation process of the *ALMMo-0* system is given in Fig. 12.2.

12.1.2 ALMMo-1 *System*

The main procedure of the learning process of the *ALMMo-1* system has the following steps [2].

Step 1. Initialize the system with the first data sample $\boldsymbol{x}_1 (K \leftarrow 1)$.
Step 2. For each newly observed data sample, $\boldsymbol{x}_K (K \leftarrow K + 1)$, generate the system output, y_K;
Step 3. Check whether $\boldsymbol{x}_K$ initializes a new *data cloud*. If yes, go to step 4, otherwise, go to step 7;
Step 4. Check whether the newly formed *data cloud* is overlapping with the nearest *data cloud* $\mathbf{C}_{n*}$, If yes, go to step 5, otherwise, go to step 6;

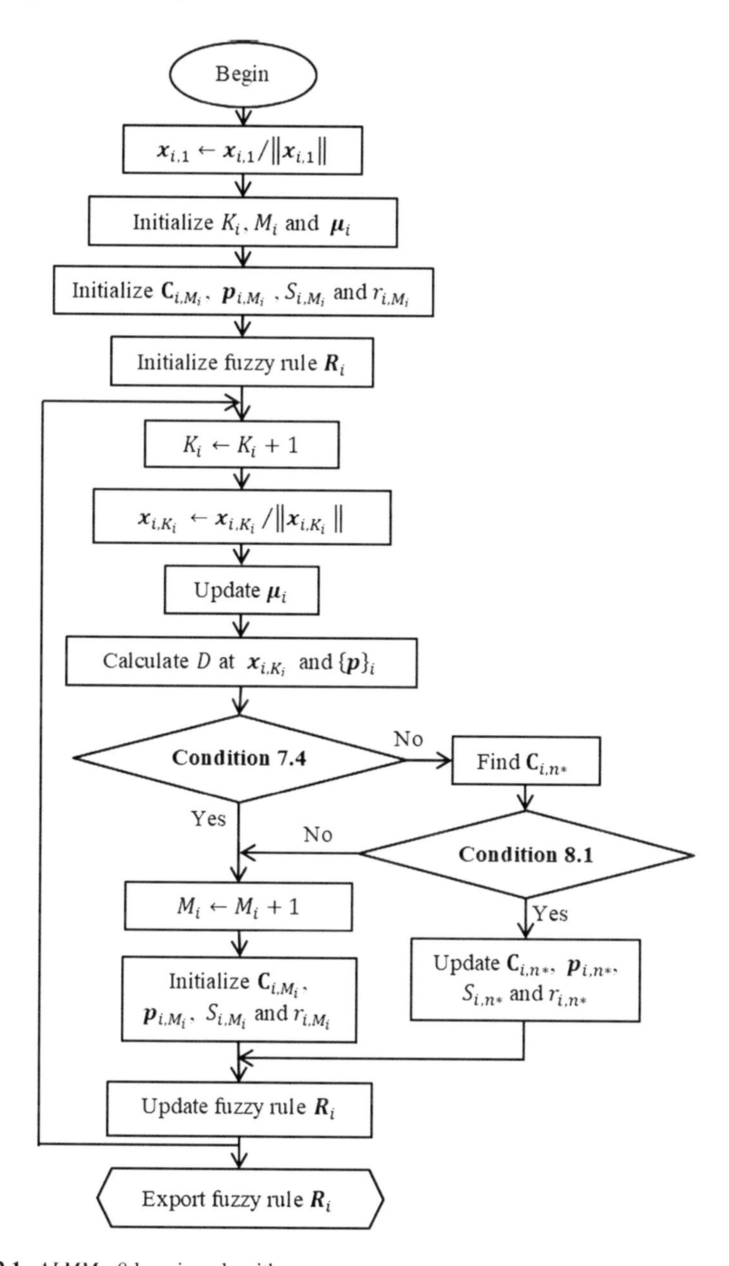

Fig. 12.1 *ALMMo-0* learning algorithm

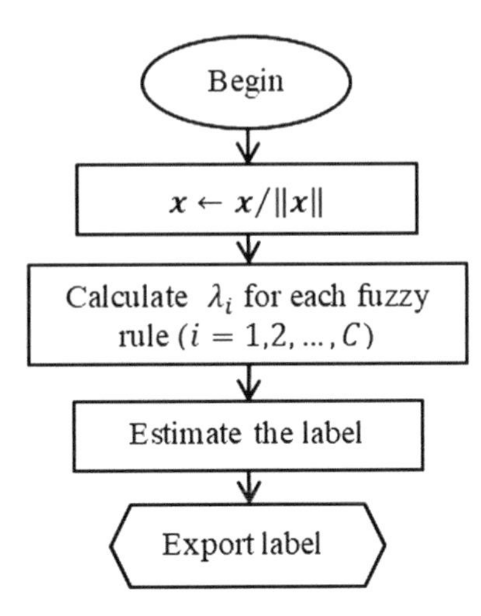

Fig. 12.2 *ALMMo-0* validation algorithm

Step 5. Replace $\mathbf{C}_{n*}$ and its meta-parameters with the *data cloud* formed around $\boldsymbol{x}_K$, replace the prototype of the corresponding fuzzy rule $\boldsymbol{R}_{n*}$ with $\boldsymbol{x}_K$, and go to step 8;
Step 6. Add the new *data cloud*, $\mathbf{C}_{M_K}$ and a new fuzzy rule $\boldsymbol{R}_{M_K}$ with $\boldsymbol{x}_K$ as the prototype $\boldsymbol{p}_{M_K}$ $(M_K \leftarrow M_K + 1)$ to the system, and go to step 8;
Step 7. Use $\boldsymbol{x}_K$ to update meta-parameters of the nearest *data cloud* $\mathbf{C}_{n*}$, and go to step 8;
Step 8. Remove the stale fuzzy rules and update the consequent parameters of the remaining fuzzy rules in the rule base;
Step 9. (Optional) Remove the attributes that have low contribution to the outcome;
Step 10. Go to step 2 if new data samples arrive; otherwise, export the rule base.

The detailed algorithmic procedure can be found in Sects. 8.3.2 and 8.3.3. The flowchart of the main procedure of the learning process for the *ALMMo-1* system is given in Fig. 12.3.

12.2 Numerical Examples of *ALMMo-0* System

In this section, the numerical examples are presented for demonstrating the performance of the *ALMMo-0* system for classification.

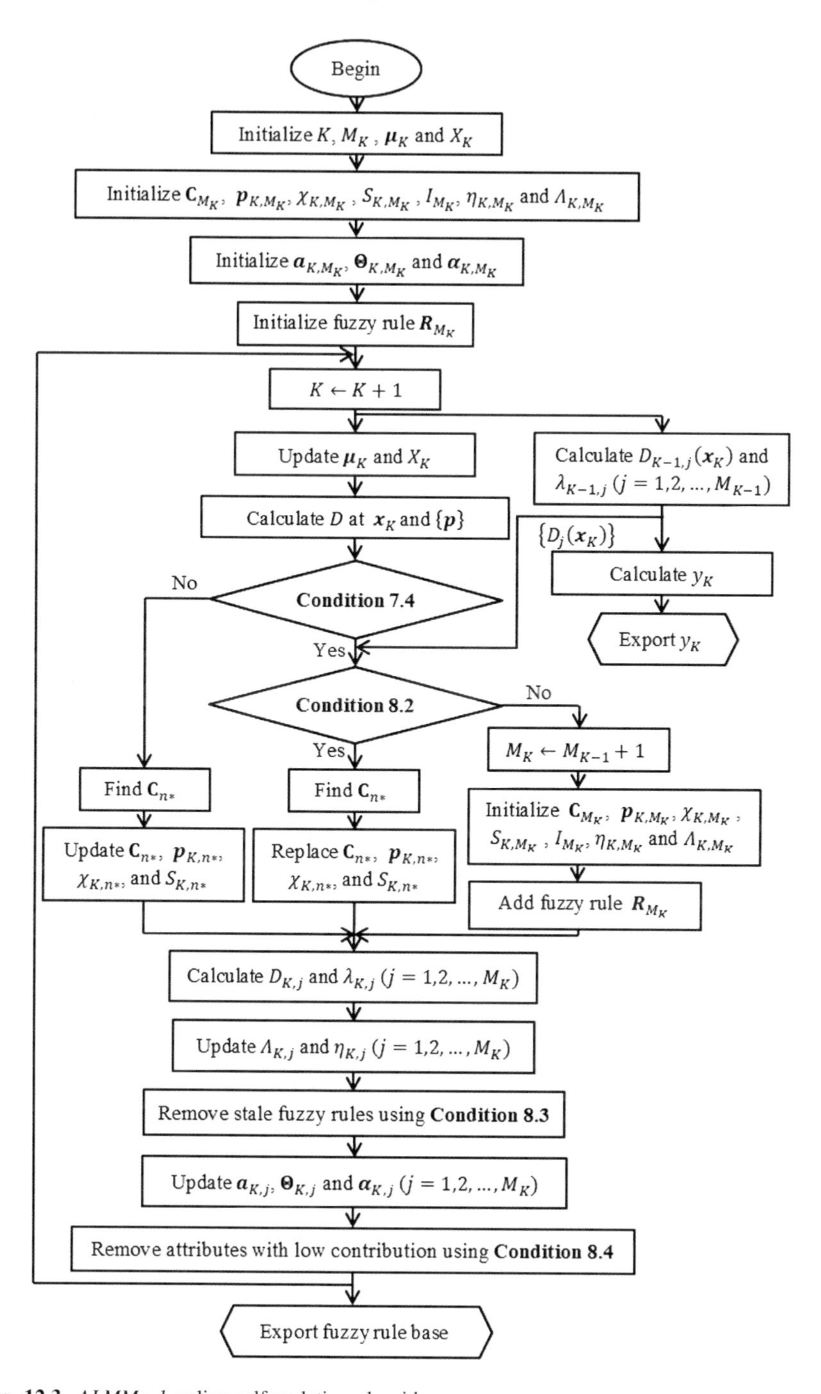

Fig. 12.3 *ALMMo-1* online self-updating algorithm

12.2.1 Datasets for Evaluation

In this section, the following benchmark datasets are considered for evaluating the performance of the *ALMMo-0* system:

(1) Optical recognition of handwritten digits dataset [3] (downloadable from [4]);
(2) Pen-based recognition of handwritten digits dataset [5] (downloadable from [6]);
(3) Multiple features dataset [7] (downloadable from [8]);
(4) Letter recognition dataset [9] (downloadable from [10]).

The details of the letter recognition dataset have been given in Sect. 11.2.1.4, while, the details of the other three benchmark datasets will be given in this subsection.

12.2.1.1 Optical Recognition of Handwritten Digits Dataset

This dataset is obtained from normalized bitmaps of handwritten digits from a preprinted form. It is composed of a training set, which has 3823 data samples, and a validation set, which contains 1797 samples, in total 5620 data samples. Each data sample is extracted from one bitmap. Each bitmap has a size of 32×32, and is divided into non-overlapping blocks of 4×4. The number of on pixels is counted for each block, which results in a 8×8 dimensional matrix, where each element is an integer from the range $[0, 16]$.

Therefore, each data sample in this dataset has 64 attributes corresponding to the 64 elements in the matrix converted from the bitmap and one label ("0" to "9") corresponding to one of the 10 digits ("0" to "9").

12.2.1.2 Pen-Based Recognition of Handwritten Digits Dataset

This dataset is obtained from the handwritten digits using a WACOM PL-100V pressure-sensitive tablet with an integrated LCD display and a cordless stylus. It is composed of a training set, which has 7494 data samples, and a validation set, which contains 3498 samples (in total, 10,992 data samples). Each data sample corresponds to one digit written inside the box of 500×500 tablet pixel resolution. Each handwritten digit is normalized to make the representation invariant to translations and scale distortions, then, it is resampled using simple linear interpolation between pairs of pixels, which results in a feature vector composed of 16 attributes.

Therefore, each data sample in this dataset is a 1×16 dimensional feature vector extracted from a handwritten digit collected using the tablet and one label ("0" to "9") corresponding to one of the 10 digits ("0" to "9").

12.2.1.3 Multiple Features Dataset

This dataset consists of different features of handwritten digits ("0" to "9") extracted from a collection of Dutch utility maps. There are 200 binary images of handwritten digits in each class (in total, 2000 handwritten digits). Each digit is represented by the following six different sets of features:

(1) 76 Fourier coefficients of the character shapes;
(2) 216 profile correlations;
(3) 64 Karhunen-Love coefficients;
(4) 240 pixel averages in 2×3 windows;
(5) 47 Zernike moments;
(6) 6 morphological features.

Therefore, in the multiple features dataset, each data sample has 649 attributes composed of the six sets of features as well as one label ("0" to "9") corresponding to one of the 10 digits ("0" to "9").

12.2.2 *Performance Evaluation*

In this subsection, the performance of the *ALMM0-0* system on benchmark classification problems is evaluated.

For the optical recognition of handwritten digits and pen-based recognition of handwritten digits datasets, we train the *ALMMo-0* system with the training set and test the trained system with the validation set. Since *ALMMo-0* system learns from the data on a sample-by-sample manner, the order of the data samples will influence its performance. Therefore, we report in Table 12.1 the overall accuracy of the classification results and the time consumption (in seconds) for training on the two benchmark datasets after 10 Monte-Carlo experiments by randomly descrambling the order of the training data.

For the multiple features and letter recognition datasets, we randomly pick out 50% of the data as the training set and use the remaining data for validation. We report in Table 12.1 the overall accuracy of the classification results and the time consumption (in seconds) for training on the two benchmark datasets after 10 Monte-Carlo experiments following the same experimental protocol.

We involve the four well-known offline classifiers as used in Sect. 11.3.2 for comparison following the same experimental protocol, which include:

(1) Support vector machine (SVM) classifier with Gaussian kernel [11];
(2) k-nearest neighbor (kNN) classifier with $k = 10$ [12];
(3) Neural network (NN) with three hidden layers of size 20 [13];
(4) Decision tree (DT) classifier [14].

Since *ALMMo-0* system is an evolving approach, we further involve the following evolving classifiers that can start "from scratch" for a fair comparison:

Table 12.1 Performance comparison between different classification approaches

Dataset	Algorithm	Overall accuracy (%)	Training time (s)
Optical recognition of handwritten digits	*ALMMo-0*	97.89	0.51
	SVM	10.13	2.16
	kNN	97.68	0.07
	NN	92.30	0.78
	DT	85.25	0.05
	eClass0	89.20	1.29
	Simpl_eClass0	89.53	1.95
	TEDAClass	91.20	1649.17
Pen-based recognition of handwritten digits	*ALMMo-0*	97.50	0.87
	SVM	10.38	7.28
	kNN	97.49	0.02
	NN	92.40	1.05
	DT	91.22	0.05
	eClass0	81.56	0.62
	Simpl_eClass0	84.53	0.93
	TEDAClass	81.46	1155.62
Multiple features	*ALMMo-0*	93.36	0.14
	SVM	10.24	0.72
	kNN	91.22	0.02
	NN	84.03	0.75
	DT	91.96	0.15
	eClass0	79.71	1.85
	Simpl_eClass0	83.63	4.10
	TEDAClass	51.54	2335.71
Letter recognition	*ALMMo-0*	92.12	1.22
	SVM	40.54	9.54
	kNN	91.84	0.02
	NN	47.19	1.93
	DT	82.29	0.07
	eClass0	46.60	0.80
	Simpl_eClass0	56.64	1.31
	TEDAClass	86.37	14,011.87

(5) eClass0 classifier [15];
(6) Simpl_eClass0 classifier [16];
(7) TEDAClass classifier [17].

The classification results obtained by the seven comparative approaches are also reported in Table 12.1.

12.2.3 Discussion and Analysis

As one can see from Table 12.1, the *ALMMo-0* classifier is able to produce highly accurate classification result after a very efficient training process. Furthermore, unlike other comparative classifiers, the *ALMMo-0* classifier can learn from streaming data on a sample-by-sample basis and be updated continuously with new observations without a full retraining. Therefore, *ALMMo-0* is a very attractive alternative classification approach for real-world applications.

12.3 Numerical Examples of *ALMMo-1* System

In this section, numerical examples are presented for demonstrating the performance of the *ALMMo-1* system for classification and regression.

12.3.1 Datasets for Evaluation

The following two benchmark datasets are used for evaluating the performance of the *ALMMo-1* system for classification problems:

(1) Pima Indians diabetes dataset [18] (downloadable from [19]);
(2) Occupancy detection dataset [20] (downloadable from [21]).

We also use the two real-world problems for evaluating the performance of the *ALMMo-1* system for regression problems:

(1) QuantQuote second resolution market dataset [22];
(2) Standard and Poor index dataset [23].

The details of the Pima Indians diabetes and occupancy detection datasets have been presented in Sects. 11.3.1.2 and 11.3.1.3. In this subsection, the detailed descriptions of QuantQuote second resolution market and Standard and Poor index datasets are presented.

12.3.1.1 QuantQuote Second Resolution Market Dataset

The QuantQuote second resolution market dataset [22] contains tick-by-tick data on all NASDAQ, NYSE, and AMEX securities from 1998 to the present moment in time. The frequency of tick data varies from one second to few minutes. This dataset contains 19144 data samples, each of which has the following five attributes:

(1) *time instance, K*;
(2) *open price, $x_{K,O}$*;

(3) *high price*, $x_{K,H}$;
(4) *low price*, $x_{K,L}$;
(5) *close price*, $x_{K,C}$.

In this book, we consider the four attributes, namely, *open price*, *high price*, *low price* and *close price* of the current time instance to predict the future values of the *open price*, 5, 10 and 15 time instances ahead, namely,

$$y_K = f\left(\boldsymbol{x}_K = \left[x_{K,O}, x_{K,H}, x_{K,L}, x_{K,C}\right]^{\mathrm{T}}\right) \tag{12.1}$$

where $y_K = x_{K+5,O}$, $x_{K+10,O}$ and $x_{K+15,O}$, respectively. As pre-processing, the data is standardized online before prediction, see Sect. 2.2.1.2.

12.3.1.2 Standard and Poor Index Dataset

The Standard and Poor index data is a more frequently used real dataset, which contains 14,893 data samples acquired from January 3, 1950 to March 12, 2009. Other prediction algorithms frequently use this dataset as a benchmark for performance because of the nonlinear, erratic and time-variant behavior of the data. The input and output relationship of the system is governed by the following equation:

$$y_K = f\left(\boldsymbol{x}_K = \left[x_{K-4}, x_{K-3}, x_{K-2}, x_{K-1}, x_K\right]^{\mathrm{T}}\right) \tag{12.2}$$

where $y_K = x_{K+1}$.

12.3.2 Performance Evaluation for Classification

For the Pima Indians diabetes dataset, we randomly pick out 90% of the data as the training set and use the remaining data for validation. For the occupancy detection dataset, similarly, we combine the two validation sets into one.

We report in Table 12.2 the overall accuracy of the classification results and the time consumption (in seconds) for training on the two benchmark datasets after 10 times Monte-Carlo experiments by randomly descrambling the order of the training data. We also compare the *ALMMo-1* system with the seven comparative approaches as used in Sect. 12.2.2, and further involve the following evolving multi-model fuzzy inference system for comparison:

(8) Fuzzily connected multi-model (FCMM) system [24].

We report the overall accuracy and training time consumption (in seconds) of the seven comparative approaches in the Table 12.2 as well.

Table 12.2 Classification performance comparison—offline scenario

Dataset	Algorithm	Overall accuracy (%)	Training time (s)
Pima Indians diabetes	*ALMMo-1*	76.04	0.08
	SVM	64.92	0.13
	kNN	72.45	0.02
	NN	73.31	0.42
	DT	69.22	0.03
	eClass0	65.39	0.03
	Simpl_eClass0	52.66	0.04
	TEDAClass	71.32	6.20
	FCMM	55.94	0.09
Occupancy detection	*ALMMo-1*	98.93	2.27
	SVM	76.07	4.77
	kNN	96.64	0.03
	NN	93.69	1.06
	DT	93.14	0.03
	eClass0	88.82	0.82
	Simpl_eClass0	96.74	0.63
	TEDAClass	96.34	416.50
	FCMM	96.97	1.02

We further conduct a comparison in an online scenario between the fully evolving algorithms that can start "from scratch", namely, *ALMMo-1*, eClass0, Simpl_eClass0 and FCMM by considering the Pima Indians diabetes and Occupancy detection datasets as data streams.

In the following numerical examples, the order of the data samples of both datasets are randomly determined, and the algorithms start classifying from the first data sample and keep updating the system structure along with the arrival of new data samples. 10 Monto Carlo experiments are conducted and the average results of the two experiments are reported in Table 12.3.

Table 12.3 Classification performance comparison—online scenario

Algorithm	Pima Indians diabetes		Occupancy detection	
	Overall accuracy (%)	Training time (s)	Overall accuracy (%)	Training time (s)
ALMMo-1	74.96	0.17	98.76	7.56
eClass0	51.98	0.08	92.57	4.93
Simpl_eClass0	58.53	0.09	95.04	2.00
TEDAClass	69.67	7.80	95.59	2943.27
FCMM	53.75	0.17	95.69	2.98

12.3.3 Performance Evaluation for Regression

In this subsection, we further consider the application of *ALMMo-1* system on real-world regression problems.

In the first numerical example, we use the QuantQuote second resolution market dataset and present the step-by-step online prediction result of using the following inputs:

(1) *open price*, $x_{K,O}$;
(2) *high price*, $x_{K,H}$;
(3) *low price*, $x_{K,L}$;
(4) *close price*, $x_{K,C}$;

of the current time instance to predict the *open price* of the five time steps in advance, namely, $y = f(\boldsymbol{x}_K)$, where $y = x_{K+5,O}$. The prediction result is depicted and compared with the true value in Fig. 12.4a and three smaller periods of the prediction result are also zoomed-in for a better illustration. The corresponding change of number of fuzzy rules/*local* models is presented in Fig. 12.4b.

In the end of the prediction process, the *ALMMo-1* system identifies six fuzzy rules, and they are given in Table 12.4. Note that the values of $x_{K,O}$, $x_{K,H}$, $x_{K,L}$ and $x_{K,C}$ presented in Table 12.4 have been destandardized using the corresponding mean and standard deviation values, and the consequent parameters obtained during the training process are kept.

For a better understanding of the performance of the *ALMMo-1* system on high frequency trading problem, we involve the following four approaches to make a comparison:

(1) Least square linear regression (LSLR) algorithm [25];
(2) Fuzzily connected multi-model (FCMM) system [24];
(3) Evolving Takagi-Sugeno (eTS) model [26];
(4) Sequential adaptive fuzzy inference system (SAFIS) [27].

The prediction performance of the algorithms is reported in Table 12.5 in terms of:

(1) non-dimensional error index (*NDEI*) [28], which is calculated by the following equation:

$$NDEI = \sqrt{\frac{\sum_{k=1}^{K} (t_k - y_k)^2}{K\sigma_t^2}} \tag{12.3}$$

where y_k is estimated value as the output of the system and t_k is the true value at the *k*th time instance; σ_t is the standard deviation of the true value.

(2) number of the identified rules (*M*);
(3) execution time (t_{exe} in seconds).

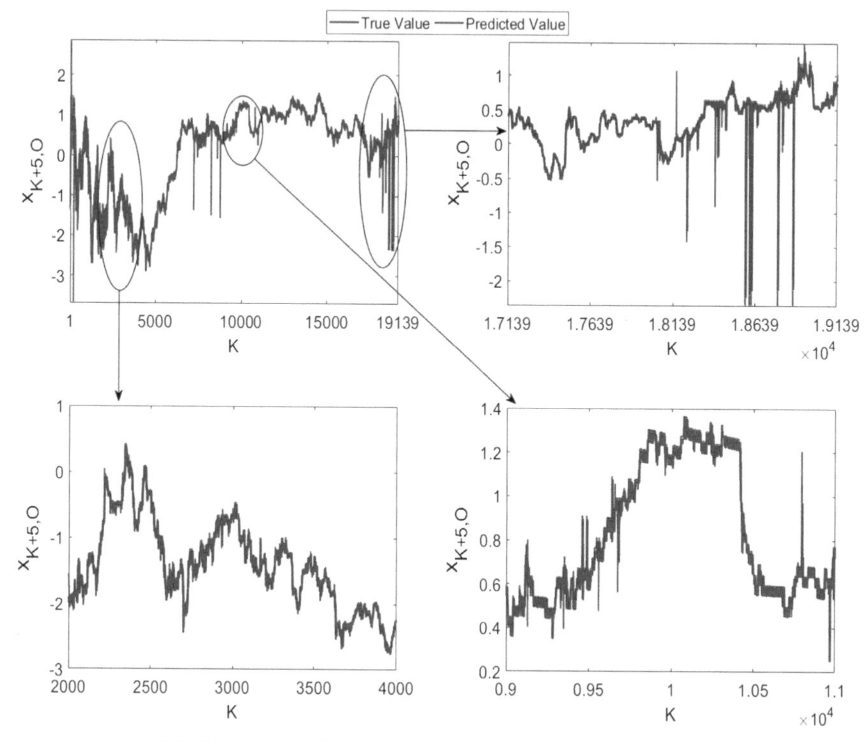

(a) Comparison between real values and the predictions

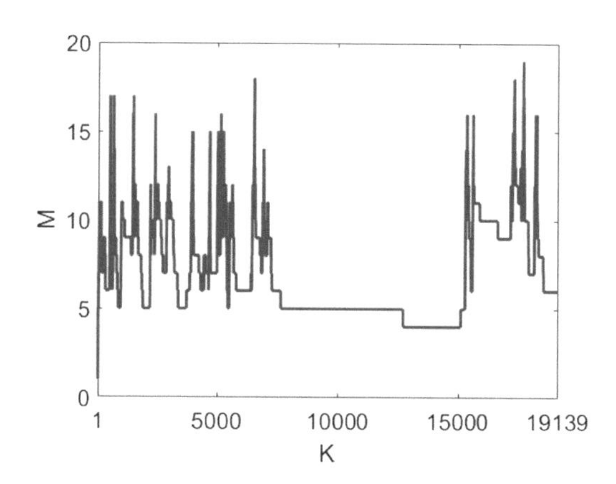

(b) The change of number of identified fuzzy rules/*local* models

Fig. 12.4 Online prediction using the QuantQuote second resolution market dataset

Table 12.4 The identified *AnYa* type fuzzy rules/*local* models

#	Fuzzy rule/*local* model
1	$IF\left([x_O, x_H, x_L, x_C]^T \sim [1405209, 1405245, 1405191, 1405220]^T\right)$ $THEN(y = 0.0251 + 0.3403x_O + 1.0163x_H + 0.1552x_L + 0.4254x_C)$
2	$IF\left([x_O, x_H, x_L, x_C]^T \sim [1404795, 1404814, 1404684, 1404750]^T\right)$ $THEN(y = 0.0130 + 0.3841x_O + 1.0980x_H + 0.1445x_L + 0.3411x_C)$
3	$IF\left([x_O, x_H, x_L, x_C]^T \sim [1408490, 1408508, 1408469, 1408488]^T\right)$ $THEN(y = 0.0057 + 0.1684x_O + 0.3636x_H + 0.2491x_L + 0.5819x_C)$
4	$IF\left([x_O, x_H, x_L, x_C]^T \sim [1407937, 1407960, 1407910, 1407936]^T\right)$ $THEN(y = 0.0079 + 0.3274x_O + 0.5950x_H + 0.4933x_L + 0.5284x_C)$
5	$IF\left([x_O, x_H, x_L, x_C]^T \sim [1407739, 1407801, 1407380, 1407423]^T\right)$ $THEN(y = 0.1171 + 0.8076x_O + 0.3803x_H + 0.5745x_L + -0.2884x_C)$
6	$IF\left([x_O, x_H, x_L, x_C]^T \sim [1407527, 1407633, 1407513, 1407613]^T\right)$ $THEN(y = 0.2725 + 0.8117x_O + 0.1449x_H + 0.5623x_L + -0.4757x_C)$

Table 12.5 Regression performance comparison on the QuantQuote second resolution market dataset

	Algorithm	*NDEI*	*M*	t_{exe}
$x_{K+5,O} = f(\boldsymbol{x}_K)$	*ALMMo-1*	0.130	6	2.17
	LSLR	0.156		6.51
	FCMM	0.159	4	5.73
	eTS	0.137	3	183.67
	SAFIS	0.742	13	6.43
$x_{K+10,O} = f(\boldsymbol{x}_K)$	*ALMMo-1*	0.159	6	2.14
	LSLR	0.184		6.51
	FCMM	0.162	4	5.55
	eTS	0.170	2	154.14
	SAFIS	0.735	14	8.45
$x_{K+15,O} = f(\boldsymbol{x}_K)$	*ALMMo-1*	0.175	6	2.14
	LSLR	0.202		6.45
	FCMM	0.176	4	5.54
	eTS	0.185	10	197.10
	SAFIS	1.038	13	8.87

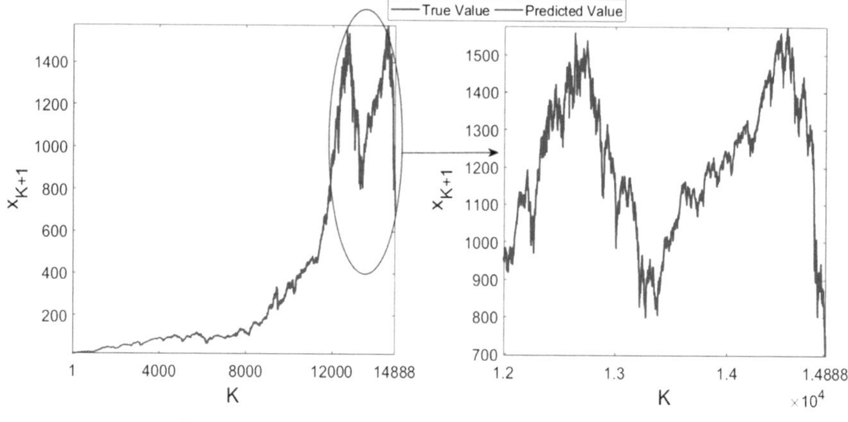

(a) Comparison between real values and the predictions

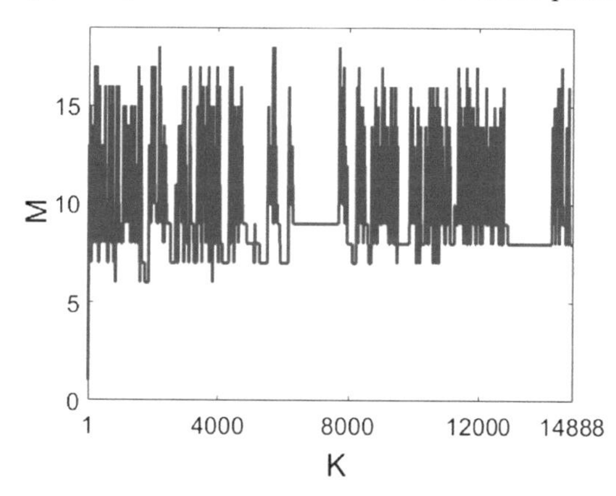

(b) The change of number of identified fuzzy rules

Fig. 12.5 Online prediction on the Standard and Poor index dataset

The prediction results using the four prices of the current time instance to predict the open price 10 and 15 time instances in advance, namely, $x_{K+10,O} = f(\boldsymbol{x}_K)$ and $x_{K+15,O} = f(\boldsymbol{x}_K)$, obtained by the five algorithms are also given in the same table.

In the following example, we use the Standard and Poor index dataset for a further evaluation. The prediction result of this dataset using *ALMMo-1* system is presented in Fig. 12.5. The prediction results obtained by the *ALMMo-1* system as well as by the other five comparative algorithms are tabulated in Table 12.6.

Table 12.6 Regression performance comparison on the Standard and Poor index dataset

Algorithm	*NDEI*	*M*
ALMMo-1	0.013	8
LSLR	0.020	
FCMM	0.014	5
eTS	0.015	14
SAFIS	0.206	6

12.3.4 Discussion and Analysis

From the previous examples we can see that, the *ALMMo-1* system is able to perform highly accurate classification and prediction results in various problems. As a new type of autonomous learning system, *ALMMo-1* touches the very foundations of the complex learning systems for data stream processing, and a wide variety of applications and extensions can be developed. For more numerical examples of using *ALMMo-1*, one can refer to [2].

12.4 Conclusions

In this chapter, the implementations of the *ALMMo* systems of zero-order and first-order are presented, and numerical examples on benchmark datasets as well as real-world problems for classification and regression/prediction are provided to demonstrate the performance of the *ALMMo* systems.

From the comparisons with the well-known approaches, one can conclude that *ALMMo* systems are able to perform highly accurate classification and regression/prediction after an efficient, nonparametric and non-iterative learning process with practically no human involvement.

References

1. P.P. Angelov, X. Gu, Autonomous learning multi-model classifier of 0-order (ALMMo-0), in *IEEE International Conference on Evolving and Autonomous Intelligent Systems*, 2017, pp. 1–7
2. P.P. Angelov, X. Gu, J.C. Principe, Autonomous learning multi-model systems from data streams. IEEE Trans. Fuzzy Syst. **26**(4), 2213–2224 (2018)
3. E. Alpaydin, C. Kaynak, Cascading classifiers. Kybernetika **34**(4), 369–374 (1998)
4. C.M. Bishop, *Pattern recognition and machine learning* (Springer, New York, 2006)
5. F. Alimoglu, E. Alpaydin, Methods of combining multiple classifiers based on different representations for pen-based handwritten digit recognition, in *Proceedings of the Fifth Turkish Artificial Intelligence and Artificial Neural Networks Symposium*, 1996, pp. 1–8

6. http://archive.ics.uci.edu/ml/datasets/Pen-Based+Recognition+of+Handwritten+Digits
7. M. van Breukelen, R.P.W. Duin, D.M.J. Tax, J.E. den Hartog, Handwritten digit recognition by combined classifiers. Kybernetika **34**(4), 381–386 (1998)
8. https://archive.ics.uci.edu/ml/datasets/Multiple+Features
9. P.W. Frey, D.J. Slate, Letter recognition using Holland-style adaptive classifiers. Mach. Learn. **6**(2), 161–182 (1991)
10. https://archive.ics.uci.edu/ml/datasets/Letter+Recognition
11. N. Cristianini, J. Shawe-Taylor, *An Introduction to Support Vector Machines and other Kernel-Based Learning Methods* (Cambridge University Press, Cambridge, 2000)
12. P. Cunningham, S.J. Delany, K-nearest neighbour classifiers. Mult. Classif. Syst. **34**, 1–17 (2007)
13. Y. LeCun, Y. Bengio, G. Hinton, Deep learning. Nat. Methods **13**(1), 35–35 (2015)
14. S.R. Safavian, D. Landgrebe, A survey of decision tree classifier methodology. IEEE Trans. Syst. Man. Cybern. **21**(3), 660–674 (1990)
15. P. Angelov, X. Zhou, Evolving fuzzy-rule based classifiers from data streams. IEEE Trans. Fuzzy Syst. **16**(6), 1462–1474 (2008)
16. R.D. Baruah, P.P. Angelov, J. Andreu, Simpl_eClass: simplified potential-free evolving fuzzy rule-based classifiers, in *IEEE International Conference on Systems, Man, and Cybernetics (SMC)*, 2011, pp. 2249–2254
17. D. Kangin, P. Angelov, J.A. Iglesias, Autonomously evolving classifier TEDAClass. Inf. Sci. (Ny) **366**, 1–11 (2016)
18. J.W. Smith, J.E. Everhart, W.C. Dickson, W.C. Knowler, R.S. Johannes, Using the ADAP learning algorithm to forecast the onset of diabetes mellitus, in *Annual Symposium on Computer Application in Medical Care*, 1988, pp. 261–265
19. https://archive.ics.uci.edu/ml/datasets/Pima+Indians+Diabetes
20. L.M. Candanedo, V. Feldheim, Accurate occupancy detection of an office room from light, temperature, humidity and CO_2 measurements using statistical learning models. Energy Build. **112**, 28–39 (2016)
21. https://archive.ics.uci.edu/ml/datasets/Occupancy+Detection+
22. https://quantquote.com/historical-stock-data
23. https://finance.yahoo.com/quote/%5EGSPC/history?p=%5EGSPC
24. P. Angelov, Fuzzily connected multimodel systems evolving autonomously from data streams. IEEE Trans. Syst. Man Cybern. Part B Cybern. **41**(4), 898–910 (2011)
25. C. Nadungodage, Y. Xia, F. Li, J. Lee, J. Ge, StreamFitter: a real time linear regression analysis system for continuous data streams, in *International Conference on Database Systems for Advanced Applications*, 2011, vol. 6588 LNCS, no. PART 2, pp. 458–461
26. P.P. Angelov, D.P. Filev, An approach to online identification of Takagi-Sugeno fuzzy models. IEEE Trans. Syst. Man Cybern. Part B Cybern. **34**(1), 484–498 (2004)
27. H.J. Rong, N. Sundararajan, G. Bin Huang, P. Saratchandran, Sequential adaptive fuzzy inference system (SAFIS) for nonlinear system identification and prediction. Fuzzy Sets Syst. **157**(9), 1260–1275 (2006)
28. M. Pratama, S.G. Anavatti, P.P. Angelov, E. Lughofer, PANFIS: a novel incremental learning machine. IEEE Trans. Neural Networks Learn. Syst. **25**(1), 55–68 (2014)

Chapter 13
Applications of Deep Rule-Based Classifiers

As a recently introduced approach for image classification with a multilayer architecture and prototype-based nature, the deep rule-based (DRB) classifiers serve as a strong alternative to the mainstream approaches, i.e. deep convolutional neural networks (DCNNs) with the following advantages:

(1) it is free from restrictive *prior* assumptions and user- and problem- specific parameters;
(2) its structure is transparent, human-interpretable and self-evolving;
(3) its training process is self-organizing, non-iterative, non-parametric, highly parallelizable and can start "from scratch".

In this chapter, the implementation and applications of the DRB classifiers are presented. The corresponding pseudo-codes are presented in Appendix B.5. The MATLAB implementation is given in Appendix C.5.

13.1 Algorithm Summary

13.1.1 Learning Process

Since the image pre-processing and feature extraction techniques involved in the DRB system are standard ones and are problem-specific, their implementation will not be specified in this section. Similar to the *ALMMo-0* algorithm, the supervised learning process can be done in parallel for each fuzzy rule within the DRB classifier, we consider the *i*th ($i = 1, 2, \ldots, C$) subsystem, and only summarize the learning process of an individual fuzzy rule. The detailed algorithmic procedure can be found in Sect. 9.4.1.

P. P. Angelov and X. Gu, *Empirical Approach to Machine Learning*, Studies in Computational Intelligence 800,
https://doi.org/10.1007/978-3-030-02384-3_13

The DRB learning algorithm for the ith subsystem consists of the following steps [1]:

Step 1. Pre-process the first image $\mathbf{I}_{i,1}$ of the ith class, extract its feature vector, $\boldsymbol{x}_{i,1}$, and initialize the ith fuzzy rule, $\boldsymbol{R}_i$ with $\mathbf{I}_{i,1}$ and $\boldsymbol{x}_{i,1}$ ($K_i \leftarrow 1$);
Step 2. For each newly arrived image, $\mathbf{I}_{i,K_i}$ of the ith class, pre-process $\mathbf{I}_{i,K_i}$ and extract its feature vector $\boldsymbol{x}_{i,K}$ ($K_i \leftarrow K_i + 1$), then, check whether $\mathbf{I}_{i,K_i}$ can be a new prototype. If yes, go to step 3; otherwise, go to step 4;
Step 3. Add a new *data cloud*, $\mathbf{C}_{i,M_i}$ ($M_i \leftarrow M_i + 1$) to the system with $\mathbf{I}_{i,K_i}$ as the prototype $\mathbf{P}_{i,M_i}$, and go to step 5;
Step 4. Use $\boldsymbol{x}_{i,K}$ to update meta-parameters of the nearest *data cloud* $\mathbf{C}_{i,n*}$, and go to step 5;
Step 5. Update $\boldsymbol{R}_i$ with the existing prototypes, and go to step 2 if new image arrives; otherwise, export $\boldsymbol{R}_i$.

The flowchart of the DRB learning algorithm of the ith subsystem in the DRB classifier is given in Fig. 13.1.

13.1.2 Validation Process

The general validation process of the DRB classifier is summarized as follows:

Step 1. For each newly arrived image, **I**, pre-process and extract its feature vector,$\boldsymbol{x}$;
Step 2. Calculate the firing strength λ_i of **I** to each fuzzy rule, $\boldsymbol{R}_i$ ($i = 1, 2, \ldots, C$);
Step 3. Decide the label of $\boldsymbol{x}$ based on the firing strength, λ_i ($i = 1, 2, \ldots, C$) following the "winner takes all" principle.

The flowchart of the DRB validation algorithm is given in Fig. 13.2.

13.2 Numerical Examples

In this section, the numerical examples are presented for demonstrating the performance of the DRB classifier on image classification.

13.2.1 Datasets for Evaluation

In this section, the following benchmark image sets are considered for evaluating the performance of the DRB classifiers, which cover four different challenging problems:

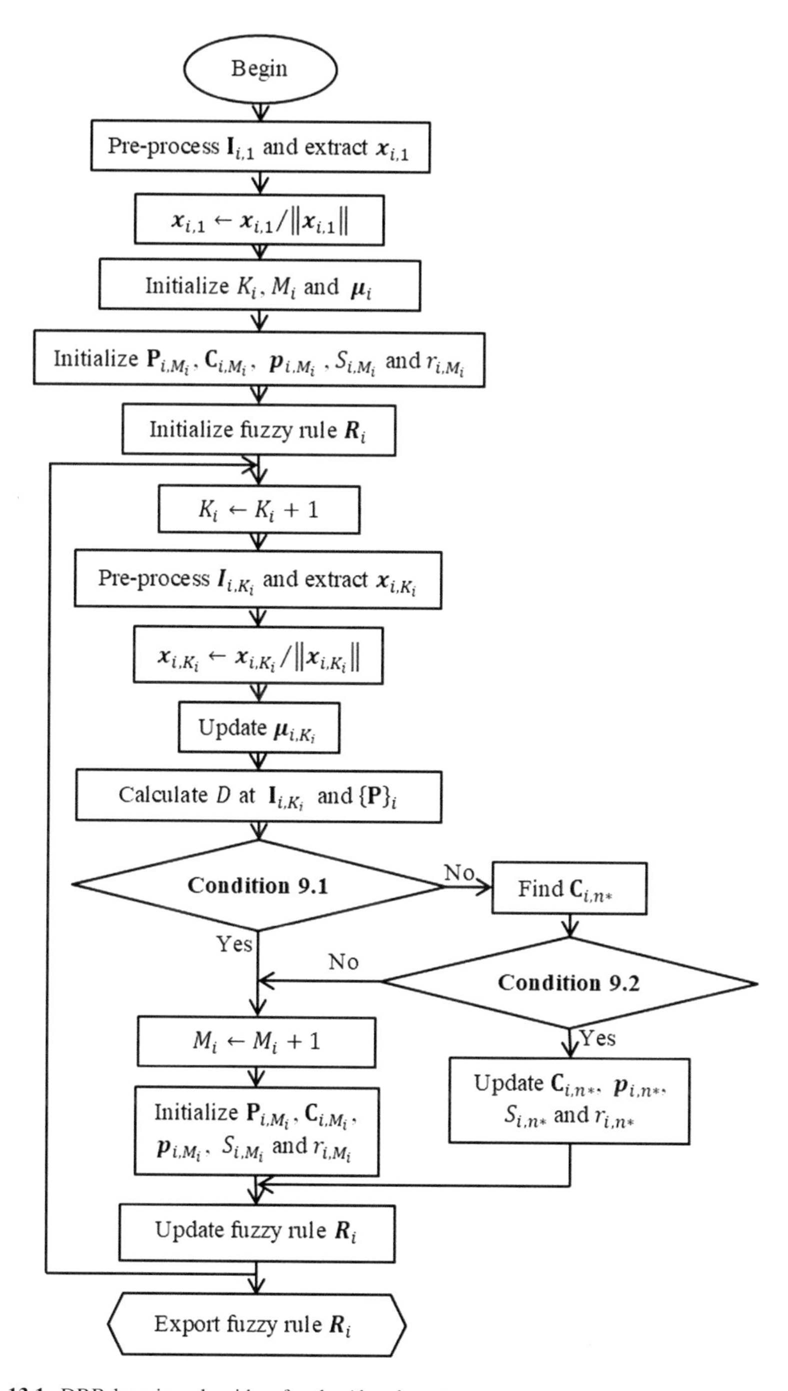

Fig. 13.1 DRB learning algorithm for the ith sub-system

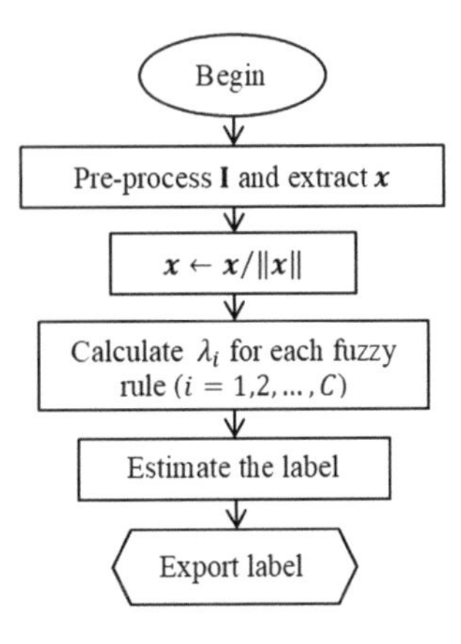

Fig. 13.2 DRB validation algorithm

A. ***Handwritten digits recognition***:

(1) Modified National Institute of Standards and Techniques (MNIST) dataset [2] (downloadable from [3]);

B. ***Face recognition***:

(2) The database of faces [4] (downloadable from [5]);

C. ***Remote sensing***:

(3) Singapore dataset [6] (downloadable from [7]);
(4) UCMerced land use dataset [8] (downloadable from [9]);

D. ***Object recognition***:

(5) Caltech101 dataset [10] (downloadable from [11]).

13.2.1.1 MNIST Dataset

The MNIST dataset is a famous benchmark database for handwritten digits recognitions. It contains 70000 grey images of handwritten digits ("0" to "9") with the size of 28 × 28. The 70000 grey images are divided into one training set, which contains 60000 images, and one validation set, which contains 10000 images. Examples of the handwritten digit images are given in Fig. 13.3 for illustration.

A lot of literature has been published to compete for the best accuracy. However, due to the fact that the dataset itself has flaws, there are a number of testing images which are unrecognizable even for humans (see Fig. 13.12 in the Sect. 13.2.2.1, for example), the testing accuracy is below 100%, although closely approaching it.

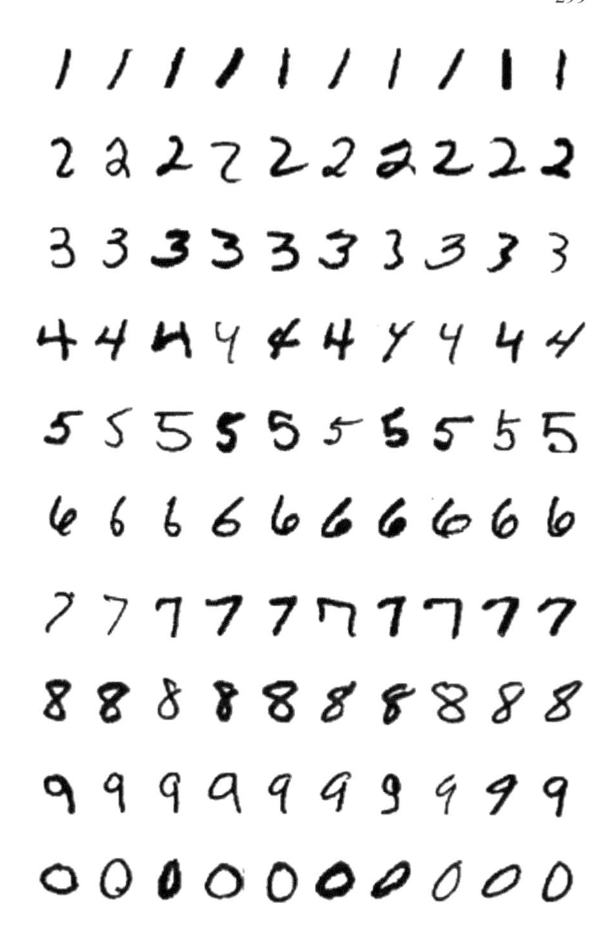

Fig. 13.3 Examples of handwritten digits

13.2.1.2 The Database of Faces

The database of faces is one of the most widely used benchmark datasets for face recognition, which contains 40 subjects with 10 different grey-level images taken with different illumination, angle, face expression and face details (glasses/no glasses, mustaches, etc.). The size of each image is 92×112 pixels. Examples of images of the database of faces are given in Fig. 13.4.

13.2.1.3 Singapore Dataset

Singapore dataset was constructed from a large satellite image of Singapore. This dataset consists of 1086 RGB images with 256×256 pixels size with nine imbalanced scene categories:

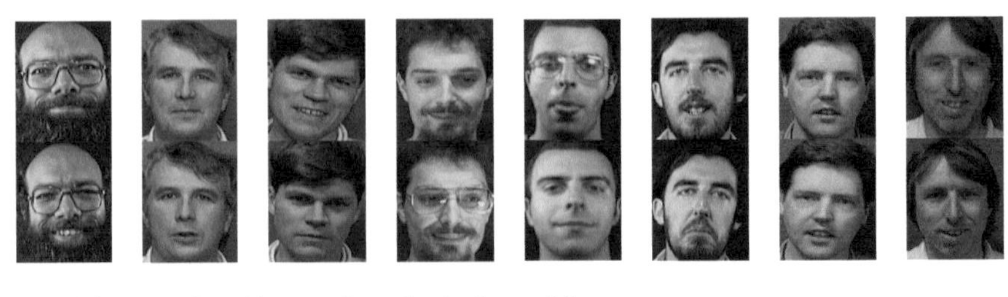

Fig. 13.4 Examples of images from the database of faces

(1) airplane (42 images);
(2) forest (141 images);
(3) harbor (134 images);
(4) industry (158 images);
(5) meadow (166 images);
(6) overpass (70 images);
(7) residential (179 images);
(8) river (84 images), and
(9) runway (112 images.

Examples of the images of the nine classes are shown in Fig. 13.5.

13.2.1.4 UCMerced Land Use Dataset

UCMerced land use dataset consists of fine spatial resolution remote sensing images of 21 challenging scene categories (including airplane, beach, building, harbor, etc.). Each category contains 100 images of the same size (256×256 pixels). Example images of the 21 classes are shown in Fig. 13.6.

13.2.1.5 Caltech101 Dataset

The Caltech 101 dataset contains 9144 pictures of objects belonging to 101 categories and one background category. The number of images per category varies

Fig. 13.5 Examples of images from the Singapore dataset

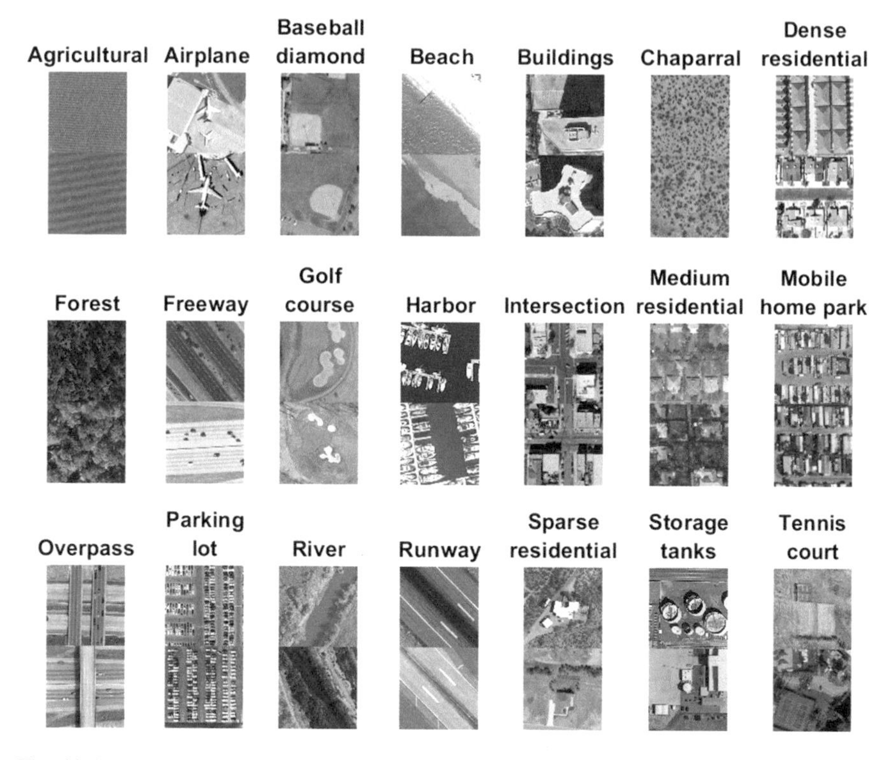

Fig. 13.6 Examples of images from UCMerced land use dataset

from 33 to 800. The size of each image is roughly 300 × 200 pixels. This data set contains categories corresponding to both rigid objects (like bikes and cars) and non-rigid objects (like animals and flowers). Therefore, the shape variance is significant. The example images of the Caltech 101 dataset are presented in Fig. 13.7.

As the four benchmark datasets are very different from each other, we will use four different, but same as in the publications [2, 6, 12, 13], experimental protocols for each dataset, respectively.

13.2.2 Performance Evaluation

In this subsection, the performance of the DRB classifier on image classification is evaluated.

Fig. 13.7 Example images of Caltech 101 dataset

13.2.2.1 Handwritten Digits Recognition

For the MNIST dataset, we use the DRB ensemble classifier is used [14]. The architecture of the DRB ensemble classifier for training is depicted in Fig. 13.8. The architecture for validation is depicted in Fig. 13.9.

As one can see from Figs. 13.8 and 13.9 that the DRB ensemble classifier consists of the following components:

(1) Normalization layer, which normalize the original value range of the pixels of the handwritten digit images from [0, 255] to [0, 1], linearly.

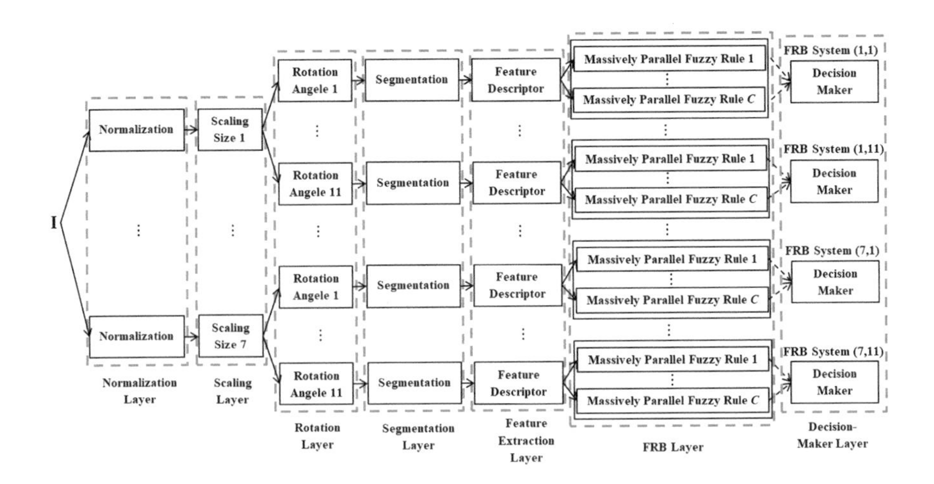

Fig. 13.8 The architecture of the DRB ensemble classifier for training

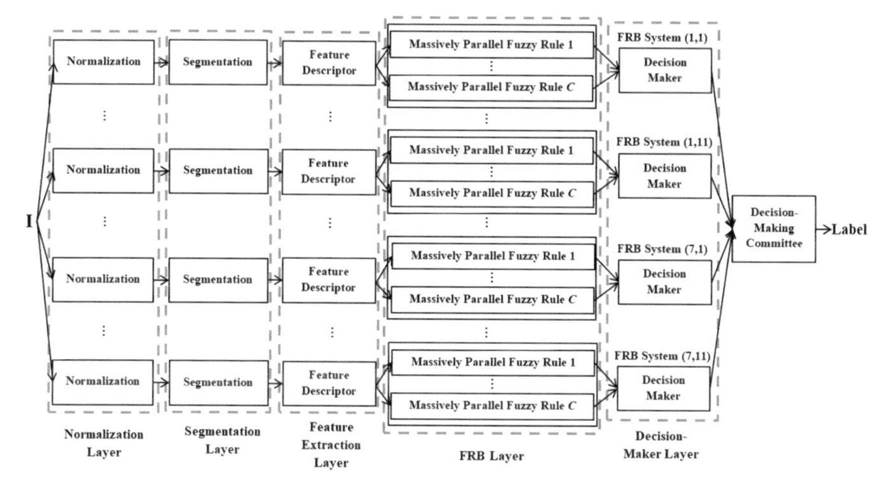

Fig. 13.9 The architecture of the DRB ensemble classifier for validation

(2) Scaling layer, which resizes the images from the original image size of 28 × 28 pixels into seven different sizes:

(i) 28 × 22;
(ii) 28 × 24;
(iii) 28 × 26;
(iv) 28 × 28;
(v) 28 × 30;
(vi) 28 × 32, and
(vii) 28 × 34.

(3) Rotation layer, which rotates each image (after the scaling operation) into 11 different angles from −15° to 15° with the interval of 3°.
Therefore, the scaling and rotation layer expands the original training set into 77 new training sets with different scaling sizes and rotation angles.

(4) Segmentation layer is for extracting the central area (22 × 22) of the training images, which discards the borders that are mostly composed of white pixels with little or no information.

(5) Feature extraction layer. In the DRB ensemble classifier for handwritten digits recognition, a low-level feature descriptor, namely GIST [15] or histogram of oriented gradient (HOG) [16], is employed, which extract a 1 × 512 dimensional GIST feature vector or a 1 × 576 dimensional HOG feature vector from each training image, respectively.

(6) Fuzzy rule-based layer, which consists of 77 fuzzy rule-based (FRB) systems. Each of them is trained with one of the two types of feature vectors from one of the 77 expanded training sets. Each FRB system has 10 FRB sub-systems (corresponding to the 10 digits “0” to “9”) and each sub-system has one

massively parallel zero-order fuzzy rule of *AnYa* type as described in Sect. 9.3. As a result, the FRB layer has 770 zero-order fuzzy rules of *AnYa* type in total. Each one of them is trained separately. The training process of these fuzzy rules is described in Sect. 9.4.1.

(7) Decision-maker layer, which is composed of 77 decision-makers. Each of them corresponds to one FRB system;
(8) Overall decision maker, which takes the partial suggestions from the 77 decision-makers and makes the final decision (class label, the digit).

During the experiment, the feature descriptor used by the DRB ensemble classifier is based on GIST or HOG features. However, due to the difference in the descriptive abilities of the two feature descriptors, the performance of the DRB ensemble classifier is somewhat different. The recognition accuracy of the proposed DRB classifier using the two different feature descriptors is tabulated in Table 13.1.

The corresponding average training times for the 10 fuzzy rules are tabulated in Table 13.2. An example of the identified massively parallel fuzzy rules is given in Fig. 13.10, from which one can appreciate its transparency and interpretability.

We further combined the DRB ensemble trained with GIST features and the DRB ensemble trained with HOG features. The architecture of this hybrid is depicted in Fig. 13.11, where the DRB ensemble classifier has the same architecture as depicted in Fig. 13.8 for training and the same as in Fig. 13.9 for validation. The decision-maker uses the Eq. (9.13) for deciding the labels of the validation images.

The combined DRB ensemble classifier achieves a better recognition performance (successfully recognized 9944 images out of 10000 validation images). Its performance is tabulated in Table 13.1 as well. The 56 mistakes made during the validation process are shown in Fig. 13.12, where one can see that most of them are, indeed, unrecognizable.

The DRB cascade classifier as described in [14] is able to achieve the best performance, which is also presented in Table 13.1. The architecture of the DRB cascade is presented in Fig. 13.13.

The DRB ensemble classifiers using GIST and HOG features have the same architecture as depicted in Fig. 13.8 for the training process and the same architecture as depicted in Fig. 13.9 for validation.

The system output integrator (see Fig. 13.13a) integrates the outputs of the 1540 massively parallel fuzzy rules of the two DRB ensemble classifiers (770 in each) into 10 overall scores of confidences (H = 154 outputs per digit) by the following equation:

$$\Lambda_i(\mathbf{I}) = \frac{1}{H}\sum_{j=1}^{H} \lambda_{i,j}(\mathbf{I}); \quad i = 1, 2, \ldots, C \tag{13.1}$$

where $\lambda_{i,j}(\mathbf{I})$ is firing strength calculated by Eq. (10.10); $C = 10$.

The conflict detector (see Fig. 13.13a) in the DRB cascade classifier will detect the rare cases in which the highest and the second highest overall scores of confidence are very close. If such cases happen, it means that there is a lack of clarity

Table 13.1 Comparison between the DRB ensembles and the state-of-the-art approaches

Algorithm	DRB-GIST	DRB-HOG	DRB-GIST + DRB-HOG	DRB Cascade
Accuracy (%)	**99.30**	**98.86**	**99.44**	**99.55**
Training time	**Less than 2 min for each part**			
PC-parameters	Core i7-4790 (3.60 GHz), 16 GB DDR3			
GPU used	**None**			
Elastic distortion	**No**			
Tuned parameters	No			
Iteration	No			
Randomness	No			
Parallelization	Yes			
Evolving ability	Yes			
Algorithm	Large Convolutional Networks [17]	Large Convolutional Networks [18]	Committee of 7 Convolutional Neural Networks [19]	Committee of 35 Convolutional Neural Networks [20]
Accuracy	99.40%	99.47%	99.73% ± 2%	99.77%
Training time	No Information	No Information	**Almost 14 h for each one of the DNNs**	
PC-parameters			Core i7-920 (2.66 GHz), 12 GB DDR3	
GPU used			**2 × GTX 480 & 2 × GTX 580**	
Elastic distortion	No	No	Yes	
Tuned parameters	Yes	Yes	Yes	
Iteration	Yes	Yes	Yes	
Randomness	Yes	Yes	Yes	
Parallelization	No	No	No	
Evolving ability	No	No	No	

Table 13.2 Computation time for the learning process per sub-system (in seconds)

Fuzzy rule #		1	2	3	4	5
Digital		"1"	"2"	"3"	"4"	"5"
Feature	GIST	32.39	41.95	45.72	37.17	34.90
	HOG	70.99	82.47	92.73	73.46	67.53
Fuzzy rule #		6	7	8	9	10
Digital		"6"	"7"	"8"	"9"	"0"
Feature	GIST	37.36	35.89	42.99	36.90	39.26
	HOG	68.48	77.93	75.83	69.90	72.03

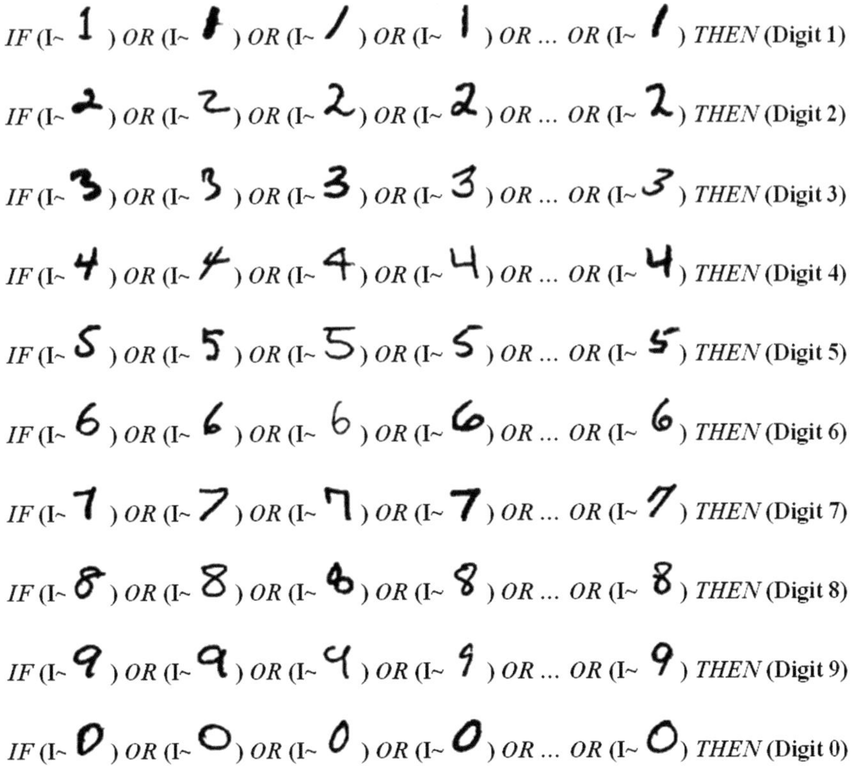

IF (I~ 1) *OR* (I~ 1) *OR* (I~ 1) *OR* (I~ 1) *OR* ... *OR* (I~ 1) *THEN* **(Digit 1)**

IF (I~ 2) *OR* (I~ 2) *OR* (I~ 2) *OR* (I~ 2) *OR* ... *OR* (I~ 2) *THEN* **(Digit 2)**

IF (I~ 3) *OR* (I~ 3) *OR* (I~ 3) *OR* (I~ 3) *OR* ... *OR* (I~ 3) *THEN* **(Digit 3)**

IF (I~ 4) *OR* (I~ 4) *OR* (I~ 4) *OR* (I~ 4) *OR* ... *OR* (I~ 4) *THEN* **(Digit 4)**

IF (I~ 5) *OR* (I~ 5) *OR* (I~ 5) *OR* (I~ 5) *OR* ... *OR* (I~ 5) *THEN* **(Digit 5)**

IF (I~ 6) *OR* (I~ 6) *OR* (I~ 6) *OR* (I~ 6) *OR* ... *OR* (I~ 6) *THEN* **(Digit 6)**

IF (I~ 7) *OR* (I~ 7) *OR* (I~ 7) *OR* (I~ 7) *OR* ... *OR* (I~ 7) *THEN* **(Digit 7)**

IF (I~ 8) *OR* (I~ 8) *OR* (I~ 8) *OR* (I~ 8) *OR* ... *OR* (I~ 8) *THEN* **(Digit 8)**

IF (I~ 9) *OR* (I~ 9) *OR* (I~ 9) *OR* (I~ 9) *OR* ... *OR* (I~ 9) *THEN* **(Digit 9)**

IF (I~ 0) *OR* (I~ 0) *OR* (I~ 0) *OR* (I~ 0) *OR* ... *OR* (I~ 0) *THEN* **(Digit 0)**

Fig. 13.10 An illustrative example of the massively parallel fuzzy rule base

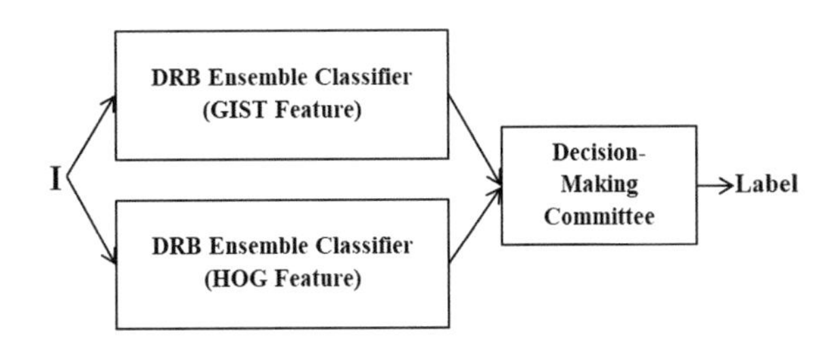

Fig. 13.11 The architecture of the combined DRB ensemble classifier

and the decision-maker will involve the conflict resolution classifier. In this case, a support vector machine (SVM) was used. Generally, however, this can be another efficient classifier which will work only on the small number of such problematic cases. Moreover, it will work as a two-class classifier (deciding only between the two most-likely cases, not between the 10 original classes). The principle for detecting a conflict is as follows:

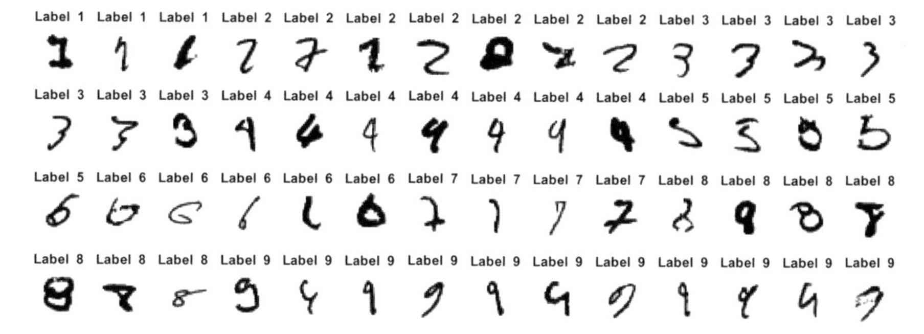

Fig. 13.12 The only 56 mistakes made by the DRB cascade out of 10000 validation images

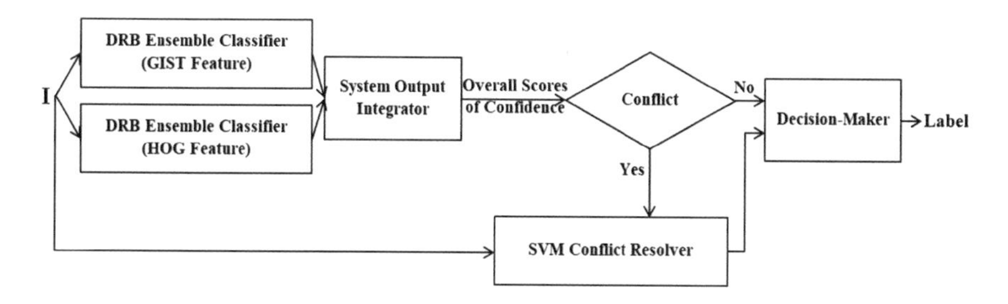

(a) The general architecture

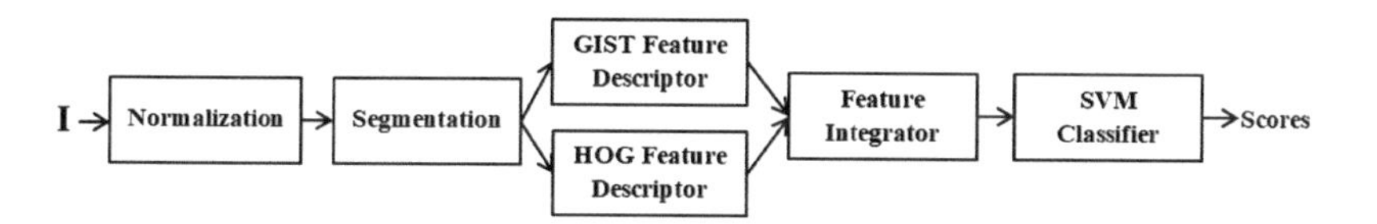

(b) The architecture of the SVM conflict resolver

Fig. 13.13 The architecture of the DRB cascade classifier [14]

Condition 13.1

$$IF\left(\Lambda_{1st}(\mathbf{I}) \leq \Lambda_{2nd}(\mathbf{I}) + \frac{\sigma_{\Lambda(\mathbf{I})}}{4}\right) THEN\ (Lack\ of\ clarity) \tag{13.2}$$

where $\Lambda_{1st}(\mathbf{I})$ and $\Lambda_{2nd}(\mathbf{I})$ are the first and second highest overall scores of confidence; $\sigma_{\Lambda(\mathbf{I})}$ is the standard deviation of the 10 overall scores of confidence of the image, $\mathbf{I}$.

The conflict resolution classifier (see Fig. 13.13) only applies to a small number (about 5%) of the validation data for which the decision-maker was not certain (there were two possible winners with close overall scores). The normalization and segmentation operations (see Fig. 13.13b) in the conflict resolution classifier is the same as used in the DRB ensemble classifiers, and the feature integrator integrates the GIST and HOG feature vectors of a handwritten digit image based on the following equation:

$$\boldsymbol{x} \leftarrow \left[\frac{\boldsymbol{x}_G^T}{\|\boldsymbol{x}_G\|}, \frac{\kappa\left(\boldsymbol{x}_H^T\right)}{\|\kappa(\boldsymbol{x}_H)\|}\right]^T \tag{13.3}$$

where $\boldsymbol{x}_{GIST}$ and $\boldsymbol{x}_{HOG}$ are the GIST and HOG features of $\mathbf{I}$; $\kappa(\cdot)$ is a nonlinear mapping function, which is used to amplify the differences [21]:

$$\kappa(x) = \mathrm{sgn}(1-x)\left(\mathrm{e}^{(1+\mathrm{sgn}(1-x)(1-x))^2} - \mathrm{e}\right) \tag{13.4}$$

In this example, without limiting the generality, the final layer of the conflict resolution classifier (see Fig. 13.13b) is a binary SVM classifier with 5th order polynomial kernel. The SVM classifier is trained with the features extracted from the original training set only (without applying rotation, scaling, etc..) because:

(1) Its training speed deteriorates significantly for large-scale datasets;
(2) Its training process cannot be parallelized;
(3) It does not support online training.

During the classification stage, the SVM conflict resolution classifier will not be functioning if Condition 13.1 is not satisfied. In such case, the decision-maker will make the decision directly based on the maximum $\Lambda_i(\mathbf{I})$ following the "winner takes all" principle.

Otherwise, the decision-maker will do a binary classification of the image between the correspondingly first and second most-likely classes with the assistance of the conflict resolver:

$$\text{Label} = \arg\max_{i=1,2}\left(\left\{\Lambda_{1st}(\mathbf{I}) + \lambda_{1st,SVM}(\mathbf{I}), \Lambda_{2nd}(\mathbf{I}) + \lambda_{2nd,SVM}(\mathbf{I})\right\}\right) \tag{13.5}$$

where $\lambda_{1st,SVM}(\mathbf{I})$ and $\lambda_{2nd,SVM}(\mathbf{I})$ are the scores obtained by the SVM classifier.

The evolving ability of the DRB classifier, as one of the most distinctive properties, plays an important role in real-world applications because there is no need for complete re-training the classifier when new data samples are coming.

To illustrate this advantage, the DRB classifier is trained with images in the form of an image stream, meanwhile, the execution time and the recognition accuracy are recorded during the process. In this example, the original training set without rescaling and rotation is used, which speeds up the process significantly. The relationship curves of the training time (the average for each of the 10 fuzzy rules) and recognition accuracy with the growing amount of the training samples are depicted in Fig. 13.14.

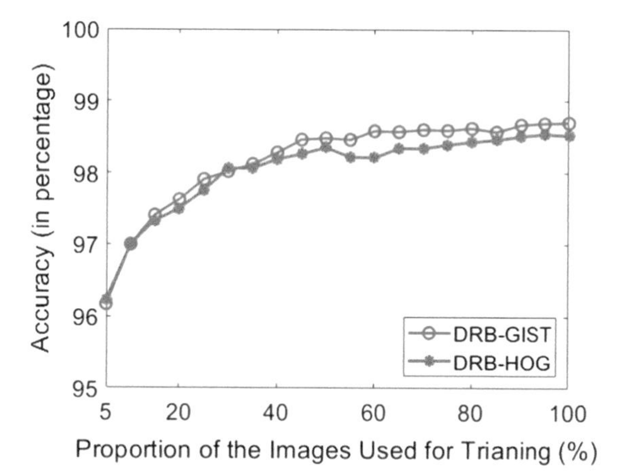

(a) Accuracy vs proportion of the images used

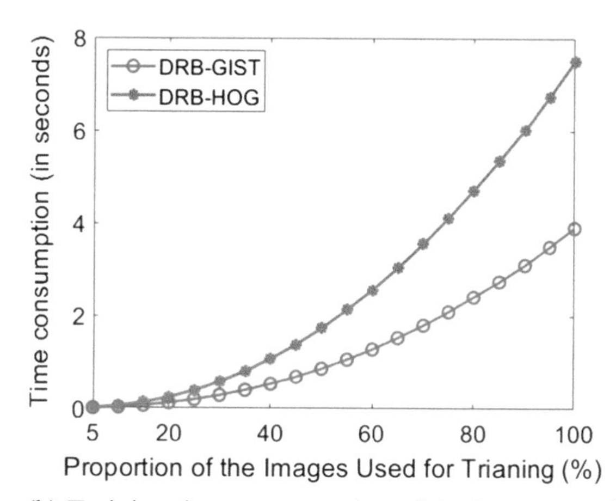

(b) Training time vs proportion of the images used

Fig. 13.14 Recognition accuracy versus the percentage of the training samples

We also compared the DRB approach with the best reported state-of-the-art techniques from the following aspects [14, 21]:

(1) **accuracy** of the image classification;
(2) **time** and complexity, namely, the time consumption for the training process;
(3) computational resources required, including the graphics processing unit (**GPU**) **used**, etc.;
(4) the use of **elastic distortion**;
(5) the requirement of **parameter tuning**;
(6) the requirement for an **iteration** process;
(7) the reproducibility (or, respectively, **randomness**) of the results;
(8) **parallelization**, namely, whether the training process is highly parallelizable;

(9) **evolving capability**, namely, whether the learning model can be updated further with new images with unseen patterns without a full retraining.

We present this comparison in Table 13.1.

13.2.2.2 Face Recognition

The architecture of the DRB classifier for face recognition does not include scaling and rotation, which is shown in Fig. 13.15. In this numerical example, the DRB classifier is composed of the following components:

(1) Normalization layer;
(2) Segmentation layer, which splits each image into smaller pieces by a 22×32 size sliding window with the step size of five pixels in both horizontal and vertical directions. This layer cuts one face image into 255 pieces.
(3) Feature extraction layer, which extracts the combined GIST and HOG features (as given by Eq. 13.3) from each segment.
(4) FRB layer, which consists of 40 massively parallel fuzzy rules, each fuzzy rule is trained based on the image segments of one of the 40 subjects.
(5) Decision-maker, which generates the labels using Eq. (13.6):

$$\text{Label} = \underset{i=1,2,\ldots,C}{\arg\max}\left(\sum_{t=1}^{T} \lambda_i(\mathbf{s}_t)\right) \tag{13.6}$$

where $\mathbf{s}_t$ denotes the tth segment of the image; T is the number of segments.

Following the commonly used experimental protocol [13], for each subject, K images are randomly selected for training and one image for testing. The average recognition accuracy (in percentage) of the DRB classifier after 50 Monte-Carlo experiments with different K ($K = 1$ to 5) is reported in Table 13.3. The DRB classifier is also compared with the state-of-the-art approaches as follows:

(1) Sparse fingerprint classification (SFC) algorithm [13];
(2) Adaptive sparse representation (ASR) algorithm [22];
(3) Sparse discriminant analysis via joint $L_{2,1}$-norm minimization ($SDAL_{21}M$) algorithm [23].

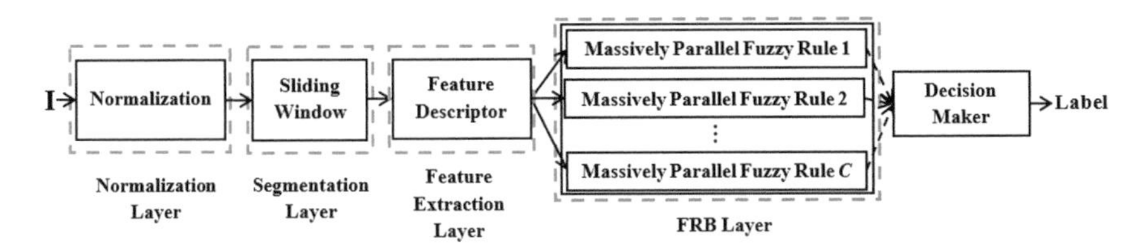

Fig. 13.15 Architecture of the DRB classifier for face recognition

Table 13.3 Comparison between the proposed approach and the state-of-the-art approaches

K	Method	Accuracy (%)
1	SFC	89
	DRB	90
2	SFC	96
	ASR	82
	DRB	97
3	SFC	98
	ASR	89
	$SDAL_{21}M$	82
	DRB	99
4	SFC	99
	ASR	93
	DRB	99
5	SFC	100
	ASR	96
	$SDAL_{21}M$	93
	DRB	100

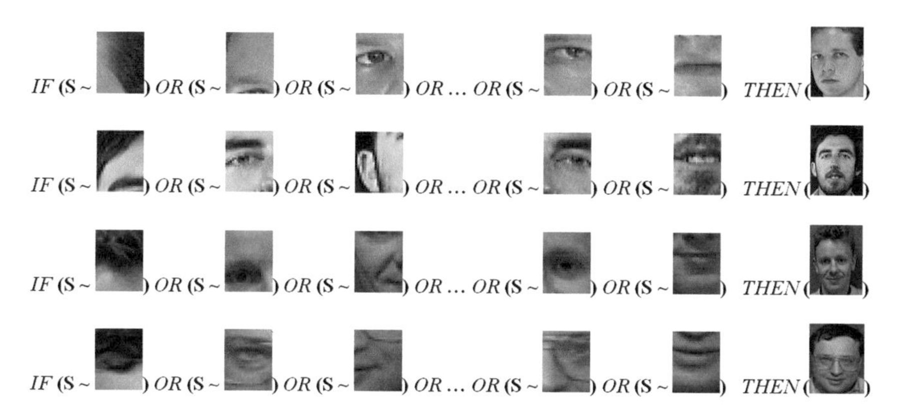

Fig. 13.16 Visual examples of the *AnYa* type fuzzy rules identified from face images

One can see from Table 13.3 that the DRB classifier can achieve higher recognition accuracy with a smaller amount of training samples. For a better illustration, four examples of the *AnYa* type fuzzy rules extracted during experiments are given in Fig. 13.16, where the segments are enlarged for visual clarity.

The recognition accuracy (in percentage) of the DRB classifier and the corresponding time consumption (t_{exe} in seconds) for each fuzzy rule in the training process with different amount of training samples is tabulated in Table 13.4, where all the values are the average after 50 Monte-Carlo experiments. As one can see

Table 13.4 Results with different amount of training samples on face recognition

K	1	2	3	4	5	6	7	8	9
Accuracy (%)	90	97	99	99	100	100	100	100	100
t_{exe} (s)	0.11	0.48	1.03	1.81	2.84	4.14	5.69	8.32	10.65

from the table, the training process is very efficient. The DRB classifier can be trained for less than 3 s per individual and achieves up to 100% accuracy in face recognition of individuals.

13.2.2.3 Remote Sensing Scene Recognition

The architecture of the DRB classifier for the remote sensing scene recognition problems is given in Fig. 13.17.

As one can see from Fig. 13.17, the DRB classifier used in the numerical examples in this subsection is composed of the following layers:

(1) Rotation layer, which rotates each remote sensing image at four different angles,

   (i) 0°;
   (ii) 90°;
   (iii) 180°, and
   (iv) 270°.

(2) Segmentation layer, which uses a sliding window to partition the remote sensing images into smaller pieces for local information extraction. By changing the size of the sliding window, the level of granularity of the segmentation result can be changed accordingly. A larger sliding window size allows the DRB to capture coarse scale spatial information at the cost of losing fine scale detail and vice versa.

   For the following numerical examples based on Singapore and UCMerced datasets, the sliding window with the window size of $\frac{6\times6}{8\times8}$ of image size and step size of $\frac{2}{8}$ width in the horizontal and $\frac{2}{8}$ length in the vertical direction is used [24].

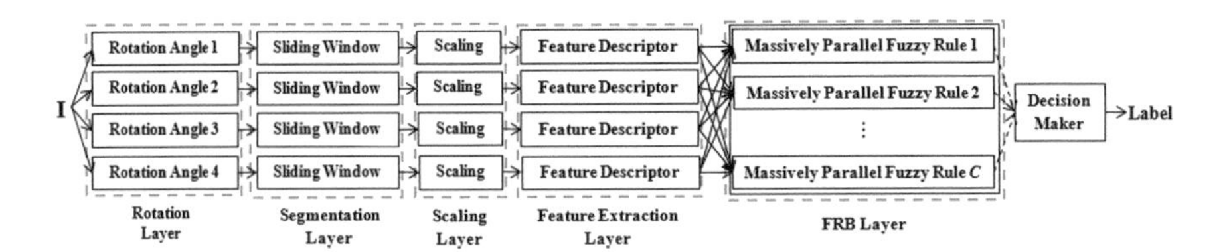

Fig. 13.17 Architecture of the DRB classifier for remote sensing scene recognition

(3) Scaling layer, which is involved in the DRB classifier to rescale the segments into the uniform size of 227×227 pixels required by the high-level feature descriptor, namely, the VGG-VD-16 model in the next layer [25];
(4) Feature extraction layer, which is the VGG-VD-16 model as described in Sect. 2.2.4.2 [25];
(5) FRB layer, which consists of C massively parallel fuzzy rules, each one of them trained based on the image segments of one of the C image categories;
(6) Decision-maker, which generates the labels using Eq. (13.6).

A. ***Numerical examples using the Singapore dataset***

Following the commonly used experimental protocol [6], the DRB classifier is trained with randomly selected 20% of the images of each land use category and uses the remaining images as a testing set. The average recognition accuracy (in percentages) is reported in Table 13.5 after five Monte-Carlo experiments.

The performance of the DRB classifier is compared with the state-of-the-art approaches as follows:

(1) Two-level feature representation (TLFP) algorithm [6];
(2) Bag of visual words (BOVW) algorithm [26];
(3) Scale invariant feature transform with sparse coding (SIFTSC) algorithm [27];
(4) Spatial pyramid matching (SPM) algorithm [28].

The recognition accuracies of the comparative approaches are reported in Table 13.5. One can see that, the DRB classifier is able to produce a significantly better recognition result than the best alternative method.

To show the evolving ability of the DRB classifier, 20% of the images of each class are randomly selected out for validation and the DRB is trained with 10, 20, 30, 40, 50, 60, 70 and 80% of the dataset. The average accuracy after five Monte-Carlo experiments is tabulated in Table 13.6. The average time for training (t_{exe} in seconds) is also reported, however, due to the unbalanced classes, the training time consumption as tabulated in Table 13.6 is the sum of the training times of the nine fuzzy rules.

B. ***Numerical examples on UCMerced dataset***

For the UCMerced land use dataset, the DRB classifier with the same architecture as presented in Fig. 13.17 is employed.

Table 13.5 Comparison between the DRB classifier and the state-of-the-art approaches on Singapore dataset

Algorithm	Accuracy (%)
TLFP	90.94
BOVW	87.41
SIFTSC	87.58
SPM	82.85
DRB	97.70

Table 13.6 Results with different amount of training samples on the Singapore dataset

Ratio (%)	10	20	30	40
Accuracy (%)	96.02	97.56	98.55	98.91
t_{exe} (s)	5.17	20.78	49.33	87.17
Ratio (%)	50	60	70	80
Accuracy (%)	99.10	99.36	99.55	99.62
t_{exe} (s)	135.00	195.57	270.89	346.14

Table 13.7 Comparison between the DRB classifier and the state-of-the-art approaches on UCMerced dataset

Algorithm	Accuracy (%)
TLFR	91.12
BOVW	76.80
SIFTSC	81.67
SPM	74.00
MUFL	88.08
RCN	94.53
DRB	96.14

Following the commonly used experimental protocol [6], 80% of the images of each class are randomly selected out for training and the remaining 20% images are used as a validation set. The average classification accuracy after five Monte-Carlo experiments is reported in Table 13.7.

The performance of the DRB classifier is also compared with the following state-of-the-art approaches:

(1) Two-level feature representation (TLFP) algorithm [6];
(2) Bag of visual words (BOVW) algorithm [26];
(3) Scale-invariant feature transform with sparse coding (SIFTSC) algorithm [27];
(4) Spatial pyramid matching (SPM) algorithm [28], [29];
(5) Multipath unsupervised feature learning (MUFL) algorithm [30];
(6) Random convolutional network (RCN) [31].

From the comparison presented in Table 13.7 one can see that, the DRB classifier, again, exhibits the best classification accuracy.

Similarly, 20% of the images of each class are selected out for validation and the DRB classifier is trained with 10, 20, 30, 40, 50, 60 and 70% of the dataset. The average accuracy and time consumption for training (per rule) are reported in Table 13.8 after five Monte-Carlo experiments. One can see that, the DRB classifier

Table 13.8 Results with different amount of training samples on the UCMerced dataset

Ratio (%)	10	20	30	40
Accuracy (%)	83.48	88.57	90.80	92.19
t_{exe} (s)	0.27	1.36	3.96	5.83
Ratio (%)	50	60	70	80
Accuracy (%)	93.48	94.19	95.14	96.10
t_{exe} (s)	10.29	11.52	15.49	18.15

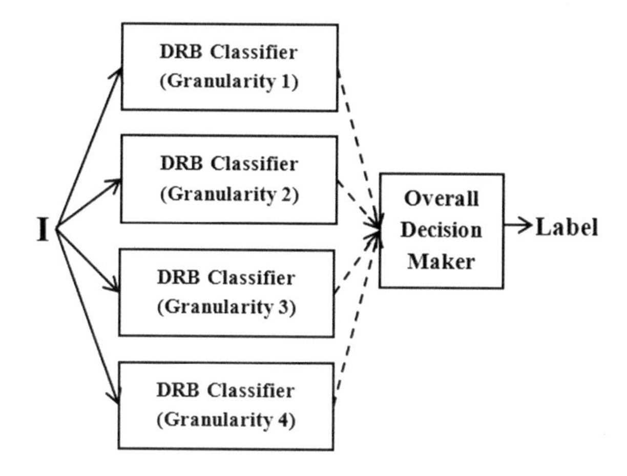

Fig. 13.18 Architecture of the DRB ensemble classifier for remote sensing scenes

can achieve over 95% classification accuracy with only less than 20 s of training for each fuzzy rule.

Furthermore, by creating an ensemble of DRB classifiers as depicted in Fig. 13.18, which consists of four DRB classifiers trained with the segments of remote sensing images at four different levels of granularity (*very small*, *small*, *medium* and *large*), which are achieved by using the sliding windows of four different sizes $\frac{4\times4}{8\times8}$,$\frac{5\times5}{8\times8}$, $\frac{6\times6}{8\times8}$ and $\frac{7\times7}{8\times8}$ with a step size of $\frac{1}{8}$ width in the horizontal and $\frac{1}{8}$ length in the vertical direction (see Fig. 13.19), the classification performance can be further improved to 97.10% [32].

13.2.2.4 Object Recognition

The architecture of the DRB classifier for object recognition is depicted in Fig. 13.20, which is the same as the latter part of the DRB classifier for remote sensing problems as presented in Fig. 13.20. The images of the Caltech 101 dataset are very uniform in presentation, aligned from left to right, and usually not occluded, therefore, the rotation and segmentation are not necessary.

Following the commonly used protocol [12], experiments are conducted by selecting 15 and 30 training images from each class for training and using the rest for validation. The average classification accuracy (in percentage) is reported in Table 13.9 after five Monte-Carlo experiments. The DRB classifier is also compared with the following state-of-the-art approaches and the results are reported in Table 13.9 as well, where one can see that the DRB classifier easily outperforms all the comparative approaches for the object recognition problem.

(1) Convolutional deep belief network (CBDN) [33];
(2) Learning convolutional feature hierarchies (LCFH) algorithm [34];
(3) Deconvolutional networks (DECN) [35];

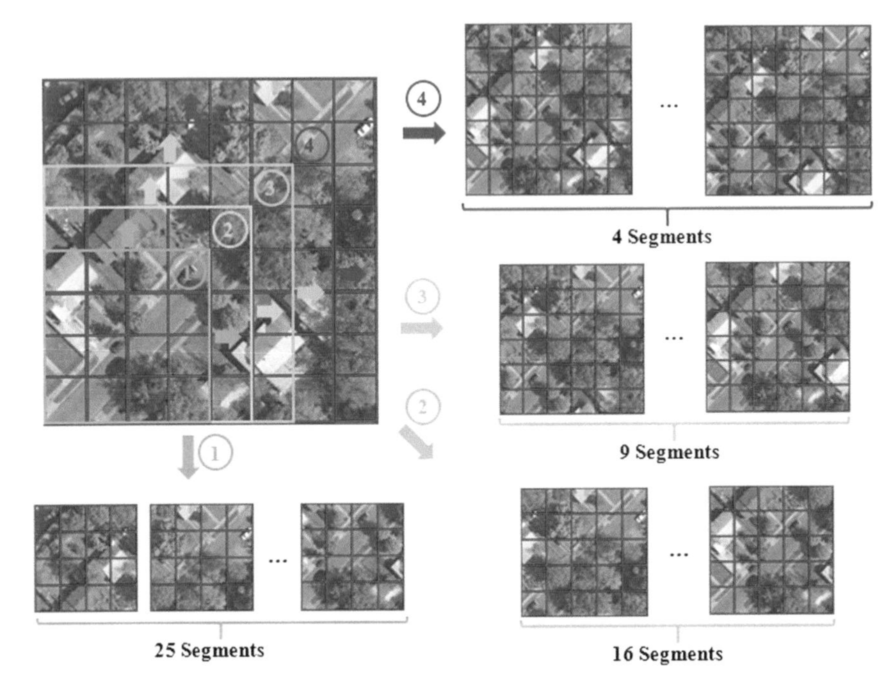

Fig. 13.19 Remote sensing image segmentation with different sliding windows

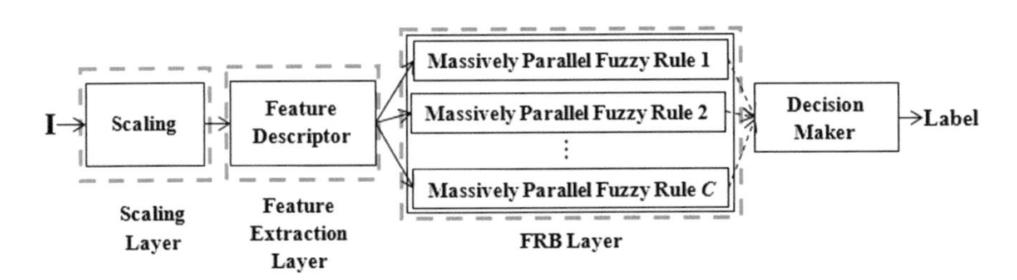

Fig. 13.20 Architecture of the DRB classifier for object recognition

(4) Linear spatial pyramid matching (LSPM) algorithm [36];
(5) Local-constraint linear coding (LCLC) algorithm [37];
(6) DEFEATnet [12];
(7) Convolutional sparse autoencoders (CSAE) [38].

Same as the previous example, we randomly select out 5, 10, 15, 20, 25, and 30 images of each class for training the DRB classifier and use the rest for validation. The average accuracy (in percentage) and time consumption (in seconds) for training (per rule) after five Monte-Carlo experiments are tabulated in Table 13.10, where one can see that, it only requires less than two seconds to train a single fuzzy rule.

Table 13.9 Comparison between the DRB classifier and the state-of-the-art approaches on Caltech 101 dataset

Algorithm	Accuracy	
	15 training images (%)	30 training images (%)
CBD	57.7	65.4
LCFH	57.6	66.3
DECN	58.6	66.9
LSPM	67.0	73.2
LCLC	65.4	73.4
DEFEATnet	71.3	77.6
CSAE	64.0	71.4
DRB	81.9	84.5

Table 13.10 Results with different amount of training samples on the Caltech 101 dataset

Training images	5	10	15	20	25	30
Accuracy (%)	76.4	80.4	81.9	83.5	83.6	84.5
t_{exe} (s)	0.14	0.39	0.99	1.02	1.25	1.42

13.2.3 *Discussion and Analysis*

As one can see from the numerical examples presented in this section that the DRB classifier is able to produce highly accurate classification result in various image classification problems including, handwritten digits recognition, remote sensing scene classification, face recognition, object recognition, etc.

Moreover, compared with the state-of-the-art approaches, the DRB classifier is able to self-organize, self-update and self-evolve its system structure and parameters in an online, autonomous and highly efficient manner. Its prototype-based nature and its semantically meaningful and transparent inner structure are very important characteristics in comparison to the existing approaches (including the best and widely used ones). For more numerical examples, please refer to [14, 21, 24, 32].

13.3 Conclusions

The DRB classifier is a general approach for various problems and serves as a strong alternative to the state-of-the-art DCNN-based approaches by providing a fully human-interpretable structure after a highly efficient, transparent, non-parametric, non-iterative training process.

In this chapter, the implementations of the DRB classifier are presented, and numerical examples based on various benchmark image sets are given to demonstrate the excellent performance of the DRB classifier.

References

1. X. Gu, P.P. Angelov, J.C. Principe, A method for autonomous data partitioning. Inf. Sci. (Ny) **460–461**, 65–82 (2018)
2. Y. LeCun, L. Bottou, Y. Bengio, P. Haffner, Gradient-based learning applied to document recognition. Proc. IEEE **86**(11), 2278–2323 (1998)
3. http://yann.lecun.com/exdb/mnist/
4. F.S. Samaria, A.C. Harter, Parameterisation of a stochastic model for human face identification, in *IEEE Workshop on Applications of Computer Vision* (1994) pp. 138–142
5. http://www.cl.cam.ac.uk/research/dtg/attarchive/facedatabase.html
6. J. Gan, Q. Li, Z. Zhang, J. Wang, Two-level feature representation for aerial scene classification. IEEE Geosci. Remote Sens. Lett. **13**(11), 1626–1630 (2016)
7. http://icn.bjtu.edu.cn/Visint/resources/Scenesig.aspx
8. Y. Yang, S. Newsam, Bag-of-visual-words and spatial extensions for land-use classification, in *International Conference on Advances in Geographic Information Systems* (2010) pp. 270–279
9. http://weegee.vision.ucmerced.edu/datasets/landuse.html
10. L. Fei-Fei, R. Fergus, P. Perona, One-shot learning of object categories. IEEE Trans. Pattern Anal. Mach. Intell. **28**(4), 594–611 (2006)
11. http://www.vision.caltech.edu/Image_Datasets/Caltech101/
12. S. Gao, L. Duan, I.W. Tsang, DEFEATnet—A deep conventional image representation for image classification. IEEE Trans. Circuits Syst. Video Technol. **26**(3), 494–505 (2016)
13. T. Larrain, J.S.J. Bernhard, D. Mery, K.W. Bowyer, Face recognition using sparse fingerprint classification algorithm. IEEE Trans. Inf. Forensics Secur. **12**(7), 1646–1657 (2017)
14. P. Angelov, X. Gu, A cascade of deep learning fuzzy rule-based image classifier and SVM, in *International Conference on Systems, Man and Cybernetics* (2017) pp. 1–8
15. A. Oliva, A. Torralba, Modeling the shape of the scene: A holistic representation of the spatial envelope. Int. J. Comput. Vis. **42**(3), 145–175 (2001)
16. N. Dalal, B. Triggs, Histograms of oriented gradients for human detection, in *IEEE Computer Society Conference on Computer Vision and Pattern Recognition* (2005) pp. 886–893
17. M. Ranzato, F.J. Huang, Y.L. Boureau, Y. LeCun, Unsupervised learning of invariant feature hierarchies with applications to object recognition, in *Proceedings of the IEEE Computer Society Conference on Computer Vision and Pattern Recognition* (2007) pp. 1–8
18. K. Jarrett, K. Kavukcuoglu, M. Ranzato, Y. LeCun, What is the best multi-stage architecture for object recognition? in *IEEE International Conference on Computer Vision* (2009) pp. 2146–2153
19. D.C. Cireşan, U. Meier, L.M. Gambardella, J. Schmidhuber, Convolutional neural network committees for handwritten character classification. Int. Conf. Doc. Analysis Recogn. **10**, 1135–1139 (2011)
20. D. Ciresan, U. Meier, J. Schmidhuber, Multi-column deep neural networks for image classification, in *Conference on Computer Vision and Pattern Recognition* (2012) pp. 3642–3649
21. P.P. Angelov, X. Gu, MICE: Multi-layer multi-model images classifier ensemble, in *IEEE International Conference on Cybernetics* (2017) pp. 436–443
22. J. Wang, C. Lu, M. Wang, P. Li, S. Yan, X. Hu, Robust face recognition via adaptive sparse representation. IEEE Trans. Cybern. **44**(12), 2368–2378 (2014)

23. X. Shi, Y. Yang, Z. Guo, Z. Lai, Face recognition by sparse discriminant analysis via joint L2,1-norm minimization. Pattern Recogn. **47**(7), 2447–2453 (2014)
24. P.P. Angelov, X. Gu, Deep rule-based classifier with human-level performance and characteristics. Inf. Sci. (Ny) **463–464**, 196–213 (2018)
25. K. Simonyan, A. Zisserman, Very deep convolutional networks for large-scale image recognition, in *International Conference on Learning Representations* (2015) pp. 1–14
26. T. Joachims, Text categorization with support vector machines: learning with many relevant features, in *European Conference on Machine Learning* (1998) pp. 137–142
27. A.M. Cheriyadat, Unsupervised feature learning for aerial scene classification. IEEE Trans. Geosci. Remote Sens. **52**(1), 439–451 (2014)
28. S. Lazebnik, C. Schmid, J. Ponce, Beyond bags of features : spatial pyramid matching for recognizing natural scene categories, in *IEEE Computer Society Conference on Computer Vision and Pattern Recognition* (2006) pp. 2169–2178
29. Y. Yang, S. Newsam, Spatial pyramid co-occurrence for image classification, in *Proceedings of the IEEE International Conference on Computer Vision* (2011) pp. 1465–1472
30. J. Fan, T. Chen, S. Lu, Unsupervised feature learning for land-use scene recognition. IEEE Trans. Geosci. Remote Sens. **55**(4), 2250–2261 (2017)
31. L. Zhang, L. Zhang, V. Kumar, Deep learning for remote sensing data. IEEE Geosci. Remote Sens. Mag. **4**(2), 22–40 (2016)
32. X. Gu, P.P. Angelov, C. Zhang, P.M. Atkinson, A massively parallel deep rule-based ensemble classifier for remote sensing scenes. IEEE Geosci. Remote Sens. Lett. **15**(3), 345–349 (2018)
33. H. Lee, R. Grosse, R. Ranganath, A.Y. Ng, Convolutional deep belief networks for scalable unsupervised learning of hierarchical representations, in *Annual International Conference on Machine Learning* (2009) pp. 1–8
34. K. Kavukcuoglu, P. Sermanet, Y.-L. Boureau, K. Gregor, M. Mathieu, Y. LeCun, Learning convolutional feature hierarchies for visual recognition, in *Advances in Neural Information Processing Systems* (2010) pp. 1090–1098
35. M. Zeiler, D. Krishnan, G. Taylor, R. Fergus, Deconvolutional networks, in *IEEE Conference on Computer Vision and Pattern Recognition* (2010) pp. 2528–2535
36. J. Yang, K. Yu, Y. Gong, T. Huang, Linear spatial pyramid matching using sparse coding for image classification, in *IEEE Conference on Computer Vision and Pattern Recognition* (2009) pp. 1794–1801
37. J. Wang, J. Yang, K. Yu, F. Lv, T. Huang, Y. Gong, Locality-constrained linear coding for image classification, in *IEEE Conference on Computer Vision and Pattern Recognition* (2010) pp. 3360–3367
38. W. Luo, J. Li, J. Yang, W. Xu, J. Zhang, Convolutional sparse autoencoders for image classification. IEEE Trans. Neural Networks Learn. Syst. **29**(7), 1–6 (2017)

Chapter 14
Applications of Semi-supervised Deep Rule-Based Classifiers

Thanks to the prototype-based nature, the deep rule-based (DRB) classifiers can be extended with fully transparent, human-interpretable and non-iterative semi-supervised learning strategies for both offline and online scenarios, namely, Semi-supervised DRB (SS_DRB) classifiers. Compared with the existing semi-supervised approaches, the SS_DRB classifier has the following distinctive features:

(1) its learning process can be conducted online on a sample-by-sample or chunk-by-chunk basis;
(2) its learning process is free from error propagation;
(3) it is able to classify out-of-sample images;
(4) it is able to learn new classes and self-evolve its system structure.

In this chapter, the implementation and applications of SS_DRB classifiers are presented. The corresponding pseudo-code is provided in Appendix B.6, and the MATLAB implementation is given in Appendix C.6.

14.1 Algorithm Summary

Since the SS_DRB classifier is extended on the basis of the DRB classifier, in this section, we will summarize the main steps of the following semi-supervised learning processes of the SS_DRB classifier:

(1) offline semi-supervised learning process;
(2) offline active semi-supervised learning process;
(3) online semi-supervised learning process;
(4) online active semi-supervised learning process.

P. P. Angelov and X. Gu, *Empirical Approach to Machine Learning*, Studies in Computational Intelligence 800,
https://doi.org/10.1007/978-3-030-02384-3_14

The validation process of the SS_DRB classifier is the same as for the DRB classifier. The detailed algorithmic procedures of the semi-supervised learning processes can be found in Sect. 9.5.

14.1.1 *Offline Semi-supervised Learning Process*

The main steps of the offline semi-supervised learning process of the DRB classifier (offline SS_DRB learning algorithm) are as follows.

Step 1. Pre-process the unlabelled training image set, $\{\mathbf{V}\}$ and extract the corresponding feature vectors $\{\boldsymbol{v}\}$;
Step 2. Calculate the vector of firing strengths for each unlabelled training image, $\boldsymbol{\lambda}(\mathbf{V}_j)\ j = 1, 2, \ldots, K$;
Step 3. Identify the unlabeled training images for which the DRB classifier is highly confident about their class labels, $\{\mathbf{V}\}_0$. If $\{\mathbf{V}\}_0$ is not empty, remove $\{\mathbf{V}\}_0$ from $\{\mathbf{V}\}$ and go to step 4, otherwise, the offline semi-supervised learning process stops and exports the rule base and the remaining images in $\{\mathbf{V}\}$ as $\{\mathbf{V}\}_2$;
Step 4. **Rank** $\{\mathbf{V}\}_0$ based on their vectors of firing strengths and obtain $\{\mathbf{V}\}_1$;
Step 5. Update the DRB classifier using the DRB learning algorithm (as described in Sects. 9.4.1 and 13.1.1) with $\{\mathbf{V}\}_1$, then go back to step 2.

The flowchart of the main procedure of the offline SS_DRB learning algorithm is given in Fig. 14.1.

14.1.2 *Offline Active Semi-supervised Learning Process*

The main steps of the offline active semi-supervised learning process of the DRB classifier (offline SS_DRB active learning algorithm) are as follows.

Step 1. Use the offline SS_DRB learning algorithm to initialize the DRB classifier, which will after that continue to learn from the unlabelled training image set, $\{\mathbf{V}\}$;
Step 2. Calculate the vector of firing strengths for each unlabelled training image within $\{\mathbf{V}\}_2$, $\boldsymbol{\lambda}(\mathbf{V}_j)\ j = 1, 2, \ldots, K$;
Step 3. Identify the unlabelled training images, $\{\mathbf{V}\}_3$ that are less similar to the prototypes within the the DRB classifier. If $\{\mathbf{V}\}_3$ is not empty, go to step 4, otherwise, the offline active semi-supervised learning process stops and exports the rule base;

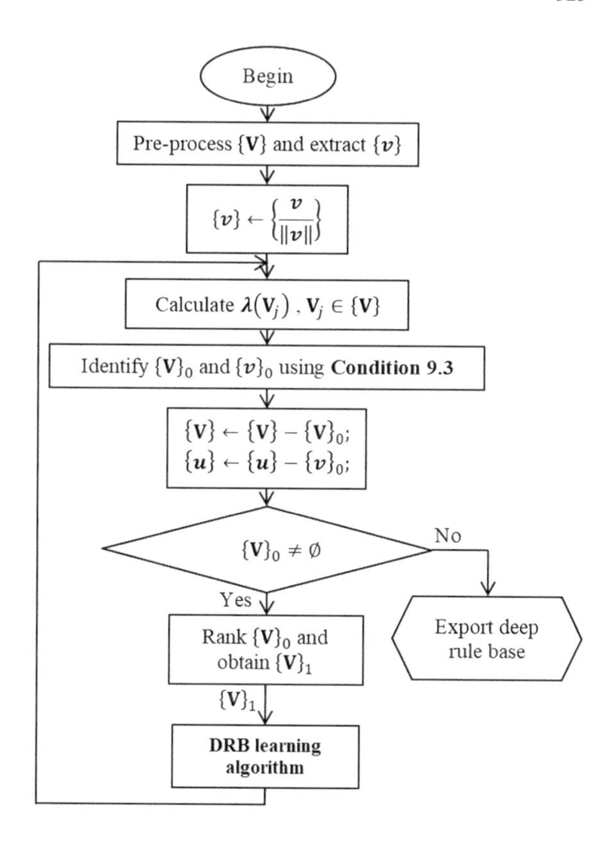

Fig. 14.1 Offline SS_DRB learning algorithm

Step 4. Identify the unlabelled training image $\mathbf{V}_{\min} \in \{\mathbf{V}\}_3$ with the lowest $\lambda_{\max}$ and add a new fuzzy rule to the rule base with $\mathbf{V}_{\min}$ as the prototype;

Step 5. Remove $\mathbf{V}_{\min}$ from $\{\mathbf{V}\}_2$ and go back to step 1.

The main procedure of the offline active semi-supervised learning process of the DRB system is summarized by the flowchart in Fig. 14.2.

14.1.3 Online (Active) Semi-supervised Learning Process

The DRB classifier is able to perform online active learning if the optional active semi-supervised learning algorithm is used. Since the online (active) semi-supervised learning algorithm can be performed on a sample-by-sample or chunk-by-chunk basis, we summarize the two operating modes of the algorithm separately.

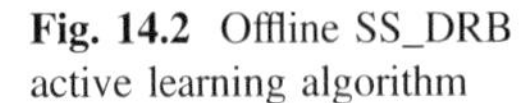
Fig. 14.2 Offline SS_DRB active learning algorithm

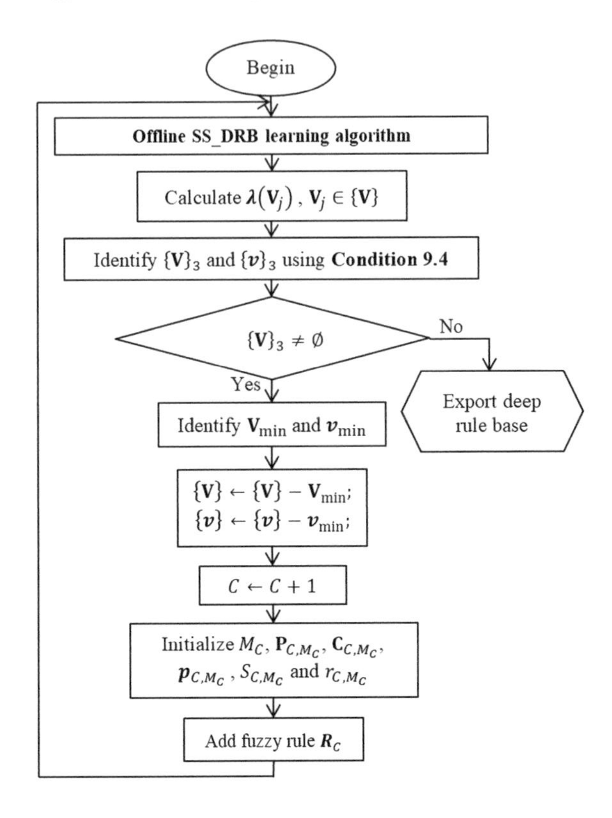

14.1.3.1 Sample-by-Sample Basis

The main steps of the online (active) semi-supervised learning process of the DRB classifier (online SS_DRB (active) learning algorithm) conducted on a sample-by-sample basis are summarized as follows. The corresponding flowchart is presented in Fig. 14.3.

Step 1. Pre-process and extract the feature vector from the available unlabeled new image $\mathbf{V}_K$;
Step 2. Calculate the vector of firing strengths, $\lambda(\mathbf{V}_K)$;
Step 3. If the DRB classifier is confident with the class of $\mathbf{V}_K$ go to step 4; otherwise, go to step 5 if the DRB classifier is required to conduct active learning or go to step 6 directly;
Step 4. Update the DRB classifier with $\mathbf{V}_K$ using the DRB learning algorithm (as described in Sects. 9.4.1 and 13.1.1) based on its estimated class, and go to step 6;
Step 5. Add a new fuzzy rule to the rule base with $\mathbf{V}_K$ as its prototype, and go to step 6;
Step 6. Go back to step 1 and processes the next image ($K \leftarrow K + 1$).

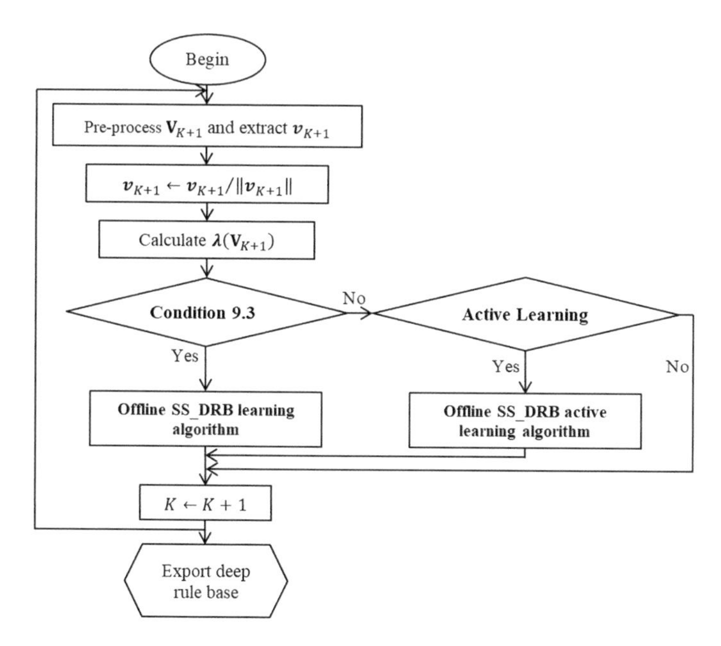

Fig. 14.3 Sample-by-sample online SS_DRB (active) learning algorithm

14.1.3.2 Chunk-by-Chunk Basis

The main steps of the online SS_DRB (active) learning algorithm conducted on a chunk-by-chunk basis are summarized as follows. The corresponding flowchart is presented in Fig. 14.4.

Step 1. Receive the next data chunk $\{\mathbf{V}\}_H^j$ $(j \leftarrow j + 1)$. Go to step 2 if the DRB classifier is required to conduct active learning; otherwise, go to step 3;

Step 2. Update the DRB classifier based on $\{\mathbf{V}\}_H^j$ using the offline SS_DRB learning algorithm, and go back to step 1;

Step 3. Update the DRB classifier based on $\{\mathbf{V}\}_H^j$ using the offline SS_DRB active learning algorithm, and go back to step 1.

14.2 Numerical Examples

In this section, the numerical examples are presented for demonstrating the performance of the SS_DRB classifier on image classification.

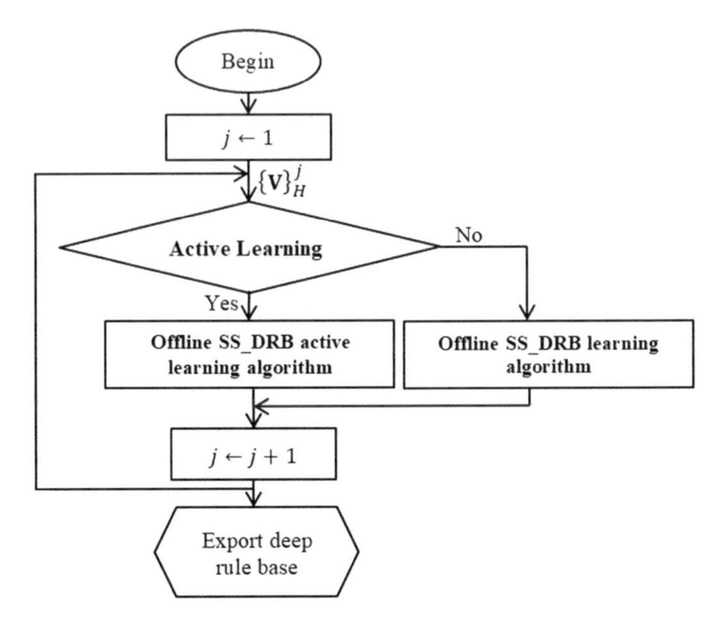

Fig. 14.4 Chunk-by-chunk online SS_DRB (active) learning algorithm

The structure of the DRB/SS_DRB classifier used for the numerical examples presented in this chapter is the same as the one used in Sect. 13.2.2.4 (Fig. 13.20).

14.2.1 Datasets for Evaluation

In this section, the following benchmark image sets are considered for evaluating the performance of the SS_DRB classifiers:

(1) Singapore dataset [1] (downloadable from [2]);
(2) UCMerced land use dataset [3] (downloadable from [4]);
(3) Caltech101 dataset [5] (downloadable from [6]).

The details of the three benchmark datasets can be found in Sects. 13.2.1.3–13.2.1.5.

14.2.2 Performance Evaluation

In this subsection, the performance of the SS_DRB classifier on image classification using the benchmark datasets is evaluated.

14.2.2.1 Illustration of the Offline Semi-supervised Learning Process

In this subsection, an illustrative example on the offline semi-supervised learning process of the DRB classifier is presented based on a subset of the Singapore dataset. In this example, $\Omega_1 = 1.2$ and $\Omega_2 = 0.5$.

For clarity, only four classes ("airplane", "forest", "industry" and "residential") are selected. The labelled training set consists of 20 images in total (five images per class), and they are presented in Fig. 14.5. The DRB classifier is trained with the labelled training set on a sample-by-sample basis using the DRB learning algorithm as given in Sect. 9.4.1. The progress of the massively parallel fuzzy rules identification and the finally identified fuzzy rules after the training process are given in Fig. 14.6.

After the system initialization, 10 images from six different classes ("airplane", "forest", "harbor", "industry", "overpass" and "residential") are used as unlabeled training images for the DRB classifier to learn in a semi-supervised manner. The 10 images are given in Fig. 14.7. As one can see, seven out of 10 images come from the previously existing classes. There is one image from the class "harbor" and two images from the class "overpass" that DRB classifier has never seen before. The firing strengths of the identified massively parallel fuzzy rules for the 10 images are tabulated in Table 14.1. The corresponding values of $(\lambda_{\max*} - \lambda_{\max**})$ of the 10 unlabeled training images are given in Fig. 14.8, where one can see that the values of the three images (ID: 5, 7 and 8) are significantly lower than the corresponding values for the others.

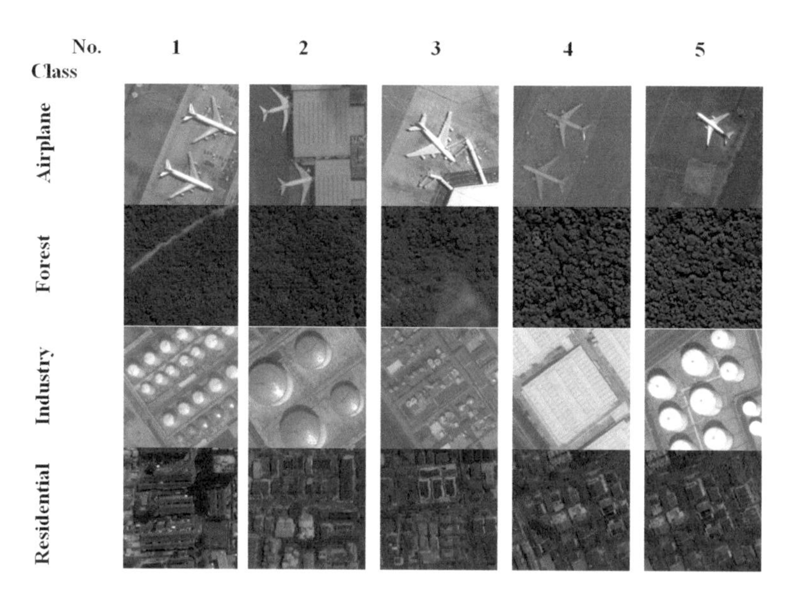

Fig. 14.5 The training images for illustration

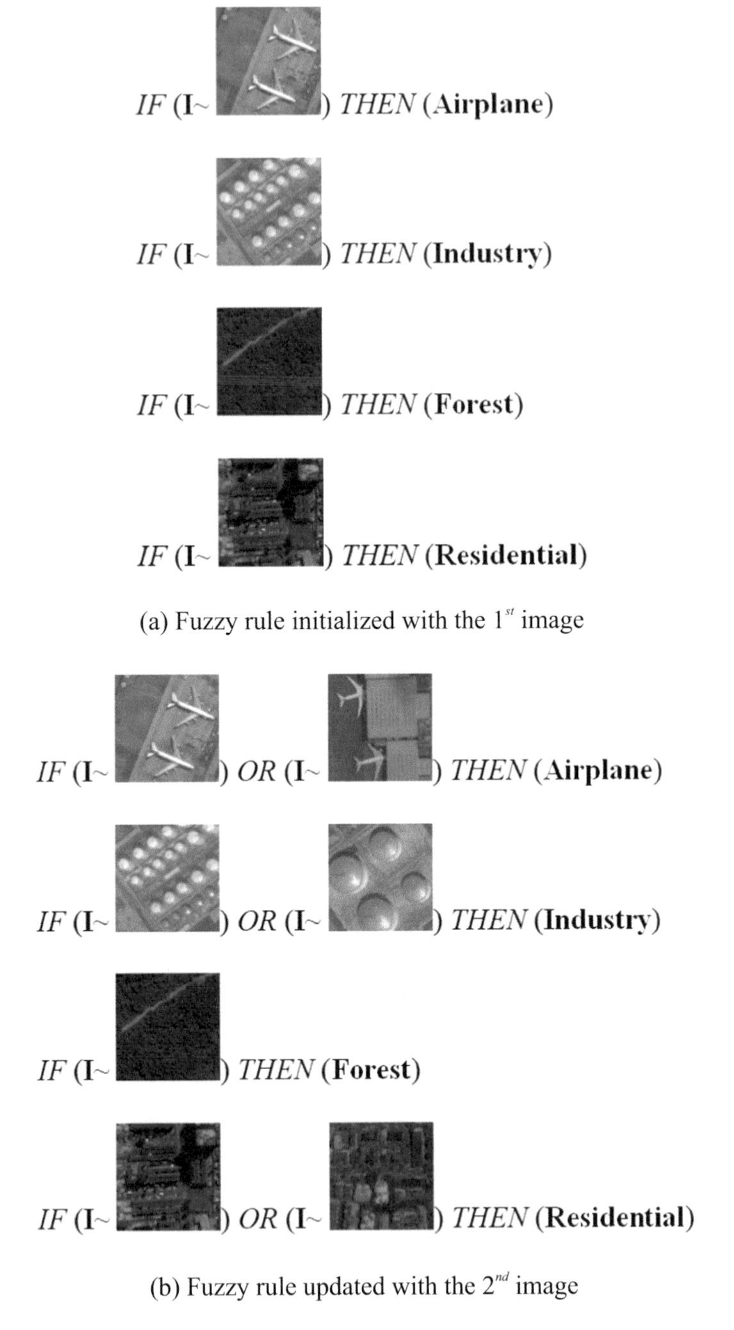

Fig. 14.6 Massively parallel fuzzy rules identified through a supervised learning process

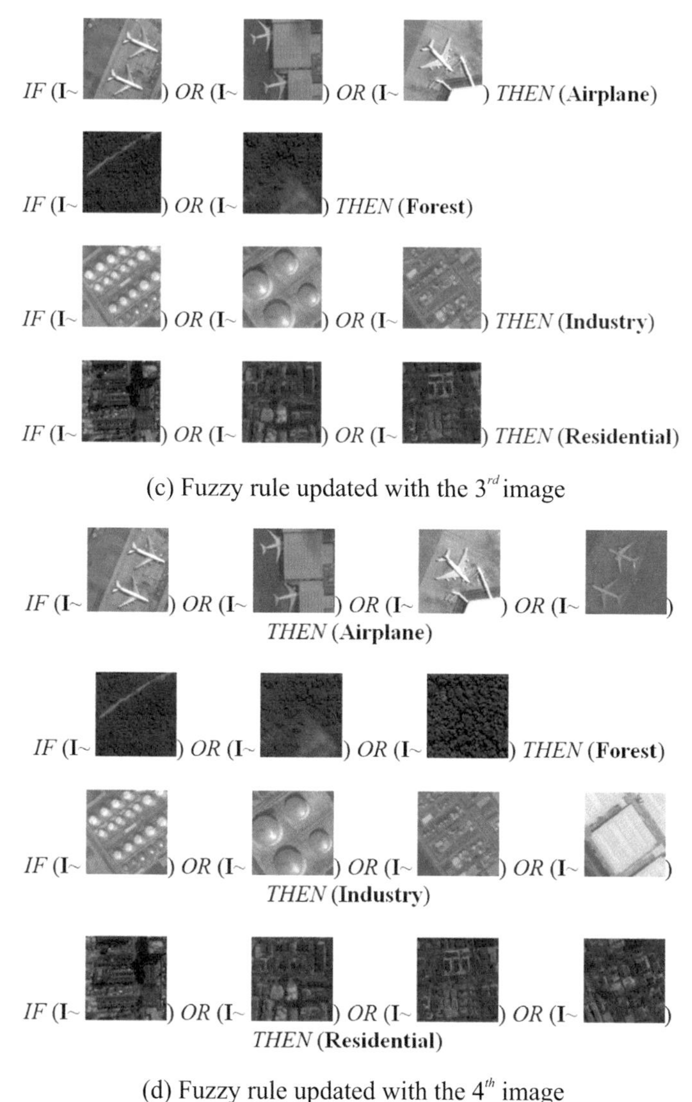

(c) Fuzzy rule updated with the 3rd image

(d) Fuzzy rule updated with the 4th image

Fig. 14.6 (continued)

The updated fuzzy rules after the semi-supervised learning process are presented in Fig. 14.9, where the newly identified prototypes during this process are highlighted with red color.

Alternatively, if the SS_DRB classifier learns from the unlabeled training set in an active manner, there are two new fuzzy rules identified from the unlabeled training set, see Fig. 14.10. However, as the algorithm cannot recognize the semantic meaning of the prototypes within the new fuzzy rules, the system automatically

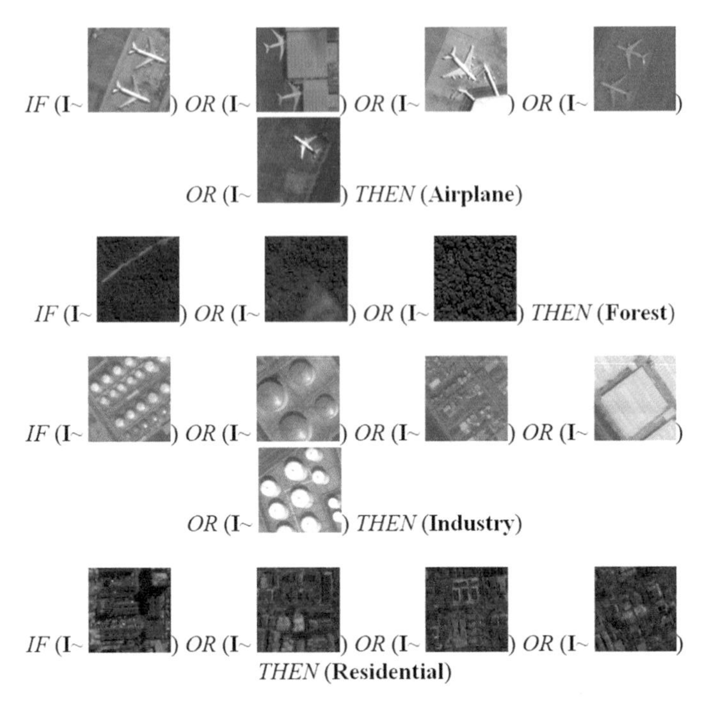

(e) Fuzzy rule updated with the 5th image

Fig. 14.6 (continued)

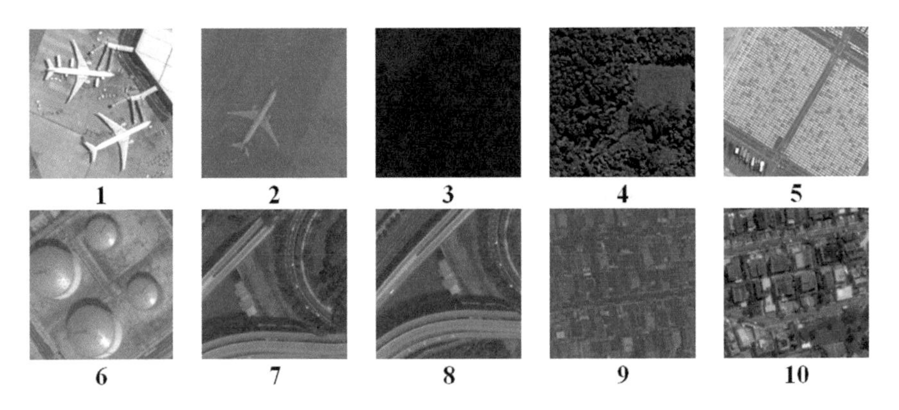

Fig. 14.7 The unlabeled training images

names the new classes as "New class 1" and "New class 2". Human experts can also examine the new fuzzy rules and give meaningful labels for them by simply checking the prototypes afterwards, i.e. "new class 1" can be renamed as "overpass" and "new class 2" can be renamed as "harbor". This process is much less laborious

Table 14.1 Firing strengths of the massively parallel fuzzy rules to the unlabeled training images

Image ID	Rule #					
	Score of confidence (λ)				Condition 10.3 (Yes/no)	Estimated label
	1	2	3	4		
1	0.6793	0.1573	0.3874	0.3206	Yes	Airplane
2	0.5916	0.2521	0.2850	0.2613	Yes	Airplane
3	0.2441	0.7165	0.1583	0.1690	Yes	Forest
4	0.2165	0.7286	0.1939	0.2090	Yes	Forest
5	0.4664	0.2008	0.5038	0.3347	No	Uncertain
6	0.2436	0.1467	0.8999	0.2735	Yes	Industry
7	0.3191	0.2045	0.3299	0.3092	No	Uncertain
8	0.2929	0.1748	0.3159	0.3086	No	Uncertain
9	0.2666	0.2204	0.3886	0.6400	Yes	Residential
10	0.3039	0.1814	0.4830	0.7456	Yes	Residential

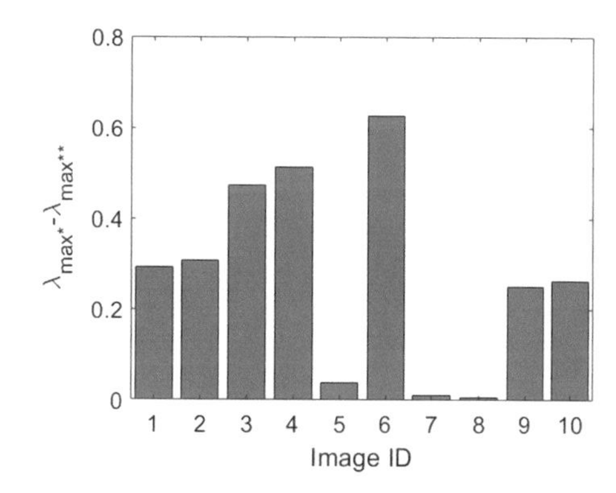

Fig. 14.8 The corresponding values of $(\lambda_{max*} - \lambda_{max**})$ of the 10 unlabeled training images

than the usual approach because it only concerns the aggregated prototypical data, not the high volume raw data, and it is more convenient for the human users.

14.2.2.2 Performance Demonstration and Analysis

In this subsection, we will conduct a series of numerical experiments based on the UCMerced dataset with the SS_DRB classifier to demonstrate its performance.

A. ***Influence of Ω_1 on the performance***

The performance of the SS_DRB classifier with different values of Ω_1 is investigated firstly. Eight images from each class are randomly selected as the labelled training set for the supervised training of the DRB classifier, and the rest of the images are used as the unlabeled training set to continue the semi-supervised learning process of the DRB classifier in both offline and online scenarios. Therefore, the labelled training set has 168 images and the unlabeled training set has 1932 images.

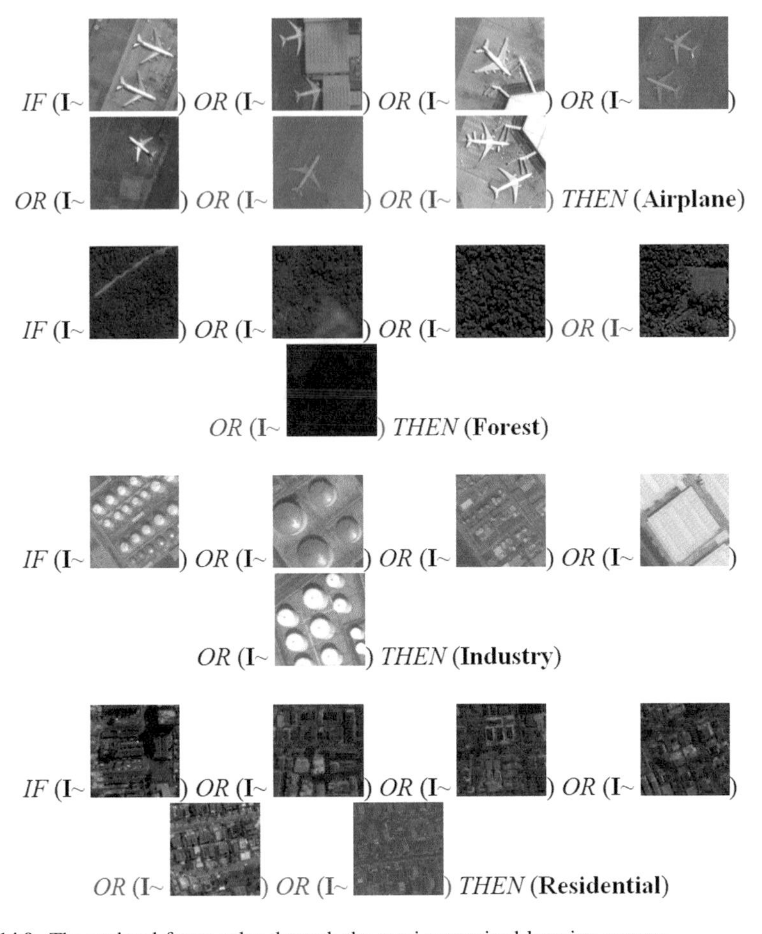

Fig. 14.9 The updated fuzzy rules through the semi-supervised learning process

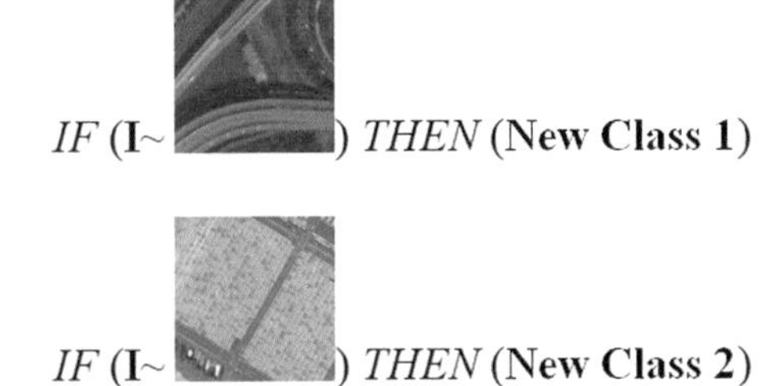

Fig. 14.10 The newly identified fuzzy rules through the active semi-supervised learning process

In the online scenario, the semi-supervised learning is conducted on both, sample-by-sample and chunk-by-chunk basis. For the former case, the order of the unlabeled images is descrambled randomly. In the latter case, we randomly divide the unlabeled training samples into two chunks, which have the same number of images.

During this numerical example, the value of Ω_1 varies from 1.05 to 1.30. The classification accuracy of the SS_DRB classifier on the unlabeled training set after 50 Monte-Carlo experiments is depicted in Fig. 14.11. The average numbers of prototypes identified (*M*) are reported in Table 14.2. The performance of the fully supervised DRB classifier is also reported as the baseline.

As one can see from Table 14.2, the higher the value of Ω_1 is, the less prototypes the SS_DRB classifier identified during the semi-supervised learning process, and, thus, the system structure is less complex and more computationally efficient. However, meanwhile, it is obvious from Fig. 14.11 that the accuracy of the classification result is not linearly correlated with the value of Ω_1. There is a certain

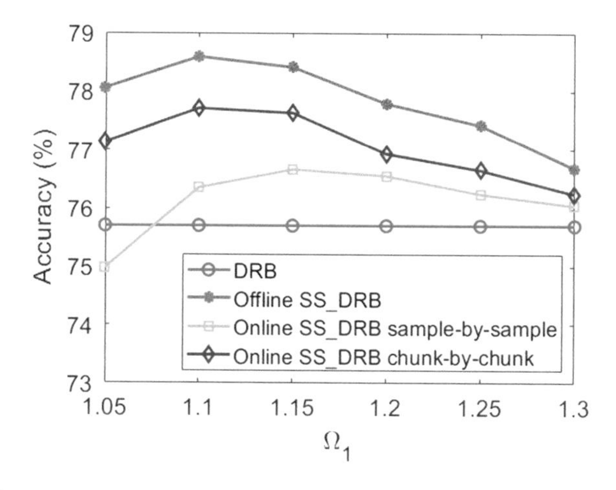

Fig. 14.11 The average accuracy curves of the SS_DRB classifiers with different values of Ω_1

Table 14.2 Performance of the SS_DRB classifier with different values of Ω_1

Algorithm	Ω_l	1.05	1.10	1.15
DRB	*M*	161.1		
Offline SS_DRB	*M*	1637	1402.8	1194.9
Online SS_DRB sample-by-sample	*M*	1432.6	1192.8	1008.2
Online SS_DRB chunk-by-chunk	*M*	1581.6	1337.8	1127.7
Algorithm	Ω_l	1.20	1.25	1.3
DRB	*M*	161.1		
Offline SS_DRB	*M*	1015.8	874.7	759.1
Online SS_DRB sample-by-sample	*M*	862.8	746.2	648.9
Online SS_DRB chunk-by-chunk	*M*	960.1	825.1	712.6

range of Ω_1 values for the SS_DRB classifier to achieve the best accuracy. Trading off the overall performance and system complexity, the best range of Ω_1 values for the experiments performed is $[1.1, 1.2]$. For consistence, $\Omega_1 = 1.2$ is used in the rest of the numerical examples presented in this chapter. However, one can also set different value for Ω_1.

B. ***Influence of the scale of labelled training set on the performance***

In this numerical example, we randomly pick out $K = 1, 2, 3, \ldots, 10$ images from each class of the UCMerced dataset as labelled training images and use the rest as the unlabeled ones to train the SS_DRB classifier in both offline and online scenarios.

Similarly to the previous experiment, the online semi-supervised learning is conducted on both, sample-by-sample and chunk-by-chunk basis. The performance in terms of accuracy classification of the SS_DRB classifiers on the unlabeled training set are reported in Fig. 14.12 after 50 Monte-Carlo experiments.

From Fig. 14.12, one can see that, with $\Omega_1 = 1.2$, the SS_DRB classifier performs best in an offline scenario, which is due to the fact that, the SS_DRB classifier is able to achieve a comprehensive representation of the ensemble properties of the static image set. In the chunk-by-chunk online learning mode, the SS_DRB classifier can only study the ensemble properties of the unlabeled images within each chunk. Its performance deteriorates further if the online semi-supervised learning is conducted on a sample-by-sample basis as unlabeled training images are isolated from each other.

C. ***Influence of Ω_2 on the active learning mechanism and performance***

In this numerical example, the influence of Ω_2 on the active learning mechanism and the performance of the SS_DRB classifier is investigated. During this experiment, the value of Ω_2 varies from 0.4 to 0.65 with a step of 0.05.

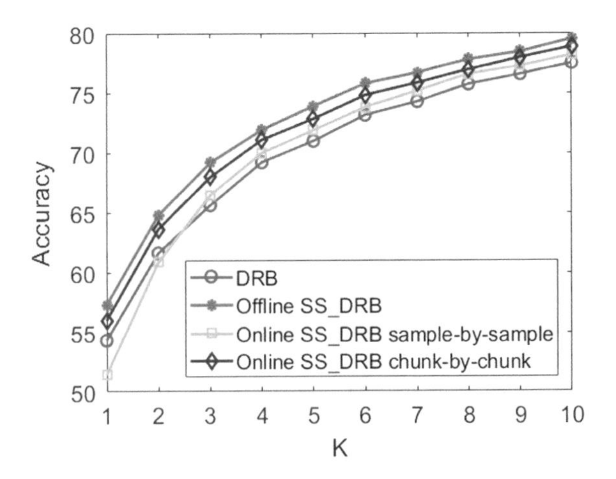

Fig. 14.12 The accuracy of the SS_DRB classifier with different number of labelled images

20 out of the 21 classes are randomly selected from the UCMerced dataset and $K = 10$ images are picked out from each of them as labelled training images. The remaining images are used as unlabeled training set to train the SS_DRB classifier in both offline and online scenarios. Similarly to the previous experiment, the semi-supervised learning is conducted on either sample-by-sample or chunk-by-chunk basis in an online scenario.

In order to calculate the classification accuracy, for each newly gained class/fuzzy rule, the class labels of its dominant members are used as the label of this class/fuzzy rule. The average classification accuracies (in percentages) of the SS_DRB classifiers on the unlabeled training set after 50 Monte-Carlo experiments are presented in Fig. 14.13. The average numbers of classes (C) after the training process during the experiment are presented in Table 14.3.

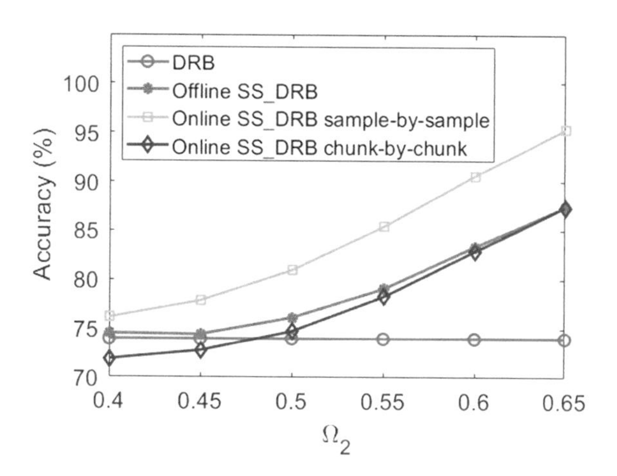

Fig. 14.13 The accuracy of the SS_DRB classifier with different values of Ω_2

Table 14.3 Performance of the SS_DRB classifier with different values of Ω_2

Algorithm	Ω_2	0.4	0.45	0.5
DRB	C	20		
Offline SS_DRB	C	20.2	32.7	74.8
Online SS_DRB sample-by-sample	C	25.2	55.3	141.6
Online SS_DRB chunk-by-chunk	C	21.3	38.6	94.3
Algorithm	Ω_2	0.55	0.6	0.65
DRB	C	20		
Offline SS_DRB	C	167.6	292.2	434.3
Online SS_DRB sample-by-sample	C	304.8	523.1	758.9
Online SS_DRB chunk-by-chunk	C	212.4	368.9	545.0

As one can see from Fig. 14.13 and Table 14.3, the higher Ω_2 is, the more new classes the SS_DRB classifier will learn from the unlabeled training images and a better classification accuracy the SS_DRB classifier exhibits.

During the experiments, many of the newly learnt fuzzy rules are actually initialized by the images from the already known classes. This is caused by the fact that images, even in the same class, usually have a wide variety of semantic content, and they can be quite distinctive from each other. The active learning algorithm allows the SS_DRB classifier to learn not only from the images from unseen classes, but also from the images of the already seen classes, which vary from other images within the same classes (that is, to reorganize the already learned class structure).

Trading off the overall performance and system complexity, the best value range of Ω_2 for the experiments performed is $[0.5, 0.6]$. For consistence, $\Omega_2 = 0.5$ is used in the rest of the numerical examples presented this chapter. However, one can also set different value for Ω_2.

14.2.3 Performance Comparison and Analysis

In this subsection, we compare the performance of the SS_DRB classifier with the following state-of-the-art approaches based on the three benchmark problems listed in Sect. 14.2.1:

A. ***Supervised classification algorithms***

(1) Support vector machine (SVM) classifier with linear kernel [7];
(2) k-nearest neighbor (kNN) classifier [8].

B. ***Semi-supervised classification algorithms***

(3) Local and global consistency (LGC) based semi-supervised classifier [9];
(4) Greedy gradient Max-Cut (GGMC) based semi-supervised classifier [10];
(5) Laplacian SVM (LapSVM) classifier [11, 12].

During the experiments, the value of k for kNN is set to be equal to K (the number of labelled training images per class). The free parameter α of the LGC classifier is set to be 0.99 as suggested in [9]. The free parameter μ of the GGMC classifier is set to be $\mu = 0.01$ as suggested in [10]. LGC and GGMC use the kNN graph with $k = K$. The LapSVM classifier uses the "one-versus-all" strategy for all the benchmark problems. Radial basis function (RBF) kernel with $\sigma = 10$ is used; the two free parameters γ_I and γ_A are set to be 1 and 10^{-6}, respectively; the number of neighbors, k, for computing the graph Laplacian is set to be 15 as suggested in [11].

For a fair comparison, we only consider the performance of the SS_DRB classifier applying offline semi-supervised learning. Therefore, all the numerical examples presented in this subsection are conducted in an offline scenario.

14.2.3.1 Performance Comparison on Singapore Dataset

For the Singapore dataset, $K = 1, 2, 3, \ldots, 10$ images from each class are randomly chosen as the labelled training images and the remaining images are used as unlabeled ones to train the SS_DRB classifier in an offline scenario. Its classification accuracy curve (in percentages) with different values of K is depicted in Fig. 14.14. The accuracy curves obtained by the five algorithms listed above are also presented in the same figure for comparison. In Fig. 14.14, each curve is the average of 50 Monte-Carlo experiments.

14.2.3.2 Performance Comparison on UCMerced Dataset

For the UCMerced dataset, $K = 1, 2, 3, \ldots, 10$ images are randomly picked out from each class as the labelled training images and use the rest as unlabeled ones. The classification performances of the SS_DRB classifier and the five comparative algorithms are depicted in Fig. 14.15 in terms of the average classification accuracy after 50 Monte-Carlo experiments on unlabeled training images.

14.2.3.3 Performance Comparison on Caltech 101 Dataset

For Caltech 101 dataset, following the commonly used experimental protocol [13], 30 images from each class are randomly picked out as the training set during the numerical experiment. The background category is left out because the images inside this category are mostly unrelated.

We further randomly choose $K = 1, 2, 3, 4, 5$ images per class from the training as labelled training set and use the rest as unlabeled training set. The classification

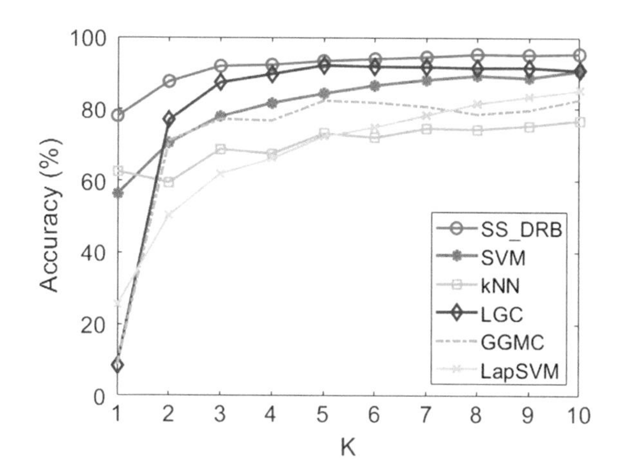

Fig. 14.14 Performance comparison on Singapore dataset

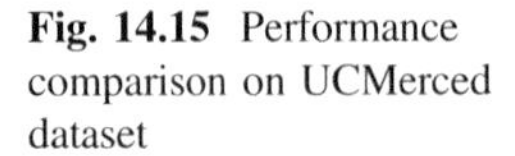

Fig. 14.15 Performance comparison on UCMerced dataset

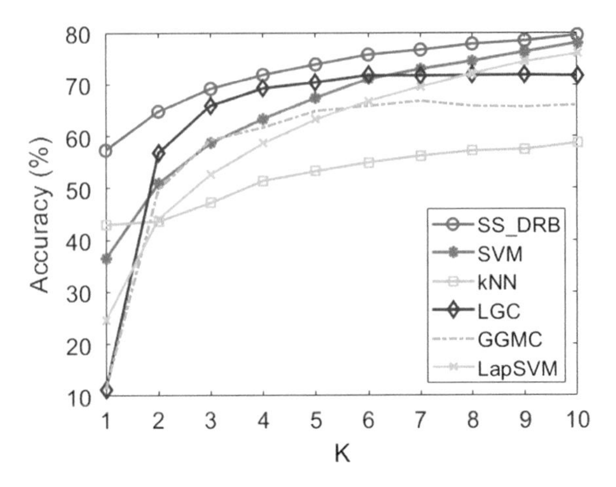

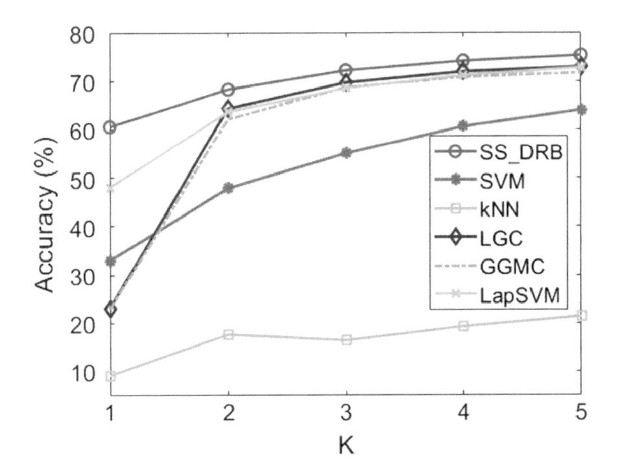

Fig. 14.16 Performance comparison on Caltech 101 dataset

performance of the SS_DRB classifier and the five comparative algorithms on the Caltech101 dataset are depicted in Fig. 14.16 in terms of average classification accuracy after 50 Monte-Carlo experiments on the unlabeled training images.

14.2.4 Discussion

From the numerical examples presented in Sect. 14.2.3 one can see that, the SS_DRB classifier is able to provide the highest classification results with only a handful of labelled training images. It consistently outperforms all the state-of-the-art classification algorithms (both, the supervised and the semi-supervised ones) on all the three popular benchmarks.

In addition, due to its non-iterative, "one pass" learning process (there is no error propagation during the process), the SS_DRB classifier is able to tolerate incorrectly pseudo-labelled images to a large degree, and, thus, it is able to consistently exhibit high performance in complex problems even with very few labelled training images. For more numerical examples, please refer to [14].

14.3 Conclusions

In this chapter, the implementation of the SS_DRB classifier, introduced as an extension of the DRB classifier, is presented. The SS_DRB classifier is able to offer very high classification rates on unlabeled images surpassing other semi-supervised approaches after a highly transparent and computationally efficient semi-supervised learning process.

Numerical examples based on large-scale benchmark image sets presented in this chapter demonstrate the strong performance of the SS_DRB classifier.

References

1. J. Gan, Q. Li, Z. Zhang, J. Wang, Two-level feature representation for aerial scene classification. IEEE Geosci. Remote Sens. Lett. **13**(11), 1626–1630 (2016)
2. http://icn.bjtu.edu.cn/Visint/resources/Scenesig.aspx
3. Y. Yang, S. Newsam, Bag-of-visual-words and spatial extensions for land-use classification, in *International Conference on Advances in Geographic Information Systems* (2010) pp. 270–279
4. http://weegee.vision.ucmerced.edu/datasets/landuse.html
5. L. Fei-Fei, R. Fergus, P. Perona, One-shot learning of object categories. IEEE Trans. Pattern Anal. Mach. Intell. **28**(4), 594–611 (2006)
6. http://www.vision.caltech.edu/Image_Datasets/Caltech101/
7. N. Cristianini, J. Shawe-Taylor, *An Introduction to Support Vector Machines and Other Kernel-Based Learning Methods* (Cambridge University Press, Cambridge, 2000)
8. P. Cunningham, S.J. Delany, K-nearest neighbour classifiers. Mult. Classif. Syst. **34**, 1–17 (2007)
9. D. Zhou, O. Bousquet, T. N. Lal, J. Weston, B. Schölkopf, Learning with local and global consistency. Adv. Neural. Inform. Process Syst., 321–328 (2004)
10. J. Wang, T. Jebara, S.F. Chang, Semi-supervised learning using greedy Max-Cut. J. Mach. Learn. Res. **14**, 771–800 (2013)
11. M. Belkin, P. Niyogi, V. Sindhwani, Manifold regularization: a geometric framework for learning from labeled and unlabeled examples. J. Mach. Learn. Res. **7**(2006), 2399–2434 (2006)
12. L. Gómez-Chova, G. Camps-Valls, J. Munoz-Mari, J. Calpe, Semisupervised image classification with Laplacian support vector machines. IEEE Geosci. Remote Sens. Lett. **5** (3), 336–340 (2008)

13. L. Fei-Fei, R. Fergus, P. Perona, Learning generative visual models from few training examples: an incremental Bayesian approach tested on 101 object categories. Comput. Vis. Image Underst. **106**(1), 59–70 (2007)
14. X. Gu, P.P. Angelov, Semi-supervised deep rule-based approach for image classification. Appl. Soft Comput. **68**, 53–68 (2018)

Chapter 15
Epilogue

In this book we systematically present the new *empirical* approach to Machine Learning. In the current state-of-the-art approaches, the starting point are the assumptions about the data distribution and generation model, amount and nature of the data, different problem- and user-specific thresholds and algorithm parameters, while in the *empirical* approach, the starting point is the data.

The first step is the data pre-processing. It includes transforming/mapping the real world problem into a M-dimensional space of features. In this book, we follow the established methods for this step of machine learning, e.g. using normalization or standardization methods, for image processing using either histogram of gradients (HOG), GIST or combination of both or a pre-trained deep convolutional neural network (DCNN), such as VGG. Data pre-processing also includes selecting the type of distance to be used. All further considerations are, however, generic, although we use more often Euclidean type of distance due to its simplicity.

We then provide a review of methods of *statistical machine learning* and methods of *computational intelligence*.

Following this review we systematically present the nonparametric quantities of the proposed new *empirical* approach, including:

(1) the *cumulative proximity*, q;
(2) the *eccentricity*, ξ and the *standardized eccentricity*, ε;
(3) the *data density*, D, and
(4) the *typicality*, τ.

We also discuss their properties and present their recursive calculation expressions. The discrete and continuous as well as unimodal and multimodal versions of the *data density* (D, D^C, D^M and D^{CM}) and the *typicality* (τ, τ^C, τ^M and τ^{CM}) are described and analyzed. The nonparametric quantities that are introduced are of great importance and serve as the theoretical foundation of the new *empirical* approach proposed in this book.

P. P. Angelov and X. Gu, *Empirical Approach to Machine Learning*, Studies in Computational Intelligence 800,
https://doi.org/10.1007/978-3-030-02384-3_15

As a fundamentally new measure in pattern recognition, the discrete version of the *typicality*, τ resembles the unimodal PDF, but is in a discrete form. The discrete multimodal typicality, τ^M resembles the PMF, but is derived from data directly and is free from the paradoxes and problems that the traditional approaches are suffering from. The continuous *typicality,* τ^C and multimodal *typicality*, τ^{CM} share many properties with the unimodal and multimodal PDFs, but is free from *prior* assumptions and user- and problem-specific parameters.

The most distinctive feature of the proposed new *empirical* approach is that it is not limited by restrictive impractical *prior* assumptions about the data generation model as the traditional probability theory and statistical learning approaches are. Instead, it is fully based on the ensemble properties and mutual distribution of the *empirically* observed discrete data. Moreover, it does not require an explicit assumption of randomness or determinism of the data, their independence, or even their number. This new *empirical* approach touches the very foundation of data analysis, and, thus, there are a wide range of applications including, but not limited to, anomaly detection, clustering/data partitioning, classification, prediction, fuzzy rule-based (FRB) system, deep rule-based (DRB) system, etc. We describe the main ones in the remainder of the book. For more details, we refer the interested readers to journal and conference publications.

Then we present the new type of fuzzy sets, named *empirical* fuzzy sets, and a new form of FRB systems, named *empirical* FRB systems, which touches the fundamental question of how to build a FRB system. Two approaches (subjective and objective) for identifying *empirical* FRB systems are presented. When compared with the traditional fuzzy sets and FRB systems, the *empirical* fuzzy sets and FRB systems have the following significant advantages:

(1) they are derived in a transparent, data-driven way without *prior* assumptions;
(2) they effectively combine the data- and human-derived models;
(3) they possess very strong interpretability and high objectiveness;
(4) they facilitate experts or even bypass the involvement of human expertise.

Then we describe the new *empirical* approach to the problem of anomaly detection. It is model-, user- and problem-specific-parameter-free and is entirely data driven. It is based on the *data de*nsity and/or on the *typicality*.

We further make a clear distinction between the global and local or contextual anomalies and propose an autonomous anomaly detection method which consists of two stages: in the first stage, we detect all potential global anomalies and in the second stage, we analyze the data pattern by forming *data clouds* and identifying possible local anomalies (in regards to these *data clouds*).

Finally, the well-known Chebyshev inequality has been simplified by using the *standardized eccentricity* to a fully autonomous approach and the problem of fault detection (FD) has been outlined. The latter approach can also be extended to a fully autonomous FD and isolation (FDI: detecting and identifying the types of faults). This is a principally new approach to the very important for practical applications problem of FDI.

Then, we introduce a new approach to partition the data autonomously into *data clouds* which form a Voronoi tessellation. This can be seen as clustering although it has some specific differences (mainly in the shapes of these *data clouds* as well as in the specific way they are being formed). The object of both, clustering and data partitioning, is transforming the large amount of raw data into a much smaller (manageable) number of more representative aggregations which can have semantic meaning.

The proposed new *empirical* data partitioning algorithm has two forms/types:

(1) offline, and
(2) online and evolving.

In addition, we formulate and propose an algorithm to guarantee **local optimality** of the structure which was derived. As a result of these proposed methods, one can start with raw data and end up with a locally optimal structure of *data clouds* represented by the focal points of each *data cloud* which are nothing else but the peaks in terms of *data density* and *typicality* (the points with locally maximum value of the *data density* and *typicality*).

This system structure is then ready to be used for analysis, building a multi-model classifier, predictor and controller or for fault isolation.

On this basis, we, then, move to address and solve these problems in a new *empirical* way (driven by actual, observed data and not by pre-defined restrictive assumptions and imposed model structures).

We then introduce the autonomous learning multi-model (*ALMMo*) systems. *ALMMo* steps upon the *AnYa* type neuro-fuzzy systems and can be seen as an universal self-developing, self-evolving, stable, locally optimal proven universal approximator. We start with the concept, and then, we go into the zero- and first-order *ALMMo* in more detail. We describe the architecture, followed by the learning methods (both, online and offline). We provide the **theoretical proof (using Lyapunov theorem) for the stability of the first-order *ALMMo* systems**, we also provide the **theoretical proof of the local optimality which satisfies Karush-Kuhn-Tucker conditions**.

Finally, we briefly review the prototype-based classifiers such as KNN, SVM, eClass0, *ALMMo-0* and move to describe the proposed deep rule-based (DRB) classifier. DRB is free from *prior* assumptions about the type of the data distribution, their random or deterministic nature, and there are no requirements to make ad hoc decisions, i.e. regarding the model structure, membership functions, number of layers, etc. Meanwhile, the prototype-based nature allows the DRB classifier to be non-parametric, non-iterative, self-organizing, self-evolving and highly parallel.

The training of the DRB classifier is a modification of the *ALMMo-0* learning algorithm, and, thus, the training process is highly, parallelizable, significantly faster and can starting "from scratch". With its prototype-based nature, the DRB classifier is able to identify a number of prototypes from the training images in a fully online, autonomous, non-iterative and nonparametric manner, and, based on this, to build a massively parallel fuzzy rule base for classification. The training and validation processes of the DRB classifier concern only the visual similarity between the identified prototypes and the unlabeled samples, and only involve the very general principles. In short, compared with the state-of-the-art approaches, the DRB classifier has the following unique features thanks mainly to its prototype-based nature:

(1) It is free from restrictive *prior* assumptions and user- and problem-specific parameters;
(2) The learning process is fully online, autonomous, non-iterative and nonparametric and can start "from scratch";
(3) The learning process is highly parallelizable and very efficient;
(4) The system structure is self-organizing, self-evolving, fully transparent and highly human-interpretable.

We further introduce the semi-supervised DRB (SS_DRB) classifier, which has the following unique features compared with the existing semi-supervised approaches in addition to the unique features of the DRB classifier:

(1) The learning process can be conducted on a sample-by-sample or chunk-by-chunk basis;
(2) It is able to perform out-of-sample classification;
(3) It is able to learn new classes in an active manner and can self-evolve its internal system structure;
(4) There are no iteration during the learning process, and, thus, no error propagation.

In this way, we introduced in this book a wholistic and systematic novel approach to machine learning which starts from the *empirically* observed data and covers all aspects of pattern recognition and data mining. It is also applicable to control (not covered in this book) as well as to problems such as anomaly detection, fault detection and isolation, predictive models.

In the last, third part of the book we provide detailed description of the various algorithms proposed in the book. A set of lecture notes, which can be used for a standalone postgraduate course or advanced machine learning undergraduate courses, are also provided.

The proposed *empirical* approach can be summarized by the overall procedure which contains five stages as listed at the end of Chap. 1, namely:

Stage 0 Pre-processing which starts by converting the real life problem into a M-dimensional vector of variables (features/attributes). They represent mathematically each data observation as a point in the feature/data space. The data is then normalized or standardized as described in Chap. 2. This equates to converting the real world problem into a hyper cube with size 1: $\boldsymbol{x} \in [0, 1]^N$. Standardization equates to converting the vast majority of the data into a hyper cube with size 6: $\boldsymbol{x} \in [-3, 3]^N$. In fact, applying first standardization followed by normalization, one can always convert the actual data space into a hyper-cube with size 1. The benefit is that in online mode ranges of the data are usually not known in advance and may change dynamically, so applying normalization in online mode is problematic, but standardization can be applied first and one can get the anomalies as a by-product easily. Then, for the already standardized data, one can apply normalization converting the hypercube of size 6 into a hypercube of size 1 (the range for each standardized variable/feature will be $[-3, 3]$ if use 3 sigma principle. This sequence of standardization followed by normalization can easily be done online.

Stage 1 essentially expands the N-dimensional data/feature space into a $(N+1)$-dimensional space whereby the additional dimension represents the *data density* and/or *typicality*. Discrete multimodal distributions as well as continuous forms of *data density* and *typicality* distributions are also introduced in this book. The peaks of these distributions play a very important role of indicating prototypes. The proposed method is constructivist, "one pass" and does not require clustering, search or optimization techniques to determine the data distribution, classification or predictive model structure.

Stage 2 includes zero- and first-order *ALMMo* Systems. Zero-order, *ALMMo-0* can be used for image processing and other classifiers. It can also be used as a predictor and even a proportional (P) controller, for fault detection and isolation, but, perhaps the most interesting application is to the DRB classifiers which are prototype based, fast, transparent and highly parallelizable. They can reach human level recognition and has been proven to surpass the mainstream Deep Neural Networks which are quite popular now. The SS_DRB classifiers are able to learn from only a small proportion of the labelled data, and after that, continue to learn in an unsupervised manner and learn new classes actively.

The first order, *ALMMo-1* are used mostly as predictors, but can also be used as classifiers as well as self-organizing controllers (fuzzy mixture of proportional-integral-derivative (PID) controllers). The classifiers based on *ALMMo-1* are more efficient for problems with relatively small number of features (tens to hundreds, not millions) and, therefore, not applied to image processing.

Stage 3 The overall output of *ALMMo* is formed by the so called fuzzily weighted recursive least square (FWRLS) algorithm. **Local optimality as well as stability and convergence of *ALMMo* models was proven theoretically**.

Stage 4 Finally, the overall output is produced as a result of a kind of fusion of partial local models through the fuzzy firing level, lambda. This opens the possibility for fusing partial outcomes of classifiers that work in different data spaces. In this way, heterogeneous classifiers which take as inputs images/video, signals (physical variables, such as temperature, pressure, location, purchase data, etc.) as well as text and produce a single output can be built based on *ALMMo* models.

In future we will develop this approach to address the following problems:

- collaborative, distributed systems, which learn autonomously and *empirically*;
- *empirical* self-evolving controllers with **proven stability**;
- *empirical* approach to dealing with heterogeneous data (heterogeneous classifier) which will fuse image, signals and text type of data;
- we will also develop new approach to organize the prototypes for fast search through them, e.g. in a hierarchical form.

Appendix A

In this appendix, the detailed derivations of the mathematical expressions of cumulative proximity using the Euclidean and Mahalanobis distances presented in Chap. 4 are provided.

A.1 Cumulative Proximity Using Euclidean Distance

Using the Euclidean distance, the recursive expressions of $q_K(\boldsymbol{x}_i)$ and $\sum_{i=1}^{K} q_K(\boldsymbol{x}_i)$ can be defined as follows:

$$\begin{aligned} q_K(\boldsymbol{x}_i) &= \sum_{j=1}^{K} d^2(\boldsymbol{x}_i, \boldsymbol{x}_j) = \sum_{j=1}^{K} (\boldsymbol{x}_i - \boldsymbol{x}_j)^T (\boldsymbol{x}_i - \boldsymbol{x}_j) \\ &= \sum_{j=1}^{K} \left(\boldsymbol{x}_i^T \boldsymbol{x}_i - \boldsymbol{x}_j^T \boldsymbol{x}_i - \boldsymbol{x}_i^T \boldsymbol{x}_j + \boldsymbol{x}_j^T \boldsymbol{x}_j \right) \\ &= K \boldsymbol{x}_i^T \boldsymbol{x}_i - K \boldsymbol{\mu}_K^T \boldsymbol{x}_i - K \boldsymbol{x}_i^T \boldsymbol{\mu}_K + K X_K \\ &= K (\boldsymbol{x}_i - \boldsymbol{\mu}_K)^T (\boldsymbol{x}_i - \boldsymbol{\mu}_K) + K X_K - K \boldsymbol{\mu}_K^T \boldsymbol{\mu}_K \\ &= K \left(\|\boldsymbol{x}_i - \boldsymbol{\mu}_K\|^2 + X_K - \|\boldsymbol{\mu}_K\|^2 \right) \end{aligned} \tag{A.1}$$

As a result, $q_K(\boldsymbol{x}_i) = K\left(\|\boldsymbol{x}_i - \boldsymbol{\mu}_K\|^2 + X_K - \|\boldsymbol{\mu}_K\|^2 \right)$.

P. P. Angelov and X. Gu, *Empirical Approach to Machine Learning*, Studies in Computational Intelligence 800,
https://doi.org/10.1007/978-3-030-02384-3

$$\begin{aligned}\sum_{i=1}^{K} q_K(\boldsymbol{x}_i) &= \sum_{i=1}^{K}\sum_{j=1}^{K} d^2(\boldsymbol{x}_i,\boldsymbol{x}_j) = \sum_{i=1}^{K}\sum_{j=1}^{K} (\boldsymbol{x}_i-\boldsymbol{x}_j)^T(\boldsymbol{x}_i-\boldsymbol{x}_j) \\ &= \sum_{i=1}^{K}\sum_{j=1}^{K}\left(\boldsymbol{x}_i^T\boldsymbol{x}_i-\boldsymbol{x}_j^T\boldsymbol{x}_i-\boldsymbol{x}_i^T\boldsymbol{x}_j+\boldsymbol{x}_j^T\boldsymbol{x}_j\right) \\ &= \sum_{i=1}^{K}\sum_{j=1}^{K}\boldsymbol{x}_i^T\boldsymbol{x}_i-\sum_{i=1}^{K}\sum_{j=1}^{K}\boldsymbol{x}_j^T\boldsymbol{x}_i-\sum_{i=1}^{K}\sum_{j=1}^{K}\boldsymbol{x}_i^T\boldsymbol{x}_j+\sum_{i=1}^{K}\sum_{j=1}^{K}\boldsymbol{x}_j^T\boldsymbol{x}_j \\ &= K\sum_{i=1}^{K}\boldsymbol{x}_i^T\boldsymbol{x}_i+K\sum_{j=1}^{K}\boldsymbol{x}_j^T\boldsymbol{x}_j-K\sum_{i=1}^{K}\boldsymbol{\mu}_K^T\boldsymbol{x}_i-K\sum_{i=1}^{K}\boldsymbol{x}_i^T\boldsymbol{\mu}_K \\ &= 2K\left(\sum_{i=1}^{K}\boldsymbol{x}_i^T\boldsymbol{x}_i-\sum_{i=1}^{K}\boldsymbol{x}_i^T\boldsymbol{\mu}_K\right) = 2K^2\left(X_K-\|\boldsymbol{\mu}_K\|^2\right)\end{aligned} \tag{A.2}$$

As a result, $\sum_{i=1}^{K} q_K(\boldsymbol{x}_i) = 2K^2\left(X_K-\|\boldsymbol{\mu}_K\|^2\right)$.

A.2 Cumulative Proximity Using Mahalanobis Distance

Using the Mahalanobis distance, the recursive expressions of $q_K(\boldsymbol{x}_i)$ and $\sum_{i=1}^{K} q_K(\boldsymbol{x}_i)$ can be derived as follows:

$$\begin{aligned}q_K(\boldsymbol{x}_i) &= \sum_{j=1}^{K} d^2(\boldsymbol{x}_i,\boldsymbol{x}_j) = \sum_{j=1}^{K}(\boldsymbol{x}_i-\boldsymbol{x}_j)^T\boldsymbol{\Sigma}_K^{-1}(\boldsymbol{x}_i-\boldsymbol{x}_j) \\ &= \sum_{j=1}^{K}\left(\boldsymbol{x}_i^T\boldsymbol{\Sigma}_K^{-1}\boldsymbol{x}_i-\boldsymbol{x}_i^T\boldsymbol{\Sigma}_K^{-1}\boldsymbol{x}_j-\boldsymbol{x}_j^T\boldsymbol{\Sigma}_K^{-1}\boldsymbol{x}_i+\boldsymbol{x}_j^T\boldsymbol{\Sigma}_K^{-1}\boldsymbol{x}_j\right) \\ &= \sum_{j=1}^{K}\boldsymbol{x}_i^T\boldsymbol{\Sigma}_K^{-1}\boldsymbol{x}_i-\sum_{j=1}^{K}\boldsymbol{x}_i^T\boldsymbol{\Sigma}_K^{-1}\boldsymbol{x}_j-\sum_{j=1}^{K}\boldsymbol{x}_j^T\boldsymbol{\Sigma}_K^{-1}\boldsymbol{x}_i+\sum_{j=1}^{K}\boldsymbol{x}_j^T\boldsymbol{\Sigma}_K^{-1}\boldsymbol{x}_j \\ &= K\boldsymbol{x}_i^T\boldsymbol{\Sigma}_K^{-1}\boldsymbol{x}_i-K\boldsymbol{x}_i^T\boldsymbol{\Sigma}_K^{-1}\boldsymbol{\mu}_K-K\boldsymbol{\mu}_K^T\boldsymbol{\Sigma}_K^{-1}\boldsymbol{x}_i+\sum_{j=1}^{K}\boldsymbol{x}_j^T\boldsymbol{\Sigma}_K^{-1}\boldsymbol{x}_j \\ &= K(\boldsymbol{x}_i-\boldsymbol{\mu}_K)^T\boldsymbol{\Sigma}_K^{-1}(\boldsymbol{x}_i-\boldsymbol{\mu}_K)+\sum_{j=1}^{K}\boldsymbol{x}_j^T\boldsymbol{\Sigma}_K^{-1}\boldsymbol{x}_j-K\boldsymbol{\mu}_K^T\boldsymbol{\Sigma}_K^{-1}\boldsymbol{\mu}_K \\ &= K\left((\boldsymbol{x}_i-\boldsymbol{\mu}_K)^T\boldsymbol{\Sigma}_K^{-1}(\boldsymbol{x}_i-\boldsymbol{\mu}_K)+X_K-\boldsymbol{\mu}_K^T\boldsymbol{\Sigma}_K^{-1}\boldsymbol{\mu}_K\right)\end{aligned} \tag{A.3}$$

As a result, $q_K(\boldsymbol{x}_i) = K\left((\boldsymbol{x}_i-\boldsymbol{\mu}_K)^T\boldsymbol{\Sigma}_K^{-1}(\boldsymbol{x}_i-\boldsymbol{\mu}_K)+X_K-\boldsymbol{\mu}_K^T\boldsymbol{\Sigma}_K^{-1}\boldsymbol{\mu}_K\right)$.

$$\begin{aligned}\sum_{i=1}^{K} q_K(\boldsymbol{x}_i) &= \sum_{i=1}^{K}\sum_{j=1}^{K}\left(\boldsymbol{x}_i-\boldsymbol{x}_j\right)^T\boldsymbol{\Sigma}_K^{-1}\left(\boldsymbol{x}_i-\boldsymbol{x}_j\right)\\
&= \sum_{i=1}^{K}\sum_{j=1}^{K}\left(\boldsymbol{x}_i^T\boldsymbol{\Sigma}_K^{-1}\boldsymbol{x}_i-\boldsymbol{x}_i^T\boldsymbol{\Sigma}_K^{-1}\boldsymbol{x}_j-\boldsymbol{x}_j^T\boldsymbol{\Sigma}_K^{-1}\boldsymbol{x}_i+\boldsymbol{x}_j^T\boldsymbol{\Sigma}_K^{-1}\boldsymbol{x}_j\right)\\
&= \sum_{i=1}^{K}\sum_{j=1}^{K}\boldsymbol{x}_i^T\boldsymbol{\Sigma}_K^{-1}\boldsymbol{x}_i-\sum_{i=1}^{K}\sum_{j=1}^{K}\boldsymbol{x}_i^T\boldsymbol{\Sigma}_K^{-1}\boldsymbol{x}_j\\
&\quad -\sum_{i=1}^{K}\sum_{j=1}^{K}\boldsymbol{x}_j^T\boldsymbol{\Sigma}_K^{-1}\boldsymbol{x}_i+\sum_{i=1}^{K}\sum_{j=1}^{K}\boldsymbol{x}_j^T\boldsymbol{\Sigma}_K^{-1}\boldsymbol{x}_j\\
&= K\sum_{i=1}^{K}\boldsymbol{x}_i^T\boldsymbol{\Sigma}_K^{-1}\boldsymbol{x}_i-K\sum_{i=1}^{K}\boldsymbol{x}_i^T\boldsymbol{\Sigma}_K^{-1}\boldsymbol{\mu}_K\\
&\quad -K\sum_{i=1}^{K}\boldsymbol{\mu}_K^T\boldsymbol{\Sigma}_K^{-1}\boldsymbol{x}_i+K\sum_{j=1}^{K}\boldsymbol{x}_j^T\boldsymbol{\Sigma}_K^{-1}\boldsymbol{x}_j\\
&= 2K^2\left(X_K-\boldsymbol{\mu}_K^T\boldsymbol{\Sigma}_K^{-1}\boldsymbol{\mu}_K\right)\end{aligned} \tag{A.4a}$$

As a result, $\sum_{i=1}^{K} q_K(\boldsymbol{x}_i) = 2K^2\left(X_K-\boldsymbol{\mu}_K^T\boldsymbol{\Sigma}_K^{-1}\boldsymbol{\mu}_K\right)$.

Alternatively, $\sum_{i=1}^{K} q_K(\boldsymbol{x}_i)$ can be further simplified by using the covariance matrix symmetricity property [1]:

$$\begin{aligned}\sum_{i=1}^{K} q_K(\boldsymbol{x}_i) &= 2\sum_{i=1}^{K}\sum_{j=1}^{K}\boldsymbol{x}_i^T\boldsymbol{\Sigma}_K^{-1}\boldsymbol{x}_i-2\sum_{i=1}^{K}\sum_{j=1}^{K}\boldsymbol{x}_j^T\boldsymbol{\Sigma}_K^{-1}\boldsymbol{x}_i\\
&= 2K\sum_{i=1}^{K}\boldsymbol{x}_i^T\boldsymbol{\Sigma}_K^{-1}\boldsymbol{x}_i-2K^2\boldsymbol{\mu}_K^T\boldsymbol{\Sigma}_K^{-1}\boldsymbol{\mu}_K\\
&= 2K\sum_{i=1}^{K}\boldsymbol{x}_i^T\left(\frac{1}{K}\sum_{j=1}^{K}\left(\boldsymbol{x}_j-\boldsymbol{\mu}_K\right)\left(\boldsymbol{x}_j-\boldsymbol{\mu}_K\right)^T\right)^{-1}\boldsymbol{x}_i\\
&\quad -2K^2\boldsymbol{\mu}_K^T\left(\frac{1}{K}\sum_{j=1}^{K}\left(\boldsymbol{x}_j-\boldsymbol{\mu}_K\right)\left(\boldsymbol{x}_j-\boldsymbol{\mu}_K\right)^T\right)^{-1}\boldsymbol{\mu}_K\\
&= \sum\left[2K\left(\sum_{i=1}^{K}\boldsymbol{x}_i\boldsymbol{x}_i^T\right)\otimes\left(\frac{1}{K}\sum_{j=1}^{K}\boldsymbol{x}_j\boldsymbol{x}_j^T-\boldsymbol{\mu}_K\boldsymbol{\mu}_K^T\right)^{-1}\right]\\
&\quad -\sum\left[2K^2\left(\boldsymbol{\mu}_K\boldsymbol{\mu}_K^T\right)\otimes\left(\frac{1}{K}\sum_{j=1}^{K}\boldsymbol{x}_j\boldsymbol{x}_j^T-\boldsymbol{\mu}_K\boldsymbol{\mu}_K^T\right)^{-1}\right]\end{aligned}$$

$$
\begin{aligned}
&= 2K^2 \sum \left[\left(\frac{1}{K} \sum_{j=1}^{K} \boldsymbol{x}_j \boldsymbol{x}_j^T - \boldsymbol{\mu}_K \boldsymbol{\mu}_K^T \right) \odot \left(\frac{1}{K} \sum_{j=1}^{K} \boldsymbol{x}_j \boldsymbol{x}_j^T - \boldsymbol{\mu}_K \boldsymbol{\mu}_K^T \right)^{-1} \right] \\
&= 2K^2 \sum \left[\left(\frac{1}{K} \sum_{j=1}^{K} \boldsymbol{x}_j \boldsymbol{x}_j^T - \boldsymbol{\mu}_K \boldsymbol{\mu}_K^T \right) \left(\frac{1}{K} \sum_{j=1}^{K} \boldsymbol{x}_j \boldsymbol{x}_j^T - \boldsymbol{\mu}_K \boldsymbol{\mu}_K^T \right)^{-1} \right] \otimes \mathbf{I}_{M \times M} \\
&= 2K^2 \text{Trace}(\mathbf{I}_{M \times M}) = 2K^2 M
\end{aligned}
\tag{A.4b}
$$

where $\otimes$ is an element-wise matrix multiplication operator; $\text{Trace}(\cdot)$ denotes the trace of the matrix.

As a result, $\sum_{i=1}^{K} q_K(\boldsymbol{x}_i) = 2K^2 M$.

Reference

1. D. Kangin, P. Angelov, and J. A. Iglesias, “Autonomously evolving classifier TEDAClass,” *Inf. Sci. (Ny).*, vol. 366, pp. 1–11, 2016.

Appendix B

In this appendix, the pseudo-codes of the *empirical* approaches presented in this book are described in detail. For simplicity of derivation, if no special declaration, the Euclidean type of distance is used by default; however, one can also use other types of distance/dissimilarity instead.

B.1 Autonomous Anomaly Detection

In this section, the pseudo-code of the algorithmic procedure of the autonomous anomaly detection (AAD) algorithm is presented. The detailed algorithmic procedure can be found in Sect. 6.3. By default, we use $n = 3$ (corresponding to "3σ" rule) and the second level of granularity, $G = 2$.

P. P. Angelov and X. Gu, *Empirical Approach to Machine Learning*, Studies in Computational Intelligence 800,
https://doi.org/10.1007/978-3-030-02384-3

AAD algorithm

Input: Static Dataset $\{x\}_K$

Algorithm Begins

1. Calculate the means, $\boldsymbol{\mu}_K$ and X_K of $\{\boldsymbol{x}\}_K$ and $\{\|\boldsymbol{x}\|^2\}_K$, respectively:

$$\boldsymbol{\mu}_K = \frac{1}{K}\sum_{i=1}^{K}\boldsymbol{x}_i ;$$

$$X_K = \frac{1}{K}\sum_{i=1}^{K}\|\boldsymbol{x}_i\|^2 ;$$

2. Identify the sorted unique data samples set $\{\boldsymbol{u}\}_L$ and the corresponding frequencies of occurrence $\{F\}_L$ from $\{\boldsymbol{x}\}_K$;
3. Calculate the multimodal *typicality* at $\{\boldsymbol{u}\}_L$:

$$\tau_K^M(\boldsymbol{u}_i) = \frac{\frac{F_i}{X_K - \|\boldsymbol{\mu}_K\|^2 + \|\boldsymbol{u}_i - \boldsymbol{\mu}_K\|^2}}{\sum_{j=1}^{L}\frac{F_j}{X_K - \|\boldsymbol{\mu}_K\|^2 + \|\boldsymbol{u}_j - \boldsymbol{\mu}_K\|^2}}; \quad \boldsymbol{u}_i \in \{\boldsymbol{u}\}_L ;$$

4. Extend τ^M to $\{\boldsymbol{x}\}_K$ and obtain $\{\tau^M(\boldsymbol{x})\}_K$;
5. Identify $\{\boldsymbol{x}\}_{PA}^1$ as $\left(\frac{1}{2n^2}\right)^{th}$ of the total number of data samples with smallest τ^M;
6. $R_{K,0} = 2\left(X_K - \|\boldsymbol{\mu}_K\|^2\right)$;

7. **FOR** $l=1$ **TO** G **:**

$$R_{K,l}=\frac{\sum_{y,z\in\{x\}_K;y\neq z;\|y-z\|^2\leq M_{K,l-1}}\|y-z\|^2}{M_{K,l}};$$

8. **END FOR**
9. Identify the unique neighbouring data samples $\{x^*\}_i$ around each $u_i\in\{u\}_L$ by **Condition 6.2**:

Condition 6.2: $IF\left(d^2\left(u_i,x_j\right)\leq R_{K,G}\right)\ \ THEN\left(\{x^*\}_i\leftarrow\{x^*\}_i+x_j\right);$

10. Calculate the means, ς_i and χ_i of $\{x^*\}_i$ and $\left\{\|x^*\|^2\right\}_i$, respectively $(i=1,2,\ldots,L)$:

$$\varsigma_i=\frac{1}{K_i}\sum_{y\in\{x^*\}_i}y;$$

$$\chi_i=\frac{1}{K_i}\sum_{y\in\{x^*\}_i}\|y\|^2;$$

11. Calculate the local *data densities* at $\{u\}_L$:

$$D_{K_i}^*\left(u_i\right)=\frac{1}{1+\frac{\|u_i-\varsigma_i\|^2}{\chi_i-\|\varsigma_i\|^2}};\quad u_i\in\{u\}_L;$$

12. Calculate the locally weighted multimodal *typicality* at $\{u\}_L$:

$$\tau_K^{M*}\left(u_i\right)=\frac{K_iD_{K_i}^*\left(u_i\right)}{\sum_{j=1}^{L}K_jD_{K_j}^*\left(u_j\right)};$$

13. Extend τ^{M*} to $\{x\}_K$ and obtain $\left\{\tau^{M*}(x)\right\}_K$;
14. Identify $\{x\}_{PA,2}$ as $\left(\frac{1}{2n^2}\right)^{th}$ of total number of data samples with smallest τ^{M*};
15. $\{x\}_{PA}\leftarrow\{x\}_{PA,1}+\{x\}_{PA,2}$;
16. Apply **ADP Algorithm** (offline version, see subsections 7.4.1 and 11.1.1) to $\{x\}_{PA}$ and obtain $\{p\}=\{p_1,p_2,\ldots,p_M\}$;
17. $i\leftarrow1$;
18. **WHILE** $\{x\}_{PA}\neq\varnothing$:

i. Identify the average radius, ω_K of the neighboring areas around each *data cloud*:

$$\gamma_K = \frac{\sum_{j=1}^{M-1}\sum_{l=j+1}^{M}\|\boldsymbol{p}_j - \boldsymbol{p}_l\|}{M(M-1)};$$

$$\varpi_K = \frac{\sum_{\substack{\boldsymbol{y},\boldsymbol{z}\in\{\boldsymbol{p}\},\boldsymbol{y}\neq\boldsymbol{z}\\ \|\boldsymbol{y}-\boldsymbol{z}\|\leq\gamma_K}}\|\boldsymbol{y}-\boldsymbol{z}\|}{M_\gamma};$$

$$\omega_K = \frac{\sum_{\substack{\boldsymbol{y},\boldsymbol{z}\in\{\boldsymbol{p}\},\boldsymbol{y}\neq\boldsymbol{z}\\ \|\boldsymbol{y}-\boldsymbol{z}\|\leq\varpi_K}}\|\boldsymbol{y}-\boldsymbol{z}\|}{M_\varpi};$$

ii. **IF** ($\boldsymbol{x}_i \in \{\boldsymbol{x}\}_{PA}$ meets **Condition 6.3**) **THEN**

Condition 6.3: $IF\left(n^* = \underset{\boldsymbol{p}\in\{\boldsymbol{p}\}_{PA}}{\arg\min}\left(d\left(\boldsymbol{x}_i, \boldsymbol{p}\right)\right)\right) AND \left(d\left(\boldsymbol{x}_i, \boldsymbol{p}_{n^*}\right) \leq \omega_K\right)$

$THEN\left(\mathbf{C}_{n^*} \leftarrow \mathbf{C}_{n^*} + \boldsymbol{x}_i\right)$

1) Add $\boldsymbol{x}_i$ to $\mathbf{C}_{n^*} \in \{\mathbf{C}\}_{PA}$:

$$\mathbf{C}_{n^*} \leftarrow \mathbf{C}_{n^*} + \boldsymbol{x}_i;$$
$$S_{n^*} \leftarrow S_{n^*} + 1;$$

iii. **ELSE**

1) $\{\boldsymbol{x}\}_A \leftarrow \{\boldsymbol{x}\}_A + \boldsymbol{x}_i$;

iv. **END IF**

v. $\{\boldsymbol{x}\}_{PA} \leftarrow \{\boldsymbol{x}\}_{PA} - \boldsymbol{x}_i$;

19. END WHILE

20. Identify $\{\mathbf{C}\}_A$ the *data clouds* satisfying **Condition 6.4** ($\mathbf{C}_i \in \{\mathbf{C}\}_{PA}$):

Condition 6.4: $IF\left(S_i < \bar{S}\right) \quad THEN\left(\{\mathbf{C}\}_A \leftarrow \{\mathbf{C}\}_A + \mathbf{C}_i\right)$

21. $\{\boldsymbol{x}\}_A \leftarrow \{\boldsymbol{x}\}_A + \{\mathbf{C}\}_A$.

Algorithm Ends

Outputs: Anomalies, $\{\boldsymbol{x}\}_A$**.**

B.2 Autonomous Data Partitioning

In this section, the implementations of the offline and evolving versions of the autonomous data partitioning (ADP) algorithm are given. The detailed algorithmic procedures can be found in Sect. 7.4.

B.2.1 Offline ADP

The pseudo code of the offline ADP algorithm is summarized as follows. For the detailed algorithmic procedure, please refer to Sect. 7.4.1.

Offline ADP algorithm

Input: Static Dataset $\{x\}_K$

Algorithm Begins

1. Calculate the means, $\boldsymbol{\mu}_K$ and X_K of $\{\boldsymbol{x}\}_K$ and $\left\{\|\boldsymbol{x}\|^2\right\}_K$, respectively:

$$\boldsymbol{\mu}_K = \frac{1}{K}\sum_{i=1}^{K}\boldsymbol{x}_i;$$

$$X_K = \frac{1}{K}\sum_{i=1}^{K}\|\boldsymbol{x}_i\|^2;$$

2. Identify the sorted unique data samples set $\{\boldsymbol{u}\}_L$ and the corresponding frequencies of occurrence $\{F\}_L$ from $\{\boldsymbol{x}\}_K$;

3. Calculate the multimodal *typicality* at $\{\boldsymbol{u}\}_L$:

$$\tau_K^M(\boldsymbol{u}_i) = \frac{\dfrac{F_i}{X_K - \|\boldsymbol{\mu}_K\|^2 + \|\boldsymbol{u}_i - \boldsymbol{\mu}_K\|^2}}{\sum\limits_{j=1}^{L}\dfrac{F_j}{X_K - \|\boldsymbol{\mu}_K\|^2 + \|\boldsymbol{u}_j - \boldsymbol{\mu}_K\|^2}}; \quad \boldsymbol{u}_i \in \{\boldsymbol{u}\}_L;$$

4. Find the unique data sample $\boldsymbol{u}_{j*}$ with maximum multimodal *typicality*:

$$j^* = \underset{\boldsymbol{u}_i \in \{\boldsymbol{u}\}_L;}{\arg\max}\left(\tau_K^M(\boldsymbol{u}_i)\right);$$

5. Initialize the indexing list $\{\boldsymbol{z}\}_L$ with $\boldsymbol{u}_{j*}$:

$$j \leftarrow 1;$$

$$\boldsymbol{z}_j \leftarrow \boldsymbol{u}_{j*};$$

$$\{\boldsymbol{z}\}_L \leftarrow \{\boldsymbol{z}_j\};$$

6. Remove $\boldsymbol{u}_{j*}$ from $\{\boldsymbol{u}\}_L$:

$$\{\boldsymbol{u}\}_L \leftarrow \{\boldsymbol{u}\}_L - \boldsymbol{u}_{j*};$$

7. Set $\boldsymbol{z}_j$ as the reference, $\boldsymbol{r}$:

$$\boldsymbol{r} \leftarrow \boldsymbol{z}_j;$$

8. **WHILE** $\{\boldsymbol{u}\}_L \neq \varnothing$:

   i. Identify the nearest unique data sample $\boldsymbol{u}_{n*}$ to the reference, $\boldsymbol{r}$:

$$n^* = \underset{u_i \in \{u\}_L;}{\arg\min}\left(\left\|u_i - r\right\|\right);$$

ii. Send u_{n*} to the indexing list $\{z\}_L$:

$$j \leftarrow j+1;$$

$$z_j \leftarrow u_{n*};$$

$$\{z\}_L \leftarrow \{z\}_L + z_j;$$

iii. Remove u_{n*} from $\{u\}_L$:

$$\{u\}_L \leftarrow \{u\}_L - u_{n*};$$

iv. Set $r \leftarrow z_j$;

9. **END WHILE**
10. Obtain the ranked multimodal *typicality* list: $\{\tau^M(z)\}_L$.
11. Identify the unique data samples corresponding to the local maxima of multimodal *typicalities* satisfying **Condition 7.1** and re-denote them as $\{z^*\} = \{z_1^*, z_2^*, \ldots, z_M^*\}$;

Condition 7.1:
$$IF\left(\tau_K^M(z_i) > \tau_K^M(z_{i-1})\right) AND \left(\tau_K^M(z_i) > \tau_K^M(z_{i+1})\right)$$
$$THEN\left(z_i \text{ is one of the local maxima}\right)$$

12. **WHILE** $\{z^*\}$ is not fixed:

i. Create *data clouds*, $\{\mathbf{C}\} = \{\mathbf{C}_1, \mathbf{C}_2, \ldots, \mathbf{C}_M\}$ around $\{z^*\}$ with $\{x\}_K$:

$$\mathbf{C}_{n*} \leftarrow \mathbf{C}_{n*} + x_j; \quad n^* = \underset{z_i^* \in \{z^*\}}{\arg\min}\left(\left\|x_j - z_i^*\right\|\right); \quad j = 1, 2, \ldots, K;$$

ii. Obtain the centers $\{c\}$ and supports $\{S\}$ of $\{\mathbf{C}\}$;

iii. Calculate the multimodal *typicality* at $\{c\}$:

$$\tau_K^M(c_i) = \frac{\dfrac{S_i}{X_K - \left\|\mu_K\right\|^2 + \left\|c_i - \mu_K\right\|^2}}{\sum_{j=1}^{M} \dfrac{S_j}{X_K - \left\|\mu_K\right\|^2 + \left\|c_j - \mu_K\right\|^2}}; \quad c_i \in \{c\}; \quad S_i \in \{S\};$$

iv. Identify the average radius, ω_K of the neighboring areas around each *data cloud*:

$$\gamma_K = \frac{\sum_{j=1}^{M-1} \sum_{l=j+1}^{M} \left\|c_j - c_l\right\|}{M(M-1)};$$

$$\varpi_K = \frac{\sum_{\substack{y,z \in \{c\}, y \neq z \\ \|y-z\| \le \gamma_K}} \|y-z\|}{M_\gamma};$$

$$\omega_K = \frac{\sum_{\substack{y,z \in \{c\}, y \neq z \\ \|y-z\| \le \varpi_K}} \|y-z\|}{M_\varpi};$$

v. Identify the neighboring *data clouds*, $\{\mathbf{C}^*\}_i$ around each *data cloud* $\mathbf{C}_i$ $(\mathbf{C}_i \in \{\mathbf{C}\})$ with the corresponding centers $\{c^*\}_i$ using **Condition 7.3;**

Condition 7.3: $IF\left(\|c_i - c_j\| \le \frac{\omega_K}{2}\right)\ THEN\left(\mathbf{C}_j \in \{\mathbf{C}^*\}_i\right)$

vi. Identify the centers of the *data clouds* representing the local maxima of multimodal *typicality* using **Condition 7.2** and re-denote them as $\{c^{**}\}$;

Condition 7.2: $IF\left(\tau_K^M(c_i) = \max\left(\tau_K^M(c_i), \{\tau^M(c^*)\}_i\right)\right)$
$THEN\,(c_i\ is\ one\ of\ the\ local\ maxima)$

vii. $\{z^*\} \leftarrow \{c^{**}\}$;

13. **END WHILE**
14. $\{p\} \leftarrow \{z^*\}$;
15. Form *data clouds* $\{\mathbf{C}\}$ around $\{p\}$ with $\{x\}_K$:

$$\mathbf{C}_{n^*} \leftarrow \mathbf{C}_{n^*} + x_j;\quad n^* = \underset{p_i \in \{p\}}{\arg\min}\left(\|x_j - p_i\|\right);\quad j = 1, 2, \ldots, K.$$

Algorithm Ends
Outputs: Prototypes, $\{p\}$ and *Data Clouds*, $\{\mathbf{C}\}$.

B.2.2 Evolving ADP

The pseudo code of the evolving ADP algorithm is summarized as follows. For the detailed algorithmic procedure, please refer to Sect. 7.4.2.

Evolving ADP algorithm

Input: Streaming Data $\{\boldsymbol{x}_1, \boldsymbol{x}_2, \boldsymbol{x}_3, \ldots\}$

Algorithm Begins

1. Initialize the global meta-parameters with $\boldsymbol{x}_1$ as:

$$K \leftarrow 1;$$
$$M \leftarrow 1;$$
$$\boldsymbol{\mu}_K \leftarrow \boldsymbol{x}_1;$$
$$X_K \leftarrow \|\boldsymbol{x}_1\|^2;$$

2. Initialize the meta-parameters of the first *data cloud* $\mathbf{C}_1$ as:

$$\mathbf{C}_M \leftarrow \{\boldsymbol{x}_1\};$$
$$\boldsymbol{c}_{K,M} \leftarrow \boldsymbol{x}_1;$$
$$S_{K,M} \leftarrow 1;$$

3. **WHILE** (new data available) **AND (**no request for interruption):

   i. $K \leftarrow K+1$;

   ii. Update $\boldsymbol{\mu}_K$ and X_K as:

$$\boldsymbol{\mu}_K \leftarrow \frac{K-1}{K}\boldsymbol{\mu}_{K-1} + \frac{1}{K}\boldsymbol{x}_K;$$
$$X_K \leftarrow \frac{K-1}{K}X_{K-1} + \frac{1}{K}\|\boldsymbol{x}_K\|^2;$$

   iii. Update γ_K as : $\gamma_K = \sqrt{2\left(X_K - \|\boldsymbol{\mu}_K\|^2\right)}$;

   iv. Calculate *data densities* D at $\boldsymbol{x}_K$ and $\boldsymbol{c}_{K-1,j}$ ($j = 1,2,\ldots,M$):

$$D_K(\boldsymbol{z}) = \frac{1}{1 + \frac{\|\boldsymbol{z} - \boldsymbol{\mu}_K\|^2}{X_K - \|\boldsymbol{\mu}_K\|^2}}; \quad \boldsymbol{z} = \boldsymbol{x}_K, \boldsymbol{c}_{K-1,1}, \boldsymbol{c}_{K-1,2}, \ldots, \boldsymbol{c}_{K-1,M}$$

   v. **IF (Condition 7.4** is satisfied) **THEN:**

$$IF\left(D_K(\boldsymbol{x}_K) > \max_{i=1,2,\ldots,M}\left(D_K\left(\boldsymbol{c}_{K-1,i}\right)\right)\right)$$

Condition 7.4: $OR\left(D_K(\boldsymbol{x}_K) < \min_{i=1,2,\ldots,M}\left(D_K\left(\boldsymbol{c}_{K-1,i}\right)\right)\right)$

$$THEN\left(\boldsymbol{x}_K \; is\; a\; new\; prototype\right)$$

   1) Add a new *data cloud* with $\boldsymbol{x}_K$ as the prototype:

$$M \leftarrow M+1;$$
$$\mathbf{C}_M \leftarrow \{\boldsymbol{x}_K\};$$
$$\boldsymbol{c}_{K,M} \leftarrow \boldsymbol{x}_K;$$
$$S_{K,M} \leftarrow 1;$$

vi. **ELSE:**

1) Find the nearest *data cloud* $\mathbf{C}_{n*}$ to $\boldsymbol{x}_K$:

$$n^* = \underset{j=1,2,\ldots,M}{\arg\min}\left(\left\|\boldsymbol{x}_K - \boldsymbol{c}_{K-1,j}\right\|\right);$$

2) **IF (Condition 7.5** is satisfied) **THEN:**

Condition 7.5: $IF\left(\left\|\boldsymbol{x}_K - \boldsymbol{c}_{K-1,n^*}\right\| \le \frac{\gamma_K}{2}\right)\ THEN\left(\boldsymbol{x}_K\ is\ assigned\ to\ \mathbf{C}_{n^*}\right)$;

* Update the meta-parameters of $\mathbf{C}_{n*}$ as:

$$\mathbf{C}_{n^*} \leftarrow \mathbf{C}_{n^*} + \boldsymbol{x}_K;$$

$$S_{K,n^*} \leftarrow S_{K-1,n^*} + 1;$$

$$\boldsymbol{c}_{K,n^*} \leftarrow \frac{S_{K-1,n^*}}{S_{K,n^*}}\boldsymbol{c}_{K-1,n^*} + \frac{1}{S_{K,n^*}}\boldsymbol{x}_K;$$

3) **ELSE:**

* Add a new *data cloud* with $\boldsymbol{x}_K$ as the prototype:

$$M \leftarrow M+1;$$

$$\mathbf{C}_M \leftarrow \{\boldsymbol{x}_K\};$$

$$\boldsymbol{c}_{K,M} \leftarrow \boldsymbol{x}_K;$$

$$S_{K,M} \leftarrow 1;$$

4) **END IF**

vii. **END IF**

viii. set $\boldsymbol{c}_{K,i} \leftarrow \boldsymbol{c}_{K-1,i}$; $S_{K,i} \leftarrow S_{K-1,i}$ for all $\mathbf{C}_i \in \{\mathbf{C}\}$ that does not receive new member;

4. END WHILE

5. $\{\boldsymbol{p}\} \leftarrow \{\boldsymbol{c}\}_K$;

6. Form *data clouds* $\{\mathbf{C}\}$ around $\{\boldsymbol{p}\}$ with $\{\boldsymbol{x}\}_K$:

$$\mathbf{C}_{n^*} \leftarrow \mathbf{C}_{n^*} + \boldsymbol{x}_j;\quad n^* = \underset{\boldsymbol{p}_i \in \{\boldsymbol{p}\}}{\arg\min}\left(\left\|\boldsymbol{x}_j - \boldsymbol{p}_i\right\|\right);\quad j = 1,2,\ldots,K.$$

Algorithm Ends

Outputs: Prototypes, $\{\boldsymbol{p}\}$ **and *Data Clouds*,** $\{\mathbf{C}\}$.

B.3 Zero-Order Autonomous Learning Multi-model System

In this section, the implementation of the zero-order autonomous learning multi-model (*ALMMo-0*) system is presented.

B.3.1 Online Self-learning Process

The pseudo code of online self-learning process of the *ALMMo-0* algorithm is presented in this subsection. By default, r_o is set to be $r_o = \sqrt{2(1 - \cos(30^{\circ}))}$. As the learning process can be done in parallel for each fuzzy rule, we consider the *i*th ($i = 1, 2, \ldots, C$) rule, and only summarize the learning process of an individual fuzzy rule. However, the same principle can be applied to all the other fuzzy rules within the rule base of the *ALMMo-0*. The detailed algorithmic procedure can be found in Sect. 8.2.2.

ALMMo-0 online self-learning algorithm

Input: Streaming Data of the ith Class $\{\boldsymbol{x}_{i,1}, \boldsymbol{x}_{i,2}, \boldsymbol{x}_{i,3}, \ldots\}$

Algorithm Begins

1. Normalize $\boldsymbol{x}_{i,1}$ by its norm $\|\boldsymbol{x}_{i,1}\|$ as:

$$\boldsymbol{x}_{i,1} \leftarrow \boldsymbol{x}_{i,1} / \|\boldsymbol{x}_{i,1}\|;$$

2. Initialize the global meta-parameters with $\boldsymbol{x}_{i,1}$ as:

$$K_i \leftarrow 1;$$
$$M_i \leftarrow 1;$$
$$\boldsymbol{\mu}_i \leftarrow \boldsymbol{x}_{i,1};$$

3. Initialize the meta-parameters of the first *data cloud* $\mathbf{C}_{i,M_i}$ as:

$$\mathbf{C}_{i,M_i} \leftarrow \{\boldsymbol{x}_{i,1}\};$$
$$\boldsymbol{p}_{i,M_i} \leftarrow \boldsymbol{x}_{i,1};$$
$$S_{i,M_i} \leftarrow 1;$$
$$r_{i,M_i} \leftarrow r_o;$$

4. Initialize the fuzzy rule:

$$\boldsymbol{R}_i: \ IF\left(\boldsymbol{x} \sim \boldsymbol{p}_{i,1}\right) \ THEN\left(Class\, i\right);$$

5. **WHILE** (new data available) **AND** (no request for interruption):

   i. $K_i \leftarrow K_i + 1;$

   ii. Normalize $\boldsymbol{x}_{i,K_i}$ by its norm $\|\boldsymbol{x}_{i,K_i}\|$ as:

$$\boldsymbol{x}_{i,K_i} \leftarrow \boldsymbol{x}_{i,K_i} / \|\boldsymbol{x}_{i,K_i}\|;$$

   iii. Update $\boldsymbol{\mu}_i$ as:

$$\boldsymbol{\mu}_i \leftarrow \frac{K_i - 1}{K_i}\boldsymbol{\mu}_i + \frac{1}{K_i}\boldsymbol{x}_{i,K_i};$$

   iv. Calculate *data densities* D at $\boldsymbol{x}_{i,K_i}$ and $\boldsymbol{p}_{i,j}$ ($j = 1, 2, \ldots, M_i$):

$$D_{K_i}(\boldsymbol{z}) = \frac{1}{1 + \frac{\|\boldsymbol{z} - \boldsymbol{\mu}_i\|^2}{1 - \|\boldsymbol{\mu}_i\|^2}}; \quad \boldsymbol{z} = \boldsymbol{x}_{i,K_i}, \boldsymbol{p}_{i,1}, \boldsymbol{p}_{i,2}, \ldots, \boldsymbol{p}_{i,M_i};$$

   v. **IF** (**Condition 7.4** is satisfied) **THEN:**

Condition 7.4:
$$IF\left(D_{K_i}\left(\boldsymbol{x}_{i,K_i}\right) > \max_{j=1,2,\ldots,M_i}\left(D_{K_i}\left(\boldsymbol{p}_{i,j}\right)\right)\right)$$
$$OR\left(D_{K_i}\left(\boldsymbol{x}_{i,K_i}\right) < \min_{j=1,2,\ldots,M_i}\left(D_{K_i}\left(\boldsymbol{p}_{i,j}\right)\right)\right)$$
$$THEN\left(\boldsymbol{x}_{i,K_i}\ is\ a\ new\ prototype\right)$$

1) Add a new *data cloud* with $\boldsymbol{x}_{i,K_i}$ as the prototype:

$$M_i \leftarrow M_i + 1;$$

$$\mathbf{C}_{i,M_i} \leftarrow \left\{\boldsymbol{x}_{i,K_i}\right\};$$

$$\boldsymbol{p}_{i,M_i} \leftarrow \boldsymbol{x}_{i,K_i};$$

$$S_{i,M_i} \leftarrow 1;$$

$$r_{i,M_i} \leftarrow r_o;$$

vi. **ELSE:**

1) Find the nearest *data cloud* $\mathbf{C}_{i,n*}$ to $\boldsymbol{x}_{i,K_i}$:

$$n* = \underset{j=1,2,\ldots,M_i}{\arg\min}\left(\left\|\boldsymbol{x}_{i,K_i} - \boldsymbol{p}_{i,j}\right\|\right);$$

2) **IF (Condition 8.1** is satisfied) **THEN:**

Condition 8.1: $IF\left(\left\|\boldsymbol{x}_{i,K_i} - \boldsymbol{p}_{i,n*}\right\| \leq r_{i,n*}\right) \quad THEN\left(\mathbf{C}_{i,n*} \leftarrow \mathbf{C}_{i,n*} + \boldsymbol{x}_{i,K_i}\right)$

* Update the meta-parameters of $\mathbf{C}_{i,n*}$ as:

$$\mathbf{C}_{i,n*} \leftarrow \mathbf{C}_{i,n*} + \boldsymbol{x}_{i,K_i};$$

$$\boldsymbol{p}_{i,n*} \leftarrow \frac{S_{i,n*}}{S_{i,n*}+1}\boldsymbol{p}_{i,n*} + \frac{1}{S_{i,n*}+1}\boldsymbol{x}_{i,K_i};$$

$$S_{i,n*} \leftarrow S_{i,n*} + 1;$$

$$r_{i,n*} \leftarrow \sqrt{\frac{1}{2}\left(r_{i,n*}^2 + \left(1 - \left\|\boldsymbol{p}_{i,n*}\right\|^2\right)\right)};$$

3) **ELSE:**

* Add a new *data cloud* with $\boldsymbol{x}_{i,K_i}$ as the prototype:

$$M_i \leftarrow M_i + 1;$$

$$\mathbf{C}_{i,M_i} \leftarrow \left\{\boldsymbol{x}_{i,K_i}\right\};$$

$$\boldsymbol{p}_{i,M_i} \leftarrow \boldsymbol{x}_{i,K_i};$$

$$S_{i,M_i} \leftarrow 1;$$

$$r_{i,M_i} \leftarrow r_o;$$

4) **END IF**

vii. **END IF**

viii. Update the fuzzy rule:

$$\boldsymbol{R}_i: \quad IF\left(\boldsymbol{x} \sim \boldsymbol{p}_{i,1}\right)OR\ldots OR\left(\boldsymbol{x} \sim \boldsymbol{p}_{i,M_i}\right) \quad THEN\left(Class\, i\right);$$

6. END WHILE

Algorithm Ends

Outputs: The ith Fuzzy Rule $\boldsymbol{R}_i$.

B.3.2 Validation Process

The pseudo code of validation process of the *ALMMo-0* algorithm is summarized as follows. The detailed algorithmic procedure can be found in Sect. 8.2.3.

ALMMo-0 validation algorithm

Input: Data Sample from Unknown Class, $\boldsymbol{x}$
Algorithm Begins

1. Normalize $\boldsymbol{x}$ by its norm $\|\boldsymbol{x}\|$ as:
$$\boldsymbol{x} \leftarrow \boldsymbol{x}/\|\boldsymbol{x}\|;$$
2. Calculate the firing strength, λ_i of each fuzzy rule ($i = 1, 2, \ldots, C$):
$$\lambda_i = \max_{\boldsymbol{p}_{i,j} \in \{\boldsymbol{p}\}_i} \left(e^{-\|\boldsymbol{x} - \boldsymbol{p}_{i,j}\|^2} \right);$$
3. Estimate the label of $\boldsymbol{x}$ as:
$$\text{Label} = \arg\max_{i=1,2,\ldots,C} (\lambda_i) \ .$$

Algorithm Ends
Outputs: The Estimated Label of $\boldsymbol{x}$**.**

B.4 First-Order Autonomous Learning Multi-Model System

In this section, the implementation of the first order autonomous learning multi-model (*ALMMo-1*) system is presented. The detailed algorithmic procedure can be found in Sect. 8.3.2, where we use the local consequent parameter learning approach as presented in Sect. 8.3.2.2 for clarity. By default, $n = 0.5$, $\Omega_o = 10$, $\eta_o = 0.1$ and $\varphi_o = 0.03$.The pseudo code of the algorithmic learning procedure of the *ALMMo-1* system is summarized as follows.

ALMMo-1 online self-learning algorithm

Input: Streaming Data $\{x_1, x_2, x_3, ...\}$

Algorithm Begins

1. Initialize the global meta-parameters with x_1 as:

$$K \leftarrow 1;$$
$$\mu_K \leftarrow x_1;$$
$$X_K \leftarrow \|x_1\|^2;$$
$$M_K \leftarrow 1;$$

2. Initialize the meta-parameters of the first *data cloud* $\mathbf{C}_{M_K}$ as:

$$\mathbf{C}_{M_K} \leftarrow \{x_1\};$$
$$p_{K,M_K} \leftarrow x_1;$$
$$\chi_{K,M_K} \leftarrow \|x_1\|^2;$$
$$S_{K,M_K} \leftarrow 1;$$
$$I_{M_K} \leftarrow K;$$
$$\eta_{K,M_K} \leftarrow 1;$$
$$\Lambda_{K,M_K} \leftarrow 1;$$

3. Initialize the consequent parameters:

$$a_{K,M_K} \leftarrow \mathbf{0}_{(N+1)\times 1};$$
$$\Theta_{K,M_K} \leftarrow \Omega_o \mathbf{I}_{(N+1)\times(N+1)};$$

4. Initialize the fuzzy rule corresponding to $\mathbf{C}_{M_K}$:

$$R_1: \ IF\left(x \sim p_{K,1}\right) \ THEN\left(y_1 = \bar{x}^T a_{K,1}\right);$$

5. **WHILE** (new data available) **AND** (no request for interruption):

   i. $K \leftarrow K+1$;

   ii. Calculate the local *data density* of x_K within each *data cloud* $\left(j = 1, 2, ..., M_{K-1}\right)$:

$$D_{K-1,j}\left(x_K\right) = \frac{1}{1 + \frac{S_{K-1,j}^2 \left\|x_K - p_{K-1,j}\right\|^2}{\left(S_{K-1,j}+1\right)\left(S_{K-1,j}\chi_{K-1,j} + \left\|x_K\right\|^2\right) - \left\|x_K + S_{K-1,j} p_{K-1,j}\right\|^2}};$$

   iii. Calculate the firing strength of each fuzzy rule ($j = 1, 2, ..., M_{K-1}$):

$$\lambda_{K-1,j} = \frac{D_{K-1,j}\left(x_K\right)}{\sum_{k=1}^{M_{K-1}} D_{K-1,k}\left(x_K\right)};$$

   iv. Calculate and **Export** the system output:

$$y_K = \sum_{i=1}^{M_{K-1}} \lambda_{K-1,i} \bar{x}_K^T a_{K-1,i};$$

   v. Update μ_K and X_K as:

$$\boldsymbol{\mu}_K \leftarrow \frac{K-1}{K}\boldsymbol{\mu}_{K-1} + \frac{1}{K}\boldsymbol{x}_K;$$

$$X_K \leftarrow \frac{K-1}{K}X_{K-1} + \frac{1}{K}\|\boldsymbol{x}_K\|^2;$$

vi. Calculate *data densities* D at $\boldsymbol{x}_K$ and $\boldsymbol{p}_{K-1,j}$ ($j = 1,2,\ldots,M_{K-1}$):

$$D_K(\boldsymbol{z}) = \frac{1}{1 + \frac{\|\boldsymbol{z}-\boldsymbol{\mu}_K\|^2}{X_K - \|\boldsymbol{\mu}_K\|^2}}; \quad \boldsymbol{z} = \boldsymbol{x}_K, \boldsymbol{p}_{K-1,1}, \boldsymbol{p}_{K-1,2}, \ldots, \boldsymbol{p}_{K-1,M_{K-1}};$$

vii. **IF** (**Condition 7.4** is satisfied) **THEN:**

$$\textbf{Condition 7.4:}\ \begin{array}{c} IF\left(D_K(\boldsymbol{x}_K) > \max\limits_{j=1,2,\ldots,M_{K-1}}\left(D_K\left(\boldsymbol{p}_{K-1,j}\right)\right)\right) \\ OR\left(D_K(\boldsymbol{x}_K) < \min\limits_{j=1,2,\ldots,M_{K-1}}\left(D_K\left(\boldsymbol{p}_{K-1,j}\right)\right)\right) \\ THEN\left(\boldsymbol{x}_K\ is\ a\ new\ prototype\right) \end{array}$$

1) **IF** (**Condition 8.2** is satisfied) **THEN:**

$$\textbf{Condition 8.2:}\ IF\left(D_{K-1,j}(\boldsymbol{x}_K) \geq \frac{1}{1+n^2}\right)\ THEN\left(\begin{array}{c}\mathbf{C}_j\ overlaps\ with \\ the\ new\ data\ cloud\end{array}\right)$$

* Find the nearest *data cloud* $\mathbf{C}_{n*}$ to $\boldsymbol{x}_K$:

$$n* = \underset{j=1,2,\ldots,M_{K-1}}{\arg\min}\left(\left\|\boldsymbol{x}_K - \boldsymbol{p}_{K-1,j}\right\|\right);$$

* Replace the meta-parameters of $\mathbf{C}_{n*}$ as ($M_K \leftarrow M_{K-1}$):

$$\mathbf{C}_{n*} \leftarrow \mathbf{C}_{n*} + \boldsymbol{x}_K;$$

$$\boldsymbol{p}_{K,n*} \leftarrow \frac{\boldsymbol{p}_{K-1,n*} + \boldsymbol{x}_K}{2};$$

$$\chi_{K,n*} \leftarrow \frac{\chi_{K-1,n*} + \|\boldsymbol{x}_K\|^2}{2};$$

$$S_{K,n*} \leftarrow \left\lceil \frac{S_{K-1,n*}+1}{2} \right\rceil;$$

2) **ELSE:**

* Add a new *data cloud* with $\boldsymbol{x}_K$ as the prototype:

$$M_K \leftarrow M_{K-1} + 1;$$

$$\mathbf{C}_{M_K} \leftarrow \{\boldsymbol{x}_K\};$$

$$\boldsymbol{p}_{K,M_K} \leftarrow \boldsymbol{x}_K;$$

$$\chi_{K,M_K} \leftarrow \|\boldsymbol{x}_K\|^2;$$

$$S_{K,M_K} \leftarrow 1;$$

$$I_{M_K} \leftarrow K;$$

$$\eta_{K-1,M_K} \leftarrow 1;$$

$$\Lambda_{K-1,M_K} \leftarrow 0;$$

* Initialize the corresponding consequent parameters:

$$\boldsymbol{a}_{K-1,M_K} \leftarrow \frac{1}{M_{K-1}} \sum_{j=1}^{M_{K-1}} \boldsymbol{a}_{K-1,j};$$

$$\boldsymbol{\Theta}_{K-1,M_K} \leftarrow \Omega_o \mathbf{I}_{(N+1)\times(N+1)};$$

* Initialize the fuzzy rule corresponding to $\mathbf{C}_{M_K}$:

$$\boldsymbol{R}_{M_K}: \ IF\left(\boldsymbol{x} \sim \boldsymbol{p}_{K,M_K}\right) \ THEN\left(y_{K,M_K} = \bar{\boldsymbol{x}}^T \boldsymbol{a}_{K,M_K}\right);$$

3) **END IF**

viii. **ELSE:**

1) Find the nearest *data cloud* $\mathbf{C}_{n^*}$ to $\boldsymbol{x}_K$:

$$n^* = \underset{j=1,2,\ldots,M_{K-1}}{\arg\max}\left(\left\|\boldsymbol{x}_K - \boldsymbol{p}_{K-1,j}\right\|\right);$$

2) $M_K \leftarrow M_{K-1}$;

3) Update the meta-parameters of $\mathbf{C}_{n^*}$ as:

$$\mathbf{C}_{n^*} \leftarrow \mathbf{C}_{n^*} + \boldsymbol{x}_K;$$

$$S_{K,n^*} \leftarrow S_{K-1,n^*} + 1;$$

$$\boldsymbol{p}_{K,n^*} \leftarrow \frac{S_{K-1,n^*}}{S_{K,n^*}} \boldsymbol{p}_{K-1,n^*} + \frac{1}{S_{K,n^*}} \boldsymbol{x}_K;$$

$$\chi_{K,n^*} \leftarrow \frac{S_{K-1,n^*}}{S_{K,n^*}} \chi_{K-1,n^*} + \frac{1}{S_{K,n^*}} \left\|\boldsymbol{x}_K\right\|^2;$$

ix. **END IF**

x. Set $\boldsymbol{p}_{K,i} \leftarrow \boldsymbol{p}_{K-1,i}$; $\chi_{K,i} \leftarrow \chi_{K-1,i}$; $S_{K,i} \leftarrow S_{K-1,i}$ for each $\mathbf{C}_i \in \{\mathbf{C}\}$ that does not receive new member;

xi. Calculate the local *data density* of $\boldsymbol{x}_K$ within each *data cloud* $(j = 1,2,\ldots,M_K)$:

$$D_{K,j}\left(\boldsymbol{x}_K\right) = \frac{1}{1 + \frac{\left\|\boldsymbol{x}_K - \boldsymbol{p}_{K,j}\right\|^2}{\chi_{K,j} - \left\|\boldsymbol{p}_{K,j}\right\|^2}};$$

xii. Calculate the firing strength of each fuzzy rule ($j = 1,2,\ldots,M_K$):

$$\lambda_{K,j} = \frac{D_{K,j}\left(\boldsymbol{x}_K\right)}{\sum_{k=1}^{M_K} D_{K,k}\left(\boldsymbol{x}_K\right)};$$

xiii. Update the accumulated firing strength $\Lambda_{K,j}$ for each fuzzy rule $(j=1,2,...,M_K)$:

$$\Lambda_{K,j} \leftarrow \Lambda_{K-1,j} + \lambda_{K,j};$$

xiv. Calculate the utility $\eta_{K,i}$ for each fuzzy rule ($j=1,2,...,M_K$):

$$\eta_{K,j} \leftarrow \frac{1}{K-I_j}\Lambda_{K,j};$$

xv. **FOR** $j=1$ **TO** M_K:

1) **IF (Condition 8.3** is satisfied) **THEN:**

Condition 8.3: $IF\left(\eta_{K,i} < \eta_o\right) \quad THEN\left(\mathbf{C}_i \text{ and } \boldsymbol{R}_i \text{ are removed }\right);$

* Remove $\mathbf{C}_j$ and $\boldsymbol{R}_j$;

* $M_K \leftarrow M_K - 1;$

2) **END IF**

xvi. **END FOR**

xvii. **FOR** $j=1$ **TO** M_K

$$1)\ \boldsymbol{\Theta}_{K,j} \leftarrow \boldsymbol{\Theta}_{K-1,j} - \frac{\lambda_{K,j}\boldsymbol{\Theta}_{K-1,j}\bar{\boldsymbol{x}}_K^T\bar{\boldsymbol{x}}_K\boldsymbol{\Theta}_{K-1,j}}{1+\lambda_{K,j}\bar{\boldsymbol{x}}_K^T\boldsymbol{\Theta}_{K-1,j}\bar{\boldsymbol{x}}_K};$$

$$2)\ \boldsymbol{a}_{K,j} \leftarrow \boldsymbol{a}_{K-1,j} + \lambda_{K,j}\boldsymbol{\Theta}_{K,j}\bar{\boldsymbol{x}}_K^T\left(y_K - \bar{\boldsymbol{x}}_K^T\boldsymbol{a}_{K-1,j}\right);$$

xviii. **END FOR**

xix. **FOR** $j=1$ **TO** M_K

$$1)\ \boldsymbol{\alpha}_{K,j} \leftarrow \boldsymbol{\alpha}_{K,j} + \left|\boldsymbol{a}_{K,j}\right|;$$

$$2)\ \bar{\boldsymbol{\alpha}}_{K,j} \leftarrow \frac{\boldsymbol{\alpha}_{K,j}}{\sum_{\alpha_{K,j,i}\in\boldsymbol{\alpha}_{K,j}} \alpha_{K,j,i}};$$

3) **IF (Condition 8.6** is satisfied) **THEN**

Condition 8.6: $IF\left(\bar{\alpha}_{K,i,j} < \frac{\varphi_o}{M_K}\sum_{l=1}^{M_K}\bar{\alpha}_{K,l,j}\right) \quad THEN\left(\text{remove the } j^{th} \text{ set from } \boldsymbol{R}_i\right)$

* Remove $a_{K,j,i}$, $\alpha_{K,j,i}$ from $\boldsymbol{a}_{K,j}$, $\boldsymbol{\alpha}_{K,j}$, respectively;

* Remove the ith row and i^{th} column from $\boldsymbol{\Theta}_{K,j}$;

4) **END IF**

xx. **END FOR**

xxi. Update the antecedent and consequent parameters of the fuzzy rules $(i=1,2,...,M_K)$:

$$\boldsymbol{R}_i:\ IF\left(\boldsymbol{x} \sim \boldsymbol{p}_{K,i}\right) \quad THEN\left(y_{K,i} = \bar{\boldsymbol{x}}^T\boldsymbol{a}_{K,i}\right);$$

6. END WHILE

Algorithm Ends

Output: Fuzzy Rule Base.

B.5 Deep Rule-Based Classifier

In this section, the implementation of the Deep Rule-Based (DRB) classifier is presented, which include:

(1) DRB learning algorithm;
(2) DRB validation algorithm.

Since the image pre-processing and feature extraction techniques involved in the DRB systems design are standard ones and are problem-specific; their implementation will not be specified in this section.

B.5.1 Supervised Learning Process

Similarly to the *ALMMo-0* algorithm, the supervised learning process can be done in parallel for each fuzzy rule within the DRB classifier, we consider the *i*th $(i = 1, 2, \ldots, C)$rule, and only summarize the learning process of an individual fuzzy rule. Nonetheless, the same principle can be applied to all the other fuzzy rules within the rule base of the DRB classifier as well. The detailed algorithmic procedure can be found in Sect. 9.4.1.

The pseudo code of online self-learning process of the *i*th subsystem of the DRB system is presented in this subsection. By default, r_o is set to be $r_o = \sqrt{2(1 - \cos(30°))}$.

DRB Learning Algorithm (for the* i*th Sub-system)

Input: Streaming Images of the ith Class $\{\mathbf{I}_{i,1}, \mathbf{I}_{i,2}, \mathbf{I}_{i,3}, \ldots\}$

Algorithm Begins

1. Pre-process $\mathbf{I}_{i,1}$ and extract its feature vector $\boldsymbol{x}_{i,1}$;
2. Normalize $\boldsymbol{x}_{i,1}$ by its norm $\|\boldsymbol{x}_{i,1}\|$ as:

$$\boldsymbol{x}_{i,1} \leftarrow \boldsymbol{x}_{i,1} / \|\boldsymbol{x}_{i,1}\|;$$

3. Initialize the global meta-parameters with $\boldsymbol{x}_{i,1}$ as:

$$K_i \leftarrow 1;$$
$$M_i \leftarrow 1;$$
$$\boldsymbol{\mu}_i \leftarrow \boldsymbol{x}_{i,1};$$

4. Initialize the meta-parameters of the first *data cloud* $\mathbf{C}_{i,M_i}$ as:

$$\mathbf{P}_{i,M_i} \leftarrow \mathbf{I}_{i,1};$$

$$\mathbf{C}_{i,M_i} \leftarrow \{\mathbf{I}_{i,1}\};$$

$$\boldsymbol{p}_{i,M_i} \leftarrow \boldsymbol{x}_{i,1};$$

$$S_{i,M_i} \leftarrow 1;$$

$$r_{i,M_i} \leftarrow r_o;$$

5. Initialize the fuzzy rule:

$$\boldsymbol{R}_1: \quad IF(\mathbf{I} \sim \mathbf{P}_{i,1}) \quad THEN(Class\, i);$$

6. **WHILE** (new image available) **AND** (no request for interruption):

i. $K_i \leftarrow K_i + 1$;

ii. Pre-process $\mathbf{I}_{i,K_i}$ and extract its feature vector $\boldsymbol{x}_{i,K_i}$;

iii. Normalize $\boldsymbol{x}_{i,K_i}$ by its norm $\|\boldsymbol{x}_{i,K_i}\|$ as:

$$\boldsymbol{x}_{i,K_i} \leftarrow \boldsymbol{x}_{i,K_i} / \|\boldsymbol{x}_{i,K_i}\|;$$

iv. Update $\boldsymbol{\mu}_i$ as:

$$\boldsymbol{\mu}_i \leftarrow \frac{K_i - 1}{K_i}\boldsymbol{\mu}_i + \frac{1}{K_i}\boldsymbol{x}_{i,K_i};$$

v. Calculate *data densities* D at $\mathbf{I}_{i,K_i}$ and $\mathbf{P}_{i,j}$ ($j = 1,2,\ldots,M_i$):

$$D_{K_i}(\mathbf{Z}) = \frac{1}{1 + \frac{\|z - \boldsymbol{\mu}_i\|^2}{1 - \|\boldsymbol{\mu}_i\|^2}}; \quad \begin{matrix} \mathbf{Z} = \mathbf{I}_{i,K_i}, \mathbf{P}_{i,1}, \mathbf{P}_{i,2}, \ldots, \mathbf{P}_{i,M_i}; \\ z = \boldsymbol{x}_{i,K_i}, \boldsymbol{p}_{i,1}, \boldsymbol{p}_{i,2}, \ldots, \boldsymbol{p}_{i,M_i}; \end{matrix}$$

v. **IF (Condition 9.1** is satisfied) **THEN:**

$$IF\left(D_{K_i}(\boldsymbol{x}_{i,K_i}) > \max_{j=1,2,\ldots,M_i}\left(D_{K_i}(\boldsymbol{p}_{i,j})\right)\right)$$

Condition 9.1: $$OR\left(D_{K_i}(\boldsymbol{x}_{i,K_i}) < \min_{j=1,2,\ldots,M_i}\left(D_{K_i}(\boldsymbol{p}_{i,j})\right)\right)$$

$$THEN\left(\mathbf{I}_{i,K_i}\ is\ a\ new\ image\ prototype\right)$$

1) Add a new *data cloud* with $\mathbf{I}_{i,K_i}$ as the prototype:

$$M_i \leftarrow M_i + 1;$$

$$\mathbf{P}_{i,M_i} \leftarrow \mathbf{I}_{i,K_i};$$

$$\mathbf{C}_{i,M_i} \leftarrow \{\mathbf{I}_{i,K_i}\};$$

$$\boldsymbol{p}_{i,M_i} \leftarrow \boldsymbol{x}_{i,K_i};$$

$$S_{i,M_i} \leftarrow 1;$$

$$r_{i,M_i} \leftarrow r_o;$$

vi. **ELSE:**

1) Find the nearest *data cloud* $\mathbf{C}_{i,n*}$ to $\mathbf{I}_{i,K_i}$:

$$n* = \underset{j=1,2,\ldots,M_i}{\arg\min}\left(\left\|\boldsymbol{x}_{i,K_i} - \boldsymbol{p}_{i,j}\right\|\right);$$

2) **IF (Condition 9.2** is satisfied) **THEN:**

Condition 9.2: $IF\left(\left\|\boldsymbol{x}_{i,K_i} - \boldsymbol{p}_{i,n*}\right\| \le r_{i,n*}\right) \quad THEN\left(\mathbf{C}_{i,n*} \leftarrow \mathbf{C}_{i,n*} + \mathbf{I}_{i,K_i}\right)$

* Update the meta-parameters of $\mathbf{C}_{i,n*}$ as:

$$\mathbf{C}_{i,n*} \leftarrow \mathbf{C}_{i,n*} + \mathbf{I}_{i,K_i};$$

$$\boldsymbol{p}_{i,n*} \leftarrow \frac{S_{i,n*}}{S_{i,n*}+1}\boldsymbol{p}_{i,n*} + \frac{1}{S_{i,n*}+1}\boldsymbol{x}_{i,K_i};$$

$$S_{i,n*} \leftarrow S_{i,n*} + 1;$$

$$r_{i,n*} \leftarrow \sqrt{\frac{1}{2}\left(r_{i,n*}^2 + \left(1 - \left\|\boldsymbol{p}_{i,n*}\right\|^2\right)\right)};$$

3) **ELSE:**

* Add a new *data cloud* with $\mathbf{I}_{i,K_i}$ as the image prototype:

$$M_i \leftarrow M_i + 1;$$

$$\mathbf{P}_{i,M_i} \leftarrow \mathbf{I}_{i,K_i};$$

$$\mathbf{C}_{i,M_i} \leftarrow \left\{\mathbf{I}_{i,K_i}\right\};$$

$$\boldsymbol{p}_{i,M_i} \leftarrow \boldsymbol{x}_{i,K_i};$$

$$S_{i,M_i} \leftarrow 1;$$

$$r_{i,M_i} \leftarrow r_o;$$

4) **END IF**

vii. **END IF**

viii. Update the fuzzy rule:

$$\boldsymbol{R}_i: \quad IF\left(\mathbf{I} \sim \mathbf{P}_{i,1}\right) OR \ldots OR\left(\mathbf{I} \sim \mathbf{P}_{i,M_i}\right) \quad THEN\left(Class\, i\right);$$

7. **END WHILE**

Algorithm Ends

Outputs: The ith Fuzzy Rule $\boldsymbol{R}_i$.

B.5.2 Validation Process

The pseudo-code of the validation process of the DRB classifier is summarized as follows. The detailed algorithmic procedure can be found in Sect. 9.4.2.

DRB Validation Algorithm

Input: Image from Unknown Class, I **Algorithm Begins**
1. Pre-process $\mathbf{I}$ and extract its feature vector $\boldsymbol{x}$; 2. Normalize $\boldsymbol{x}$ by its norm $\lVert\boldsymbol{x}\rVert$ as: $\boldsymbol{x} \leftarrow \boldsymbol{x}/\lVert\boldsymbol{x}\rVert$; 3. Calculate the firing strength, $\lambda_i(\mathbf{I})$ of each fuzzy rule ($i = 1,2,\ldots,C$): $$\lambda_i(\mathbf{I}) = \max_{j=1,2,\ldots,M_i}\left(e^{-\lVert x-p_{i,j}\rVert^2}\right);$$ 4. Estimate the label of $\mathbf{I}$ as: $$\text{Label} = \arg\max_{i=1,2,\ldots,C}\left(\lambda_i(\mathbf{I})\right).$$
Algorithm Ends **Outputs: The Estimated Label of I.**

B.6 Semi-supervised Deep Rule-Based Classifier

In this section, the implementation of the Semi-Supervised Deep Rule-Based (SS_DRB) classifier is presented, which include:

(3) Offline SS_DRB learning algorithm;
(4) Offline SS_DRB active learning algorithm;
(5) Online SS_DRB (active) learning algorithms.

The detailed algorithmic procedure of SS_DRB learning algorithms can be found in Sect. 9.5.

B.6.1 Offline Semi-supervised Learning Process

The offline semi-supervised learning process of the SS_DRB system is summarized below.

Offline SS_DRB Learning Algorithm

Input: ***1)*** **Unlabeled Images,** $\{\mathbf{V}\}$
2) **Threshold** Ω_1 **;**
Algorithm Begins

1. Pre-process $\{\mathbf{V}\}$ and extract their feature vectors $\{\boldsymbol{v}\}$;
2. Normalize each $\boldsymbol{v}_j \in \{\boldsymbol{v}\}$ by its norm $\|\boldsymbol{v}_i\|$ as: $\boldsymbol{v}_j \leftarrow \boldsymbol{v}_j / \|\boldsymbol{v}_j\|$;
3. Calculate the vector of firing strengths, $\boldsymbol{\lambda}(\mathbf{V}_j) = [\lambda_1(\mathbf{V}_j), \lambda_2(\mathbf{V}_j), \ldots, \lambda_C(\mathbf{V}_j)]$ for each unlabeled training image $\mathbf{V}_j \in \{\mathbf{V}\}$:
$$\lambda_k(\mathbf{V}_j) = \max_{l=1,2,\ldots,M_k}\left(e^{-\|\boldsymbol{v}_j - \boldsymbol{p}_{k,l}\|^2}\right); \quad k = 1,2,\ldots,C$$
4. Identify $\{\mathbf{V}\}_0$ by **Condition 9.3;**
Condition 9.3: $IF\left(\lambda_{\max*}(\mathbf{V}_j) > \Omega_1 \cdot \lambda_{\max**}(\mathbf{V}_j)\right) \quad THEN\left(\mathbf{V}_j \in \{\mathbf{V}\}_0\right)$
5. Find the feature vectors $\{\boldsymbol{v}\}_0$ of $\{\mathbf{V}\}_0$;
6. $\{\mathbf{V}\} \leftarrow \{\mathbf{V}\} - \{\mathbf{V}\}_0$;
7. $\{\boldsymbol{v}\} \leftarrow \{\boldsymbol{v}\} - \{\boldsymbol{v}\}_0$;
8. **WHILE** $\{\mathbf{V}\}_0 \neq \varnothing$ **:**
   i. Rank $\{\mathbf{V}\}_0$ in a descending order based on the value of $\left(\lambda_{\max*}(\mathbf{V}) - \lambda_{\max**}(\mathbf{V})\right)$ and obtain $\{\mathbf{V}\}_1$;
   ii. **Execute** the **DRB learning algorithm** with $\{\mathbf{V}\}_1$ as the input to update the deep rule base;
   iii. Calculate the vector of firing strengths, $\boldsymbol{\lambda}(\mathbf{V}_j) = [\lambda_1(\mathbf{V}_j), \lambda_2(\mathbf{V}_j), \ldots, \lambda_C(\mathbf{V}_j)]$ for each unlabeled training image $\mathbf{V}_j \in \{\mathbf{V}\}$;
   iv. Identify $\{\mathbf{V}\}_0$ by **Condition 9.3**;
   v. $\{\mathbf{V}\} \leftarrow \{\mathbf{V}\} - \{\mathbf{V}\}_0$;
   vi. $\{\boldsymbol{v}\} \leftarrow \{\boldsymbol{v}\} - \{\boldsymbol{v}\}_0$;
9. **END WHILE**

Algorithm Ends
Outputs: Deep Rule Base;

B.6.2 Offline Active Semi-supervised Learning Process

The main procedure of offline active semi-supervised learning process of the DRB system is summarized by the following pseudo code.

Offline SS_DRB active learning algorithm

Input: ***1)*** **Unlabeled Images,** $\{\mathbf{V}\}$ **;**

2) **Threshold** Ω_1 **;**

3) **Threshold** Ω_2 **;**

Algorithm Begins

1. $\{\mathbf{V}\}_2 \leftarrow \{\mathbf{V}\}$;

i. **WHILE** $\{\mathbf{V}\}_2 \neq \varnothing$:

ii. **Execute** the **offline SS_DRB learning algorithm**;

iii. Calculate the vector of firing strengths, $\boldsymbol{\lambda}(\mathbf{V}_j) = [\lambda_1(\mathbf{V}_j), \lambda_2(\mathbf{V}_j), \ldots, \lambda_C(\mathbf{V}_j)]$ for $\mathbf{V}_j \in \{\mathbf{V}\}$:

$$\lambda_k(\mathbf{V}_j) = \max_{l=1,2,\ldots,M_k}\left(e^{-\|\boldsymbol{v}_j - \boldsymbol{p}_{k,l}\|^2}\right); \quad k = 1,2,\ldots,C;$$

iv. Identify $\{\mathbf{V}\}_2$ from $\{\mathbf{U}\}$ by **Condition 9.4;**

Condition 9.4: $IF\left(\lambda_{\max^*}(\mathbf{V}_j) \leq \Omega_2\right) \quad THEN\left(\mathbf{V}_j \in \{\mathbf{V}\}_2\right)$

v. Identify $\mathbf{V}_{\min} \in \{\mathbf{V}\}_2$ as:

$$\mathbf{V}_{\min} \leftarrow \mathbf{V}_j; \quad j = \underset{\mathbf{V}_k \in \{\mathbf{V}\}_2}{\arg\min}\left(\lambda_{\max^*}(\mathbf{V}_k)\right);$$

vi. $\{\mathbf{V}\} \leftarrow \{\mathbf{V}\} - \mathbf{V}_j$;

vii. $\{\boldsymbol{v}\} \leftarrow \{\boldsymbol{v}\} - \boldsymbol{v}_j$;

viii. Add a new class: $C \leftarrow C + 1$;

ix. Initialize the *data cloud* of the new class:

$$M_C \leftarrow 1;$$

$$\mathbf{P}_{C,M_C} \leftarrow \mathbf{V}_{\min};$$

$$\mathbf{C}_{C,M_C} \leftarrow \{\mathbf{V}_{\min}\};$$

$$\boldsymbol{p}_{C,M_C} \leftarrow \boldsymbol{v}_{\min};$$

$$S_{C,M_C} \leftarrow 1;$$

$$r_{C,M_C} \leftarrow r_o;$$

x. Add a new fuzzy rule corresponding to $\mathbf{C}_{C,M_C}$:

$$\boldsymbol{R}_C: \quad \begin{array}{c} IF\left(\mathbf{I} \sim \mathbf{P}_{C,1}\right) \\ THEN\left(New Class\, C\right); \end{array}$$

2. **END WHILE**

Algorithm Ends

Outputs: Deep Rule Base

B.6.3 Online (Active) Semi-supervised Learning Process

The online semi-supervised learning processes of the SS_DRB system are summarized below.

A. SS_DRB (active) learning algorithm (sample-by-sample)

Input: ***1)*** **Unlabeled Streaming Images,** $\{\mathbf{V}_{K+1}, \mathbf{V}_{K+2}, \mathbf{V}_{K+3}, \ldots\}$ **;**

2) **Threshold** Ω_1 **;**

3) **Threshold** Ω_2 **(Optional, for Active Learning Only);**

Algorithm Begins

1. **WHILE** (new image available) **AND (**no request for interruption):

   i. Pre-process $\mathbf{V}_{K+1}$ and extract its feature vector $\boldsymbol{u}_{K+1}$;

   ii. Normalize $\boldsymbol{v}_{K+1}$ by its norm $\|\boldsymbol{u}_{K+1}\|$ as:

   $$\boldsymbol{v}_{K+1} \leftarrow \boldsymbol{v}_{K+1} / \|\boldsymbol{v}_{K+1}\|;$$

   iii. Calculate the vector of firing strengths, $\boldsymbol{\lambda}(\mathbf{V}_{K+1}) = [\lambda_1(\mathbf{V}_{K+1}), \lambda_2(\mathbf{V}_{K+1}), \ldots, \lambda_C(\mathbf{V}_{K+1})]$ for $\mathbf{V}_{K+1}$:

   $$\lambda_k(\mathbf{V}_{K+1}) = \max_{j=1,2,\ldots,N_i}\left(e^{-\|\boldsymbol{v}_{K+1} - \boldsymbol{p}_{i,j}\|^2}\right); \quad k = 1,2,\ldots,C;$$

   iv. **IF (Condition 9.3 is satisfied) THEN :**

   Condition 9.3: $IF\left(\lambda_{\max*}(\mathbf{V}_{K+1}) > \Omega_1 \cdot \lambda_{\max**}(\mathbf{V}_{K+1})\right) \quad THEN\left(\mathbf{V}_{K+1} \in \{\mathbf{V}\}_0\right)$

   1) **Execute** the **offline SS_DRB learning algorithm** to learn from $\mathbf{V}_{K+1}$;

   v. **ELSE (Optional) ;**

   1) **Execute** the **offline SS_DRB active learning algorithm** to learn from $\mathbf{V}_{K+1}$;

   vi. **END IF**

   vii. $K \leftarrow K+1$;

2. **END WHILE**

Algorithm Ends

Outputs: Deep Rule Base

B. SS_DRB (active) learning algorithm (chunk-by-chunk)

Input: ***1)*** **Unlabeled Streaming Image Chunks** $\{\mathbf{V}\}_H^1, \{\mathbf{V}\}_H^2, \{\mathbf{V}\}_H^3, \ldots$ **;**

2) **Threshold** Ω_1 **;**

3) **Threshold** Ω_2 **(Optional, for Active Learning Only);**

Algorithm Begins

1. Set $j \leftarrow 1$;
2. **WHILE** (new image chunk available) **AND (**no request for interruption):
   i. **IF** (**Active learning** is performed) **THEN**
   1) **Execute** the **offline SS_DRB learning algorithm** to learn from $\{\mathbf{V}\}_H^j$;
   ii. **ELSE**
   1) **Execute** the **offline SS_DRB active learning algorithm** to learn from $\{\mathbf{V}\}_H^j$;
   iii. **END IF**
   iv. $j \leftarrow j+1$**;**
3. **END WHILE**

Algorithm Ends

Outputs: Deep Rule Base

Appendix C

In this appendix, the MATLAB implementations and examples of the *empirical* approaches presented in this book are given for illustration.

C.1 Autonomous Anomaly Detection

C.1.1 MATLAB Implementation

The autonomous anomaly detection (AAD) algorithm is implemented within MATLAB environment by the following function:

$$[Output] = AAD(Input)$$

The input (*Input*) of the function *AAD* is a static dataset.

Input	Description
Input.Data	The input data matrix; each row is a data sample

The outputs (*Output*) of the function *AAD* consist of two parts as follows.

Output	Description
Output.IDX	Indices of anomalous data samples; each row is the index of the data sample in the same row as in the *Output.Anomaly;*
Output. Anomaly	Anomalous data samples; each row is an anomalous data sample identified from the input dataset

P. P. Angelov and X. Gu, *Empirical Approach to Machine Learning*, Studies in Computational Intelligence 800,
https://doi.org/10.1007/978-3-030-02384-3

C.1.2 Simple Illustrative Example

In this subsection, an illustrative example based on the Iris dataset [1] (downloadable from [2]) is presented to demonstrate the concept. Iris dataset is one of the most classical ones in the field and is still frequently referenced. It contains three classes ("setosa", "versicolor" and "virginica") with 50 samples in each (150 samples in total). Each data sample has four attributes:

(i) sepal length;
(ii) sepal width;
(iii) petal length, and
(iv) petal width.

To identify the anomalous data samples from the dataset, one can type the following MATLAB commands into the Command Window:

```
load fisheriris
Input.data=data;
[Output]=AAD(Input);
```

After the execution of the above commands, one can type the following commands into the Command Window to visualize the result, and a similar figure to Fig. C.1 will appear.

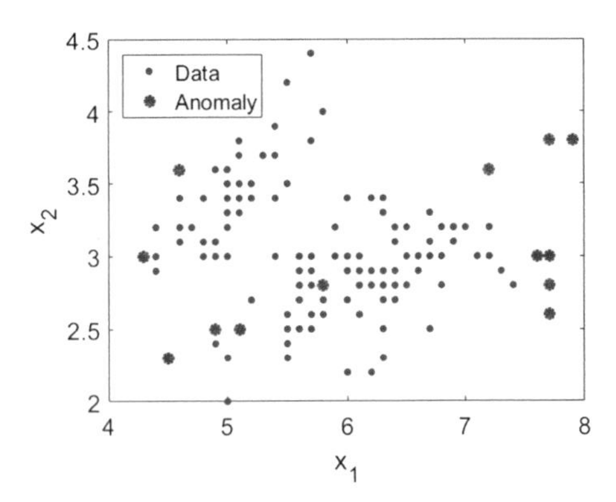

Fig. C.1 A simple demonstration of the AAD algorithm

```
plot(meas(:,1),meas(:,2),'b.','markersize',16); hold on;

plot(meas(Output.IDX,1),meas(Output.IDX,2),'r*','linewidth',2,'markersize',8);

xlabel('x_1');

ylabel('x_2');

legend('Data','Anomaly','location','northwest');

set(gca,'fontsize',16);
```

The identified anomalies are listed in *Output.Anomaly*, and the corresponding indices are given in *Output.IDX*.

As one can see from *Output* that the AAD algorithm recognizes 13 data samples as anomalies, which are tabulated in Table C.1.

Table C.1 The detected anomalies from Iris dataset

#	Detected anomalies
1	$\boldsymbol{x}_{14} = [4.30, 3.00, 1.10, 0.10]^T$
2	$\boldsymbol{x}_{23} = [4.60, 3.60, 1.00, 0.20]^T$
3	$\boldsymbol{x}_{42} = [4.50, 2.30, 1.30, 0.30]^T$
4	$\boldsymbol{x}_{99} = [5.10, 2.50, 3.00, 1.10]^T$
5	$\boldsymbol{x}_{106} = [7.60, 3.00, 6.60, 2.10]^T$
6	$\boldsymbol{x}_{107} = [4.90, 2.50, 4.50, 1.70]^T$
7	$\boldsymbol{x}_{110} = [7.20, 3.60, 6.10, 2.50]^T$
8	$\boldsymbol{x}_{115} = [5.80, 2.80, 5.10, 2.40]^T$
9	$\boldsymbol{x}_{118} = [7.70, 3.80, 6.70, 2.20]^T$
10	$\boldsymbol{x}_{119} = [7.70, 2.60, 6.90, 2.30]^T$
11	$\boldsymbol{x}_{123} = [7.70, 2.80, 6.70, 2.00]^T$
12	$\boldsymbol{x}_{132} = [7.90, 3.80, 6.40, 2.00]^T$
13	$\boldsymbol{x}_{136} = [7.70, 3.00, 6.10, 2.30]^T$

C.2 Autonomous Data Partitioning

C.2.1 MATLAB Implementation

The autonomous data partitioning (ADP) algorithm is implemented within MATLAB environment by the following function:

$$[Output] = ADP(Input, Mode)$$

As the ADP algorithm has two versions, namely, evolving and offline. The function *ADP* supports three different operating modes (*Mode*):

Mode	Description
'*Offline*'	Executes the offline ADP algorithm with static data
'*Evolving*'	Executes the evolving ADP algorithm with the streaming data "from scratch". In this mode, the algorithm learns from the data on a sample-by-sample basis
'*Updating*'	Updates the trained model using the evolving ADP algorithm with the newly observed data on a sample-by-sample basis, and later performs data partitioning in regards to all the data

The inputs (*Input*) of the function *ADP* can be different under different operating mode.

Mode	Input	Description
'*Offline*' '*Evolving*' '*Updating*'	*Input.Data*	The observed data matrix; each row is a data sample
'*Updating*'	*Input.HistData*	The data matrix consisting of all the previously processed data; each row is a data sample; this is for producing the ultimate data partitioning
'*Updating*'	*Input.SysParms*	The trained model See also *Output.SysParms*

The outputs (*Output*) of the function *ADP* consists of three parts as follows.

Output	Description
Output.C	Centers/prototypes of the identified *data clouds*; each row is a center/prototype
Output.IDX	*Data cloud* indices of data the samples; each row is the index of the data sample in the same row in the input data matrix
Output.SysParms	Structure and meta-parameters of the model identified through the learning process, with which the model can continue self-evolving and self-updating without a full retraining when new data samples are observed

C.2.2 Simple Illustrative Examples

In this subsection, an illustrative example based on the banknote authentication dataset [3] (downloadable from [4]) is presented to demonstrate the concept. The details of the banknote authentication dataset have been given in Sect. 11.2.1.1.

C.2.2.1 Offline ADP Algorithm

To conduct data partitioning of the dataset with ADP algorithm offline, one can type the following MATLAB commands into the Command Window:

```
load BanknoteAuthentication_data

Input.Data=data;

[Output]=ADP(Input, 'Offline');
```

After the execution of the function *ADP*, one can visualize the data partitioning result with the following commands, and a similar figure to Fig. C.2 will appear, where dots "." in different color stand for data samples in different *data clouds*; black asterisks "*" present for the identified prototypes.

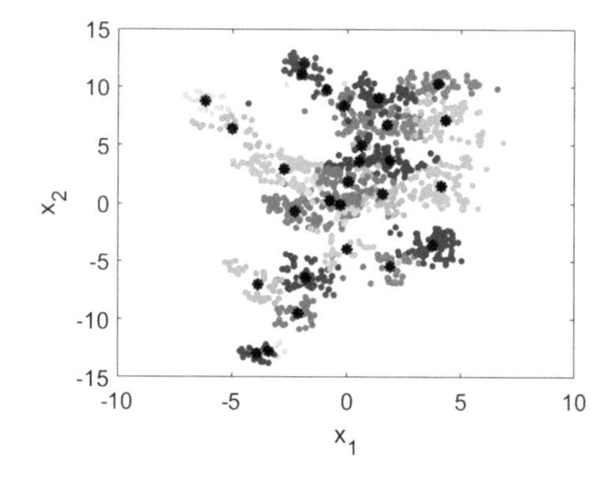

Fig. C.2 A simple demonstration of the offline ADP algorithm

```
for ii=1:1:size(Output.C,1)

   plot(data(Output.IDX==ii,1),data(Output.IDX==ii,2),...
      '.','markersize',16); hold on;

end

plot(Output.C(:,1),Output.C(:,2),'k*','linewidth',2,'markersize',8);

xlabel('x_1'); ylabel('x_2');

set(gca,'fontsize',16);
```

The prototypes are given in *Output.C*, and the corresponding *data cloud* indexes of the data are given in *Output.IDX*. In this example, the offline ADP algorithm identified 28 prototypes and, thus, 28 *data clouds* are formed around them.

C.2.2.2 Evolving ADP Algorithm

To conduct data partitioning on the dataset with ADP algorithm on a sample-by-sample basis, one can simply change the second line of the MATLAB commands from *Mode='Offline'* to *Mode='Evolving'*,

```
load BanknoteAuthentication_data

Input.Data=data;

[Output]=ADP(Input,'Evolving');
```

By using the same MATLAB commands provided in Sect. C.2.2.1 for visualizing after the successful execution of the evolving ADP algorithm, a similar figure to Fig. C.3 will appear. In this example, the evolving ADP algorithm identified 27 prototypes and 27 *data clouds* are formed.

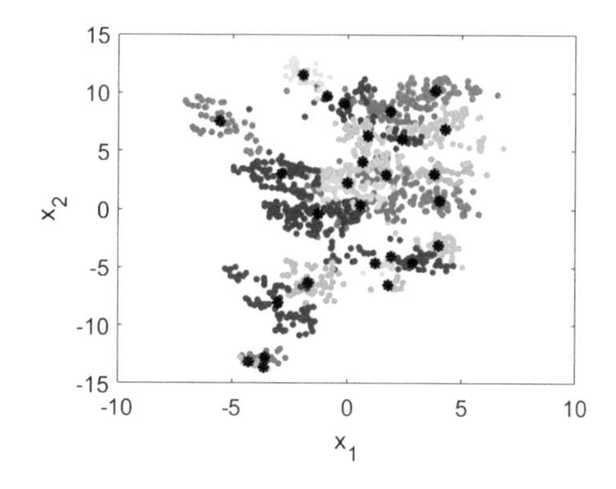

Fig. C.3 A simple demonstration of the evolving ADP algorithm

C.3 Zero-Order Autonomous Learning Multi-model System

C.3.1 MATLAB Implementation

The zero-order autonomous learning multi-model (*ALMMo-0*) system is implemented within MATLAB environment by the following function:

$$[Output] = ALMMo0(Input, Mode)$$

The function *ALMMo0* supports three different operating modes (*Mode*):

Mode	Description
'*Learning*'	Trains the *ALMMo-0* system with training data on a sample-by-sample basis
'*Updating*'	Updates the previously trained *ALMMo-0* system with new training data on a sample-by-sample basis
'*Validation*'	Validates the trained *ALMMo-0* system on the testing data

The respective inputs (*Input*) of the function *ALMMo0* in the three operating modes are as follows:

Mode	Input	Description
'*Learning*'	*Input. Data*	The data matrix of the training data; each row is a data sample
	Input. Labels	The labels of the data; each row is the label of the data sample in the corresponding row in *Input.Data*
'*Updating*'	*Input. Data*	The matrix of new training data; each row is a data sample
	Input. Labels	The labels of the new training data; each row is the label of the data sample in the corresponding row in *Input.Data*

(continued)

(continued)

Mode	Input	Description
	Input. SysParms	The trained *ALMMo-0* system Also see *Output.SysParms*
'*Validation*'	*Input. Data*	The data matrix of the testing data; each row is a data sample
	Input. Labels	The labels of the testing data; each row is the label of the data sample in the corresponding row in *Input.Data*. This input will be used for calculating the confusion matrix and classification accuracy only
	Input. SysParms	The trained *ALMMo-0* system See also *Output.SysParms*

The respective outputs (*Output*) of the function *ALMMo0* in the three operating modes are as follows:

Mode	Output	Description
'*Learning*' and '*Updating*'	*Output. SysParms*	Structure and meta-parameters of the model identified through the learning process, with which the model can continue self-evolving and self-updating without a full retraining when new data samples are observed
'*Validation*'	*Output. EstLabs*	The estimated labels of the testing data; each row is an estimated label of the corresponding row in *Input.Data*
	Output. ConfMat	The confusion matrix calculated by comparing the estimation and the ground truth
	Output. ClasAcc	The classification accuracy

The *ALMMo-0* classifier is highly parallelizable and its training process can be speeded up using multiple workers, but in the function *ALMMo0* provided in this book is implemented for a single worker without losing generality. Nonetheless, users can easily modify the code if they prefer.

C.3.2 Simple Illustrative Example

In this subsection, an illustrative example based on the seeds dataset [5] (downloadable from [6]) is presented to demonstrate the concept. Seeds dataset is one of the most popular benchmark datasets for classification, which contains three classes (1, 2 and 3) with 70 samples per class. Each data sample has seven attributes:

(i) area;
(ii) perimeter;
(iii) compactness;

(iv) length of the kernel;
(v) width of the kernel;
(vi) asymmetry coefficient, and
(vii) length of the kernel groove.

First of all, the following commands are used to randomly divide the three subsets with 70 samples in each (two for training and one for validation):

```
load seeds_data
load seeds_labels
seq=randperm(size(data,1));
data=data(seq,:);
label=label(seq);
TrainingData1=data(1:size(data,1)/3,:);
TrainingLabels1=label(1:size(data,1)/3);
TrainingData2=data(1+size(data,1)/3:2*size(data,1)/3,:);
TrainingLabels2=label(1+size(data,1)/3:2*size(data,1)/3);
ValidationData=data(1+2*size(data,1)/3:end,:);
ValidationLabels=label(1+2*size(data,1)/3:end);
```

Firstly, the *ALMMo-0* classifier is trained with the first training set, namely, *TrainingData1* and *TrainingLabels1*. The training process can be completed using the following MATLAB commands:

```
Input0.Data=TrainingData1;
Input0.Labels=TrainingLabels1;
[Output0]=ALMMo0(Input0,'Learning');
```

The trained *ALMMo-0* classifier can be tested with the validation data, namely *ValidationData*, using the following commands,

```
Input1=Output0;
Input1.Data=ValidationData;
Input1.Labels=ValidationLabels;
[Output1]=ALMMo0(Input1,'Validation');
Output1.ConfMat
Output1.ClasAcc
```

and a confusion matrix regarding the classification result should appear in the Command Window afterwards.

```
ans =

   22    2    3
    0   21    0
    1    0   21

ans =

   0.9143
```

Then, the trained *ALMMo-0* classifier is updated with the second training set, namely, *TrainingData2* and *TrainingLabels2*. The training process with the new training set can be completed using the following MATLAB commands:

```
Input2=Output0;
Input2.Data=TrainingData2;
Input2.Labels=TrainingLabels2;
[Output2]=ALMMo0(Input2,'Updating');
```

We can test the updated *ALMMo-0* classifier with the validation data again using the same commands provided above. Then, a new confusion matrix calculated based on the new classification result will appear.

```
Input3=Output2;

Input3.Data=ValidationData;

Input3.Labels=ValidationLabels;

[Output3]=ALMMo0(Input3,'Validation');

Output3.ConfMat

Output3.ClasAcc

ans =

   27   0   0
    0  21   0
    1   0  21

ans =

   0.9857
```

C.4 First-Order Autonomous Learning Multi-model System

C.4.1 MATLAB Implementation

The first-order autonomous learning multi-model (*ALMMo-1*) system is implemented within MATLAB environment by the following function:

$$[Output] = ALMMo1(Input, Mode, Functionality)$$

The function *ALMMo1* supports three different operating modes (*Mode*) and two different functionalities (*Functionality*):

Mode	Description
'*Learning*'	Trains the *ALMMo-1* system with training data on a sample-by-sample basis
'*Updating*'	Updates the previously trained *ALMMo-1* system with new training data on a sample-by-sample basis
'*Validation*'	Validates the trained *ALMMo-1* system on the testing data

Functionality	Description
'*Classification*'	Trains the *ALMMo-1* system for classification purpose
'*Regression*'	Trains the *ALMMo-1* system for regression purpose

C.4.1.1 Classification

For classification purpose (*'Classification'*), the respective input (*Input*) of the function *ALMMo1* in the three operating modes are as follows:

Mode	Input	Description
'*Learning*'	*Input.X*	The data matrix of the training data; each row is a data sample
	Input.Y	The labels of the data; each row is the label of the data sample in the corresponding row in *Input.X*
'*Updating*'	*Input.X*	The matrix of new data; each row is a data sample
	Input.Y	The labels of the new data; each row is the label of the data sample in the corresponding row in *Input.X*
	Input. SysParm	The trained *ALMMo-1* system See also *Output.SysParms*
'*Validation*'	*Input.X*	The data matrix of the testing data; each row is a data sample
	Input.Y	The labels of the testing data; each row is the label of the data sample in the corresponding row in *Input.X*. This input will be used for calculating the confusion matrix and classification accuracy only
	Input. SysParm	The trained *ALMMo-1* system See also *Output.SysParms*

The respective output (*Output*) of the function *ALMMo1* in the three operating modes are as follows.

Mode	Output	Description
'*Learning*' and '*Updating*'	*Output. SysParms*	Structure and meta-parameters of the model identified through the learning process, with which the model can continue self-evolving and self-updating without a full retraining when new data samples are observed

(continued)

(continued)

Mode	Output	Description
	Output.EstLabs	The estimated labels of the input data; each row is the estimated label of the corresponding row in *Input.X*
	Output.ConfMat	The confusion matrix calculated by comparing the estimation and the ground truth
	Output.ClasAcc	The classification accuracy
'Validation'	*Output.EstLabs*	The estimated labels of the testing data; each row is the estimated label of the corresponding row in *Input.X*
	Output.ConfMat	The confusion matrix calculated by comparing the estimation and the ground truth
	Output.ClasAcc	The classification accuracy

C.4.1.2 Regression

For classification purpose (*'Regression'*), the respective input (*Input*) of the function *ALMMo1* in the three operating modes are as follows.

Mode	Input	Description
'Learning'	*Input.X*	The data matrix of the training data; each row is a data sample
	Input.Y	The desired system output corresponding to the input data; each row is the desired output of the system to the data sample in the corresponding row in *Input.X*
'Updating'	*Input.X*	The matrix of new data; each row is a data sample
	Input.Y	The desired system outputs corresponding to the input data; each row is the desired output of the system to the data sample in the corresponding row in *Input.X*
	Input.SysParms	The trained *ALMMo-1* system See also *Output.SysParms*
'Validation'	*Input.X*	The data matrix of the testing data; each row is a data sample
	Input.Y	The desired system outputs corresponding to the input data; each row is the desired output of the system to the data sample in the corresponding row in *Input.X*. This input will be used for calculating the root mean square error (RMSE) and non-dimensional error index (NDEI) only
	Input.SysParms	The trained *ALMMo-1* system See also *Output.SysParms*

The respective output (*Output*) of the function *ALMMo1* in the three operating modes are as follows:

Mode	Output	Description
'*Learning*' and '*Updating*'	*Output. SysParms*	Structure and meta-parameters of the model identified through the learning process, with which the model can continue self-evolving and self-updating without a full retraining when new data samples are observed
	Output. PredSer	The system outputs based on the input data; each row is the system output of the corresponding row in *Input.X*
	Output. RMSE	The root mean square error (RMSE) of the prediction
	Output. NDEI	The non-dimensional error index (NDEI) of the prediction
'*Validation*'	*Output. PredSer*	The system outputs based on the testing data; the testing data; each row is the estimated label of the corresponding row in *Input.X*
	Output. RMSE	The root mean square error (RMSE) of the prediction
	Output. NDEI	The non-dimensional error index (NDEI) of the prediction

C.4.2 Simple Illustrative Examples

C.4.2.1 Classification

In this subsection, a numerical example based on the banknote authentication dataset [3] (downloadable from [4]) is presented for illustration. Details of the banknote authentication dataset have been given in Sect. 11.2.2.1.

```
load BanknoteAuthentication_data

load BanknoteAuthentication_label

seq0=randperm(size(data,1));

label=label(seq0);

data=data(seq0,1:1:4);

TrainingData=data(1:size(data,1)*0.5,:);

ValidationData=data(1+size(data,1)*0.5:end,:);

TrainingLabels=label(1:size(data,1)*0.5);

ValidationLabels=label(1+size(data,1)*0.5:end);
```

Firstly, we divide the banknote authentication dataset into two subsets randomly using the following commands. Each of the two subsets contains 50% of the data.

To train the *ALMMo-1* system based on the training samples, one can type the following MATLAB commands into the Command Window:

```
Input0.X=TrainingData;

Input0.Y=TrainingLabels;

[Output0]=ALMMo1(Input0,'Learning','Classification')
```

and a similar output will appear in the Command Window as follows:

```
Output0 =

  struct with fields:

    SysParm: [1×1 struct]
    EstLabs: [686×1 double]
    ConfMat: [2×2 double]
    ClasAcc: 0.9665
```

After the *ALMMo-1* system is trained, one can test it on the validation data using the following commands:

```
Input1.SysParm=Output0.SysParm;

Input1.X=ValidationData;

Input1.Y=ValidationLabels;

[Output1]=ALMMo1(Input1,'Validation','Classification')
```

and a similar output as the one given below will appear in the Command Window.

```
Output1 =

  struct with fields:

    EstLabs: [686×1 double]
    ConfMat: [2×2 double]
    ClasAcc: 0.9854
```

The *ALMMo-1* system can also be updated using new data samples after the training without a full retraining. One can easily achieve this by using the *'Updating'* mode of the function *ALMMo1*. By using the following commands, the *ALMMo-1* system is further trained with the validation data.

```
Input2.SysParm=Output0.SysParm;

Input2.X=ValidationData;

Input2.Y=ValidationLabels;

[Output2]=ALMMo1(Input2,'Updating','Classification')

Output2 =

  struct with fields:

    SysParm: [1×1 struct]
    EstLabs: [686×1 double]
    ConfMat: [2×2 double]
    ClasAcc: 0.9825
```

C.4.2.2 Regression

In this subsection, a numerical example based on the well-known regression benchmark dataset triazines [7] (downloadable from [8]) is given to demonstrate the performance of *ALMMo-1* system for regression. The goal of the triazines dataset is to predict the inhibition of dihydrofolate reductase by triazines. This dataset is composed of 186 data samples in total. There are 100 data samples in the training set and 86 in the validation set. Each data sample has 60 attributes as the input vector and one attribute as the desired output.

First of all, we use the following MATLAB commands to read the data and generate the training and validation sets.

```
data=load('triazines_train');
TraY=data(:,61);
TraX=data(:,1:60);
data=load('triazines_test');
TesY=data(:,61);
TesX=data(:,1:60);
```

Then, the *ALMMo-1* system can be trained using the training set by the following commands:

```
Input0.X=TraX;
Input0.Y=TraY;
[Output0]=ALMMo1(Input0,'Learning','Regression')
```

and the output will appear in the Command Window:

```
Output0 =
struct with fields:
  SysParm: [1×1 struct]
  PredSer: [100×1 double]
     RMSE: 0.0891
     NDEI: 0.3378
```

The trained *ALMMo-1* system can be validated with the validation data using the following commands:

```
Input1.SysParm=Output0.SysParm;

Input1.X=TesX;

Input1.Y=TesY;

[Output1]=ALMMo1(Input1,'Validation','Regression')
```

and the system output will appear in the Command Window:

```
Output1 =

  struct with fields:

    PredSer: [86×1 double]
       RMSE: 0.0078
       NDEI: 0.0313
```

One can also plot the predicted time sequence (system output) and compare it with the real time sequence (expected system output) by the following commands:

```
figure

plot(1:length(TesY),TesY,'r-', 1:length(TesY),Output1.PredSer,'b-','linewidth',2)

axis([1 length(TesY) 0 1])

xlabel('K')

ylabel('y')

legend('Real Value','Predicted Value')

set(gca,'fontsize',16)
```

Then, a figure similar to Fig. C.4 will appear.

Similarly, one can further update the *ALMMo-1* system with the validation data:

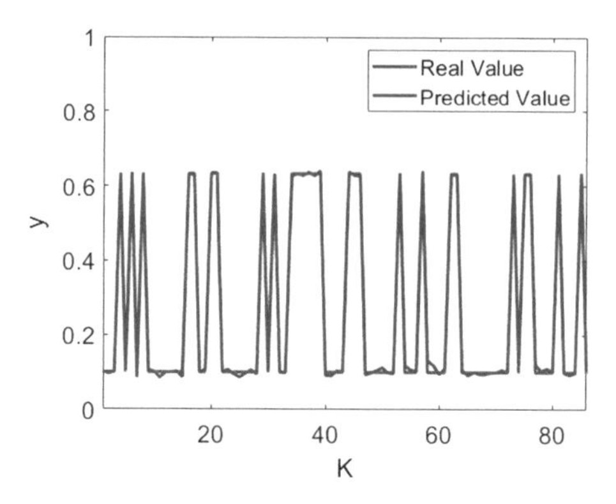

Fig. C.4 Comparison between the predicted values and the real values

```
Input2.SysParm=Output0.SysParm;
Input2.X=TesX;
Input2.Y=TesY;
[Output2]=ALMMo1(Input2,'Updating','Regression')
```

C.5 Deep Rule-Based Classifier

C.5.1 MATLAB Implementation

The Deep Rule-Based (DRB) classifier is implemented within MATLAB environment by the following function:

$$[Output] = DRB(Input, Mode)$$

The function *DRB* supports three different operating modes (*Mode*):

Mode	Description
'*Learning*'	Trains the DRB classifier with training images on a sample-by-sample basis
'*Updating*'	Updates the previously trained DRB classifier with new training images on a sample-by-sample basis
'*Validation*'	Validates the trained DRB classifier on the testing images

The respective inputs (*Input*) of the function *DRB* in the three operating modes are as follows.

Mode	Input	Description
'*Learning*'	*Input. Images*	The training images; each cell is an image
	Input. Features	The feature vectors of the training images; each row is the feature vector of the training image stored in the corresponding cell in *Input.Images*
	Input. Labels	The labels of the training images; each row is the label of the training image stored in the corresponding cell in *Input.Images*
'*Updating*'	*Input. Images*	The training images; each cell is an image
	Input. Features	The feature vectors of the training images; each row is the feature vector of the training image stored in the corresponding cell in *Input.Images*
	Input. Labels	The labels of the training images; each row is the label of the training image stored in the corresponding cell in *Input.Images*
	Input. DRBParms	The trained DRB classifier See also *Output.DRBParms*
'*Validation*'	*Input. Images*	The testing images; each cell is an image
	Input. Features	The feature vectors of the testing images; each row is the feature vector of the testing image stored in the corresponding cell in *Input.Images*
	Input. Labels	The real labels of the testing images; each row is the label of the unlabeled image stored in the corresponding cell in *Input. Images*. This input will be used for calculating the confusion matrix and classification accuracy only
	Input. DRBParms	The trained DRB classifier Also see *Output.DRBParms*

The respective outputs (*Output*) of the function *DRB* in the three operating modes are as follows.

Mode	Output	Description
'*Learning*' and '*Updating*'	*Output. DRBParms*	Structure and meta-parameters of the DRB classifier identified through the learning process, with which the model can continue to self-evolve and self-update without a full retraining when new images are given
'*Validation*'	*Output. EstLabs*	The estimated labels of the testing image; each row is the estimated label of the corresponding cell in *Input.Data*
	Output. ConfMat	The confusion matrix calculated by comparing the estimation and the ground truth

(continued)

(continued)

Mode	Output	Description
	Output. ClasAcc	The classification accuracy
	Output. Scores	The matrix of scores of confidence of the testing images; Each row is the scores of confidence of the testing images in the corresponding cell given by the massively parallel fuzzy rules

Similarly to the *ALMMo-0* classifier, the DRB classifier is highly parallelizable and its training process can be speeded up using multiple workers, but in the function DRB provided by this book is implemented on a single worker without losing generality. Nonetheless, users can easily modify the code if they prefer.

C.5.2 Simple Illustrative Example

In this section, an illustrative example based on a subset of MNIST (Modified National Institute of Standards and Technique) dataset [9] (downloadable from [10]) is presented to demonstrate the concept. MNIST dataset is the most well-known and widely studied image set for handwritten digits recognition. The details of the MNIST dataset can be found in Sect. 13.2.1.1.

For illustration, we randomly picked out 10 images from each class, 100 images in total, as the labelled training set to train the DRB classifier. All the images are given in Fig. C.5. The 512×1 dimensional GIST feature vectors are extracted from the images (one image per feature vector) after normalization, which transforms the pixel value range from $[0, 255]$ to $[0, 1]$.

Fig. C.5 Training images

Then, we can use the following MATLAB commands to train the DRB classifier. One may notice that the training images, their feature vectors and labels are stored in variables *TrainingImages, TrainingFeatures* and *TrainingLabels* respectively, where *TrainingImages* is a 1×100 cell, each cell is an image with size 28×28 pixels; *TrainingFeatures* is a 100×512 matrix, each row of which is a GIST feature of the image in the corresponding cell in *TrainingImages*; *TrainingLabels* is a 100×1 vector, each row is the label of the image in the corresponding cell of *TrainingImages*.

```
Input0.Images=TrainingImages;

Input0.Features=TrainingFeatures;

Input0.Labels=TrainingLabels;

Mode='Learning';

[Output0]=DRB(Input0,Mode);
```

The identified prototypes after the training process are tabulated in Table C.2. The massively parallel fuzzy rules built upon these prototypes are given in Fig. C.6 for illustration.

We also selected out randomly 20 images per class, 200 images in total, as the validation set. The selected images are given in Fig. C.7. By applying the same pre-processing and feature extraction techniques to the testing images, we obtain the GIST feature vectors of the images and use the following MATLAB commands to conduct the validation process. One may notice that the unlabeled images, their feature vectors and labels are stored in variables *UnlabelledImages*, *UnlabelledFeatures* and *GroundTruth*, respectively.

Table C.2 The identified prototypes

Digit	Identified prototypes
1	1 1 1 1 1
2	2 2 2 2 2 2 2
3	3 3 3 3 3 3 3 3
4	4 4 4 4 4 4 4
5	5 5 5 5 5 5 5 5
6	6 6 6 6 6 6
7	7 7 7 7 7 7
8	8 8 8 8 8 8 8 8
9	9 9 9 9 9
0	0 0 0 0 0 0

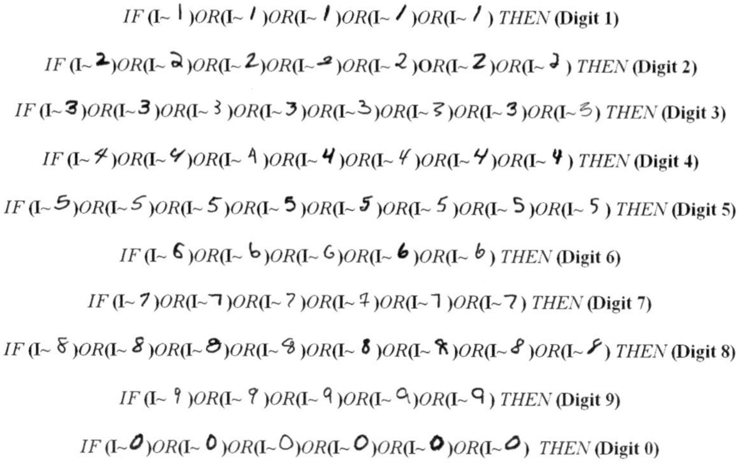

IF (I~ 1)*OR*(I~ 1)*OR*(I~ 1)*OR*(I~ 1)*OR*(I~ 1) *THEN* **(Digit 1)**

IF (I~ 2)*OR*(I~ 2)*OR*(I~ 2)*OR*(I~ 2)*OR*(I~ 2)*OR*(I~ 2)*OR*(I~ 2) *THEN* **(Digit 2)**

IF (I~ 3)*OR*(I~ 3)*OR*(I~ 3)*OR*(I~ 3)*OR*(I~ 3)*OR*(I~ 3)*OR*(I~ 3)*OR*(I~ 3)*OR*(I~ 3) *THEN* **(Digit 3)**

IF (I~ 4)*OR*(I~ 4)*OR*(I~ 4)*OR*(I~ 4)*OR*(I~ 4)*OR*(I~ 4)*OR*(I~ 4) *THEN* **(Digit 4)**

IF (I~ 5)*OR*(I~ 5)*OR*(I~ 5)*OR*(I~ 5)*OR*(I~ 5)*OR*(I~ 5)*OR*(I~ 5)*OR*(I~ 5) *THEN* **(Digit 5)**

IF (I~ 6)*OR*(I~ 6)*OR*(I~ 6)*OR*(I~ 6)*OR*(I~ 6) *THEN* **(Digit 6)**

IF (I~ 7)*OR*(I~ 7)*OR*(I~ 7)*OR*(I~ 7)*OR*(I~ 7)*OR*(I~ 7) *THEN* **(Digit 7)**

IF (I~ 8)*OR*(I~ 8)*OR*(I~ 8)*OR*(I~ 8)*OR*(I~ 8)*OR*(I~ 8)*OR*(I~ 8)*OR*(I~ 8) *THEN* **(Digit 8)**

IF (I~ 9)*OR*(I~ 9)*OR*(I~ 9)*OR*(I~ 9)*OR*(I~ 9) *THEN* **(Digit 9)**

IF (I~ 0)*OR*(I~ 0)*OR*(I~ 0)*OR*(I~ 0)*OR*(I~ 0)*OR*(I~ 0) *THEN* **(Digit 0)**

Fig. C.6 The identified massively parallel fuzzy rules

Fig. C.7 The randomly selected images

1 1 1 1 1 1 1 1 1 1
1 1 1 1 1 1 1 1 1 1
2 2 2 2 2 2 2 2 2 2
2 2 2 2 2 2 2 2 2 2
3 3 3 3 3 3 3 3 3 3
3 3 3 3 3 3 3 3 3 3
4 4 4 4 4 4 4 4 4 4
4 4 4 4 4 4 4 4 4 4
5 5 5 5 5 5 5 5 5 5
5 5 5 5 5 5 5 5 5 5
6 6 6 6 6 6 6 6 6 6
6 6 6 6 6 6 6 6 6 6
7 7 7 7 7 7 7 7 7 7
7 7 7 7 7 7 7 7 7 7
8 8 8 8 8 8 8 8 8 8
8 8 8 8 8 8 8 8 8 8
9 9 9 9 9 9 9 9 9 9
9 9 9 9 9 9 9 9 9 9
0 0 0 0 0 0 0 0 0 0
0 0 0 0 0 0 0 0 0 0

```
Input1.DRBParms=Output0.DRBParms;

Input1.Images=UnlabelledImages;

Input1.Features=UnlabelledFeatures;

Input1.Labels=GroundTruth;

Mode='Validation';

[Output1]=DRB(Input1,Mode);
```

After the execution of the commands above, one can check the accuracy and confusion matrix of the classification result on the unlabeled images by typing the following two lines in the Command Window:

```
Output1.ClasAcc

Output1.ConfMat
```

and the results will appear as follows:

```
ans =

  0.8550

ans =

  19   0   0   0   0   0   1   0   0   0
   0  15   1   0   0   0   0   1   1   2
   0   1  16   0   2   0   0   1   0   0
   0   0   0  16   0   0   1   0   3   0
   0   0   0   0  18   0   0   2   0   0
   0   0   0   0   0  20   0   0   0   0
   1   0   0   0   0   0  17   0   2   0
   0   0   1   0   1   0   0  18   0   0
   0   0   1   1   0   0   3   1  13   1
   0   0   1   0   0   0   0   0   0  19
```

One can further update the trained DRB classifier with the testing images if the labels are provided by using the following MATLAB commands.

```
Input2.DRBParms=Output0.DRBParms;

Input2.Images=UnlabelledImages;

Input2.Features=UnlabelledFeatures;

Input2.Labels=GroundTruth;

Mode='Updating';

[Output2]=DRB(Input2,Mode);
```

C.6 Semi-supervised Deep Rule-Based Classifier

C.6.1 MATLAB Implementation

The Semi-Supervised Deep Rule-Based (SS_DRB) classifier as described in Sects. 10.5.1 and 10.5.2 is implemented within MATLAB environment by the following function:

$$[Output] = SSDRB(Input, Mode)$$

The SS_DRB classifier with the active learning capability as described in Sect. 10.5.3 is implemented by the following function:

$$[Output] = ASSDRB(Input, Mode)$$

Both functions, *SSDRB* and *ASSDRB*, support two different operating modes (*Mode*):

Mode	Description
'*Offline*'	Train the (active) SS_DRB classifier with the static image set
'*Online*'	Train the (active) SS_DRB classifier with the streaming images. In this mode, the classifier learns from the images on a sample-by-sample basis

The required inputs (*Input*) of the functions *SSDRB* and *ASSDRB* in the two operating modes are as follows.

Input	Description
Input.Images	The unlabeled training images; each cell is an image
Input. Features	The feature vectors of the unlabeled training images; each row is the feature vector of the image stored in the corresponding cell in *Input.Images*
Input.Labels	The real labels of the unlabeled images; each row is the label of the training image stored in the corresponding cell in *Input.Images.* This input will be used for calculating the confusion matrix and classification accuracy only in the later steps. If the ground truth is missing, one can replace it with all zeros, however, one needs to spend more efforts in manually identifying the real labels later
Input. DRBParms	The trained *DRB* classifier, which can be the output of functions *DRB*, *SSDRB* or *ASSDRB*. See also *Output.DRBParms*
Input. Omega1	The threshold $\Omega_1$in **Condition 9.3**
Input. Omega2	The threshold Ω_2 in **Condition 9.4**

It has to be stressed that SS_DRB classifier is built upon the DRB classifier, and both functions *SSDRB* and *ASSDRB*, require the system parameters of a trained DRB classifier as the system input. After a DRB classifier has been primed with the labelled training images, one can use the functions *DRB, SSDRB* and *ASSDRB* to learn (actively) from the images in a supervised or semi-supervised manner.

The outputs (*Output*) of the functions *SSDRB* and *ASSDRB* are the system structure and parameters stored in *Output.DRBParms*.

Output	Description
Output. DRBParms	Structure and meta-parameters of the DRB classifier identified through the learning process, with which the model can continue to self-evolve and self-update without a full retraining when new images are given

One can use the function *DRB* to conduct validation on the trained SS_DRB classifier using the validation mode (*Mode=‘Validation’*).

C.6.2 Simple Illustrative Example

In this subsection, we will use the same image set as the one used in Sect. 13.1.2 for illustration.

First of all, the MATLAB commands for SS_DRB classifier to learn from the unlabeled training images are given, where we use the offline semi-supervised learning mode to train the DRB classifier primed by the labeled training images, and Ω_1 is set to be $\Omega_1 = 1.1$.

```
Input0.Images=TrainingImages;
Input0.Features=TrainingFeatures;
Input0.Labels=TrainingLabels;
Mode='Learning';
[Output0]=DRB(Input0,Mode);
Input1.DRBParms=Output0.DRBParms;
Input1.Images=UnlabelledImages;
Input1.Features=UnlabelledFeatures;
Input1.Labels=GroundTruth
Input1.Omega1=1.1;
Mode='Offline';
[Output1]=SSDRB(Input1,Mode);
```

After the offline semi-supervised learning process is finished, one can visualize the identified prototypes and test the classification accuracy using the same commands as given in Sect. 13.1.2. The prototypes identified from both, the labelled and unlabeled training images are given in Table C.3, where the identified visual prototypes from the unlabeled training images are presented as the ones with the inverse colors.

To train the SS_DRB classifier in the online mode, one can modify the above commands by simply replacing *Mode* = '*Offline*' with *Mode* = '*online*'.

Then, the following example is presented to illustrate how the SS_DRB classifier can learn actively from the unlabeled training images.

Firstly, we remove the labelled training images of digit "0" and train the DRB classifier with nine massively parallel rules corresponding to digits "1" to "9" using the following commands.

Table C.3 The prototypes identified after the offline semi-supervised learning process

Digit	Identified prototypes
1	
2	
3	
4	
5	
6	
7	
8	
9	
0	

```
for ii=1:1:90
    Input0.Images{ii}=TrainingImages{ii};
end
Input0.Features=TrainingFeatures(1:1:90,:);
Input0.Labels=TrainingLabels(1:1:90,:);
Mode='Learning';
[Output0]=DRB(Input0,Mode);
```

Then, we can use the following commands to enable the primed DRB classifier to learn actively from the unlabeled training images in a semi-supervised manner. In this example, the learning process is conducted on a sample-by-sample basis, Ω_1 is set to be $\Omega_1 = 1.1$ and Ω_2 is set to be $\Omega_2 = 0.75$. One can also replace "*Mode* = '*Online*'" with "*Mode* = '*Offline*'" to allow the SS_DRB classifier to learn in offline mode.

```
Input1.DRBParms=Output0.DRBParms;

Input1.Images=UnlabelledImages;

Input1.Features=UnlabelledFeatures;

Input1.Labels=GroundTruth;

Input1.Omega1=1.1;

Input1.Omega2=0.75;

Mode='Online';

[Output1]=ASSDRB(Input1,Mode);
```

The nine massively parallel fuzzy rules after the semi-supervised active learning process are depicted in Fig. C.8, where the newly added prototypes during the process are presented with inverse colors.

There are 11 new fuzzy rules added to the rule base during the semi-supervised active learning process as well, which are presented in Fig. C.9, and the consequent parts of these fuzzy rules are presented as "New Class 1", "New Class 2", … , and "New Class 11".

The SS_DRB classifier will automatically provide the label for each new fuzzy rule based on the real label of the images of the dominant class associated with this particular rule if ground truth is provided. Otherwise, human expertise is required to be involved later to assign the labels for the newly added rules.

After the semi-supervised learning process, one can further test the performance of the trained SS_DRB classifier on the unlabeled images using the following MATLAB commands:

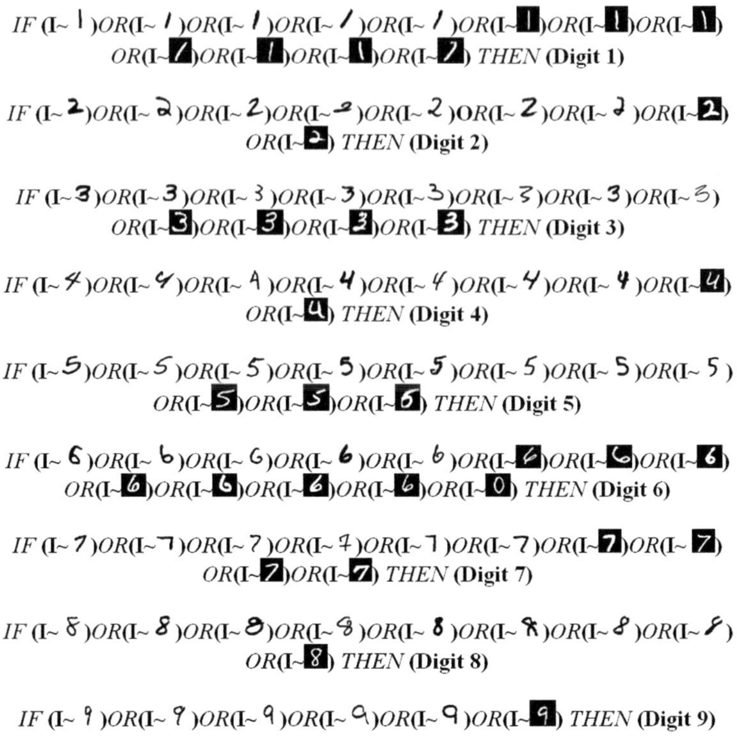

IF (I~)*OR*(I~)*OR*(I~)*OR*(I~)*OR*(I~)*OR*(I~)*OR*(I~)*OR*(I~)
OR(I~)*OR*(I~)*OR*(I~)*OR*(I~) *THEN* **(Digit 1)**

IF (I~)*OR*(I~)*OR*(I~)*OR*(I~)*OR*(I~)*OR*(I~)*OR*(I~)*OR*(I~)
OR(I~) *THEN* **(Digit 2)**

IF (I~)*OR*(I~)*OR*(I~)*OR*(I~)*OR*(I~)*OR*(I~)*OR*(I~)*OR*(I~)
OR(I~)*OR*(I~)*OR*(I~)*OR*(I~) *THEN* **(Digit 3)**

IF (I~)*OR*(I~)*OR*(I~)*OR*(I~)*OR*(I~)*OR*(I~)*OR*(I~)*OR*(I~)
OR(I~) *THEN* **(Digit 4)**

IF (I~)*OR*(I~)*OR*(I~)*OR*(I~)*OR*(I~)*OR*(I~)*OR*(I~)*OR*(I~)
OR(I~)*OR*(I~)*OR*(I~) *THEN* **(Digit 5)**

IF (I~)*OR*(I~)*OR*(I~)*OR*(I~)*OR*(I~)*OR*(I~)*OR*(I~)*OR*(I~)
OR(I~)*OR*(I~)*OR*(I~)*OR*(I~)*OR*(I~) *THEN* **(Digit 6)**

IF (I~)*OR*(I~)*OR*(I~)*OR*(I~)*OR*(I~)*OR*(I~)*OR*(I~)*OR*(I~)
OR(I~)*OR*(I~) *THEN* **(Digit 7)**

IF (I~)*OR*(I~)*OR*(I~)*OR*(I~)*OR*(I~)*OR*(I~)*OR*(I~)*OR*(I~)
OR(I~) *THEN* **(Digit 8)**

IF (I~)*OR*(I~)*OR*(I~)*OR*(I~)*OR*(I~)*OR*(I~) *THEN* **(Digit 9)**

Fig. 8 The updated massively parallel fuzzy rules

```
Input2.DRBParms=Output1.DRBParms;

Input2.Images=UnlabelledImages;

Input2.Features=UnlabelledFeatures;

Input2.Labels=GroundTruth;

Mode='Validation';



[Output2]=DRB(Input2,Mode);

Output2.ConfMat
```

and the confusion matrix of the classification result will appear in the Command Window:

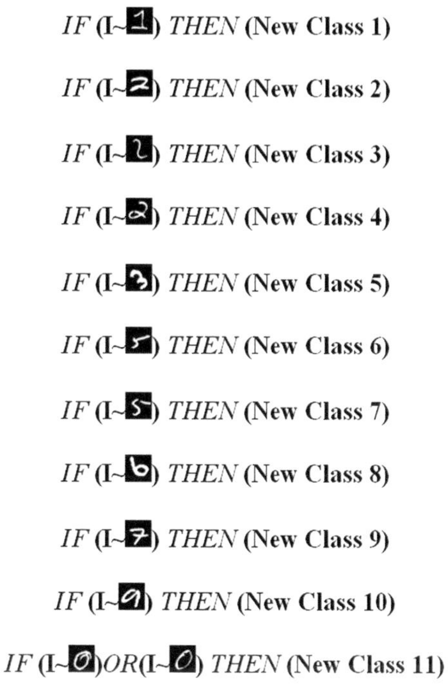

Fig. C.9 The identified massively parallel fuzzy rules

IF **(I~) ***THEN* **(New Class 1)**

IF **(I~) ***THEN* **(New Class 2)**

IF **(I~) ***THEN* **(New Class 3)**

IF **(I~) ***THEN* **(New Class 4)**

IF **(I~) ***THEN* **(New Class 5)**

IF **(I~) ***THEN* **(New Class 6)**

IF **(I~) ***THEN* **(New Class 7)**

IF **(I~) ***THEN* **(New Class 8)**

IF **(I~) ***THEN* **(New Class 9)**

IF **(I~) ***THEN* **(New Class 10)**

IF **(I~)***OR***(I~) ***THEN* **(New Class 11)**

```
ans =

  20   0   0   0   0   0   0   0   0   0
   0  18   0   0   0   0   0   1   1   0
   0   0  18   0   1   0   0   1   0   0
   0   0   0  17   0   0   0   0   3   0
   0   0   0   0  20   0   0   0   0   0
   0   0   0   0   0  20   0   0   0   0
   1   0   0   0   0   0  17   0   2   0
   0   0   1   0   0   0   0  19   0   0
   0   0   1   1   0   0   1   0  17   0
   1   0   2   0   0   6   0   1   0  10
```

If more unlabeled training images are provided after the learning process, one can further use the functions *SSDRB* and *ASSDRB* to update the trained SS_DRB classifier. In the following example, after the DRB classifier is primed by the labelled training images, we use the function *SSDRB* to train the *DRB* classifier in a semi-supervised manner (obtaining a SS_DRB classifier) on a sample-by-sample basis (namely, *Mode* = '*Online*') with half of the unlabeled training images, and then, use the function *ASSDRB* to allow the SS_DRB classifier to learn actively on the remaining unlabeled images as a static image set (namely, *Mode* = '*Offline*').

```
Input0.Images=TrainingImages;
Input0.Features=TrainingFeatures;
Input0.Labels=TrainingLabels;
Mode='Learning';
[Output0]=DRB(Input0,Mode);
Input1.DRBParms=Output0.DRBParms;
for ii=1:1:100
   Input1.Images{ii}=UnlabelledImages{ii};
end
Input1.Features=UnlabelledFeatures(1:1:100,:);
Input1.Labels=GroundTruth(1:1:100,:);
Input1.Omega1=1.1;
Mode='Online';
[Output1]=SSDRB(Input1,Mode);
Input2.DRBParms=Output1.DRBParms;
for ii=1:1:100
   Input2.Images{ii}=UnlabelledImages{ii+100};
end
Input2.Features=UnlabelledFeatures(101:1:200,:);
Input2.Labels=GroundTruth(101:1:200,:);
Input2.Omega1=1.1;
Input2.Omega2=0.75;
Mode='Online';
```

```
[Output2]=ASSDRB(Input2,Mode);

Input3.DRBParms=Output2.DRBParms;

Input3.Images=UnlabelledImages;

Input3.Features=UnlabelledFeatures;

Input3.Labels=GroundTruth;

Mode='Validation';

[Output3]=DRB(Input3,Mode);

Output3.ConfMat
```

and the confusion matrix of the classification result will appear in the Command Window:

```
ans =

   19   0   0   0   0   0   1   0   0   0
    1  17   0   0   0   0   0   1   1   0
    0   1  17   0   1   0   0   1   0   0
    0   0   0  17   0   0   0   0   3   0
    0   0   0   0  20   0   0   0   0   0
    0   0   0   0   0  20   0   0   0   0
    1   0   0   0   0   0  17   0   2   0
    0   0   1   0   0   0   0  19   0   0
    0   0   1   1   0   0   1   0  17   0
    0   0   1   0   0   0   0   0   0  19
```

References

1. R. A. Fisher, "The use of multiple measurements in taxonomic problems," *Ann. Eugen.*, vol. 7, no. 2, pp. 179–188, 1936.
2. http://archive.ics.uci.edu/ml/datasets/Iris

3. V. Lohweg, J. L. Hoffmann, H. Dörksen, R. Hildebrand, E. Gillich, J. Hofmann, and J. Schaede, "Banknote authentication with mobile devices," in *Media Watermarking, Security, and Forensics*, 2013, p. 866507.
4. https://archive.ics.uci.edu/ml/datasets/banknote+authentication
5. M. Charytanowicz, J. Niewczas, P. Kulczycki, P. A. Kowalski, S. Lukasik, and S. Zak, "A complete gradient clustering algorithm for features analysis of X-ray images," *Adv. Intell. Soft Comput.*, vol. 69, pp. 15–24, 2010.
6. https://archive.ics.uci.edu/ml/datasets/seeds
7. L. J. Layne, "Prediction for compound activity in large drug datasets using efficient machine learning approaches," in *Infomration and Resources Management Association International Conference*, 2005, pp. 57–62.
8. https://www.csie.ntu.edu.tw/~cjlin/libsvmtools/datasets/regression.html
9. Y. LeCun, L. Bottou, Y. Bengio, and P. Haffner, "Gradient-based learning applied to document recognition," *Proc. IEEE*, vol. 86, no. 11, pp. 2278–2323, 1998.
10. http://yann.lecun.com/exdb/mnist/

Index

P. P. Angelov and X. Gu, *Empirical Approach to Machine Learning*, Studies in Computational Intelligence 800,
https://doi.org/10.1007/978-3-030-02384-3

Q

R

Printed in the United States
By Bookmasters